# Peyresq Lectures on
# Nonlinear
# Phenomena

## Volume 3

# Peyresq Lectures on

# Nonlinear Phenomena

## Volume 3

Editors

## Freddy Bouchet
Ecole Normale Supérieure de Lyon and CNRS, France

## Basile Audoly
Institut Jean Le Rond d'Alembert, CNRS and
Université Pierre et Marie Curie, France

## Jacques Alexandre Sepulchre
Institut Non Linéaire de Nice, France

 **World Scientific**

NEW JERSEY · LONDON · SINGAPORE · BEIJING · SHANGHAI · HONG KONG · TAIPEI · CHENNAI

*Published by*

World Scientific Publishing Co. Pte. Ltd.

5 Toh Tuck Link, Singapore 596224

*USA office:* 27 Warren Street, Suite 401-402, Hackensack, NJ 07601

*UK office:* 57 Shelton Street, Covent Garden, London WC2H 9HE

**British Library Cataloguing-in-Publication Data**

A catalogue record for this book is available from the British Library.

**PEYRESQ LECTURES ON NONLINEAR PHENOMENA**
**(Volume 3)**

ISBN 978-981-4440-58-5

Printed in Singapore.

# Preface

This is the third volume of notes compiled from lectures delivered at the yearly *Peyresq non-linear meetings* (rencontres non-linéaires de Peyresq) over the period 2003–2009. The first and second volume were published earlier, spanning the periods 1998–1999 and 2000–2002 respectively [1].

These meetings cover a wide range of topics in non-linear sciences, from mathematical to mechanical, physical and biological aspects. Their aim is to bring together researchers coming from various branches of non-linear sciences, expose them to problems that are different from — but ultimately related to — their everyday themes, and provide them with a set of general methods and concepts which they can adapt and apply to their own problems. The benefit of mixing communities has been a strong and constant motivation for starting and continuing these meetings. It makes the atmosphere of these gatherings lively and extremely refreshing. This unique atmosphere is hopefully reflected by the present book, whose aim is to make these stimulating lectures accessible to a wide audience.

The audience ranges from Ph-d students and post-docs, to young or senior researchers. Addressing such a wide audience was made possible by the format of the lectures. Unlike communications in specialized conferences, they were given in 4 to 5 block of 2 hours spread over one week, leaving enough time for an in-depth introduction to the field followed by a detailed presentation of some current research themes. All lectures have been designed from scratch for accessibility. The present lecture notes clearly benefit from this commitment. As a result, this book should be of interest to any researcher (whether theoretician, numerician or experimentalist) or engineer working in a field connected to non-linear sciences, such as: dynamical systems, Hamiltonian systems and partial differential equations; stochastic processes; hydrodynamics and complex fluids; instabilities and turbulence;

---

[1] See *Peyresq Lecture Notes on Nonlinear Phenomena*, R. Kaiser and J. Montaldi, editors. World Scientific, 2000, and *Peyresq Lectures on Nonlinear Phenomena, Vol. 2*, J.-A. Sepulchre and J.-L. Beaumont, editors. World Scientific, 2003

optics; quantum mechanics; biology and physiology; physics and mechanics of interfaces and slender bodies; physics of out-of-equilibrium system; statistical physics; numerical analysis. Methods in non-linear sciences are often generic and can be exported from one field to the next. However, being relatively new and spread over different scientific communities, these methods have become taught more widely at the University only recently; as a result many researchers lack a comprehensive knowledge of this field. The aim of this book is to teach the general methods of non-linear sciences to those that are new in the field, and to get those that are already in the field interested into new problems.

The different chapters in this book cover some problems that have emerged or have been reconsidered recently by combining ideas and concepts from different fields under the general heading of non-linear sciences. In his contribution, Argentina considers the question of locomotion with a physicist's eye, explaining how biological systems can take advantage of fluid mechanics and elasticity to move. Bouchet and Venaille approach geophysical flows using a combination of methods from classical physics of fluids and 2D turbulence, statistical mechanics and thermodynamics. De Bouard studies the influence of noise, representing for instance thermal agitation, on the properties of dynamical systems such as the formation of spatial structures, which are classically studied at zero temperature. Josserand, Lagrée and Lhuillier's contribution concerns the flow of granular material; their presentation borrows ideas from statistical physics to characterize phenomena that lie in-between fluid and solid mechanics. Le Bellac's review on the general theory of relativity (in French) emphasizes the geometrical aspects of the theory; his smooth introduction to modern differential geometry is extremely valuable given the number of areas where geometry plays a key role, such as the theory of dynamical systems, classical mechanics, numerical analysis. Perez and Pincet investigate the opposite concept of cell adhesion using statistical physics, electrostatics and elasticity. Pocheau's in-depth review on the non-linear growth dynamics and instabilities starts from the classical Saffman-Taylor instability in fluid mechanics, presenting ideas which have found applications in physics, material sciences and biology.

The meetings are held in late may in the wonderful village of Peyresq. This village at the foot of the Alps was restored by a team of volunteering Belgian academics; its isolated situation makes it a wonderful place for scientific exchanges, and is becoming known more and more widely. The foundation *Peyresq foyer d'humanisme*, run by Mady Smets, makes the place available for scientific — and more broadly, humanistic — meetings for a very reasonable cost. The foundation and the staff, composed of delightful and devoted people, are all warmly thanked. We would also like to thank those who organized the meetings throughout the years, and made it a lasting achievement. The current form of the meetings started in 1998, building on a previous form of the meeting initiated by Pierre Coullet from the Institut non-linéaire de Nice. He was later joined by Robin Kaiser and James Montaldi. In a second step, two labs joined the organizing committee (the Institut de Recherche sur les Phénomènes Hors d'Équilibre in Marseille and the Laboratoire de Physique Statistique of the École Normale Supérieure in Paris). The meetings have been made possible by the long-standing support of the CNRS and by contributions from the laboratories. They have been living for so many years thanks to the generous work of Jean-Luc Beaumont, Jacques-Alexandre Sepulchre, Freddy Bouchet and Yves Pomeau.

Basile Audoly, Freddy Bouchet, and Jacques Alexandre Sepulchre.

Figure 1. A view close to Peyresq village.

# Contents

# Chapter 1

# Some Examples of Animal Locomotion in Fluids

Médéric Argentina

*Université de Nice Sophia-Antipolis, LJAD*
*Parc Valrose, 06108 Nice, France*

In these lectures, we present some models describing the locomotion of animals. We first propose some examples of dimensional analysis that provide good prediction of locomotion velocity. We then focus our discussions on organisms moving into fluids. The locomotion velocity is computed in the low and high Reynolds number limit.

## Contents

## 1. Introduction and some scaling laws

For physicists, it may appear curious to consider that animal locomotion is an interesting subject. Moreover, the complexity of a moving organism like those associated to its shape, its internal or external composition, and others biological peculiarities could represent an obstacle for a profound analysis of locomotion. Nevertheless one can feel that the physicist approach may give some insights, and some answers to natural questions. For example: what is the maximum velocity that a human can walk? This question may become tricky if one wants to include the effect of complexity of muscles, borne. In fact it is possible to give some insights. To be more precise, with a simple and physical argument, an estimation of the walking speed can be presented. When an organism is walking, it moves its legs, that are linked to the body via a rotating articulation. One would expect that locomotion velocity to be of order of the leg velocity. What is the typical leg velocity? To move the leg, it is necessary to work against the gravitational force, and the comparison with a pendulum then becomes evident. The characteristic velocity of a pendulum is $\sqrt{lg}$, with $l$ length of the leg, and $g$ the gravitational acceleration. A human leg is of order 1 meter, and this estimation gives $3$ m.s$^{-1}$. So this rough argument gives the good order of magnitude. A nice survey on locomotion can be found in the book of Mc. Neil Alexander[1] who treated this example with great attention. In fact, Galileo-Galilei was maybe one of the first scientists to try to understand scalings in animals: why elephants have such large legs? Because living animals are composed of water, the weight scales as the cube of the size of the animal, and legs must be bigger to counter-balance the pressure induced by the weight.

Another illustration of the physicist approach: why most of the car just go to 200 km.h$^{-1}$, or why most of the cars does not run over 200 km.h$^{-1}$? By inquiring the function, power of the motor $P$ vs the maximum velocity $U$ cars can reach, a very simple answer appear: $P \sim U^3$. The complexity of a motor and cars in general could prevent us to present a simple law, but physics can give some insights: for big vehicles the drag stress is proportional to $\rho_a U^2$, with $\rho_a$ the air density. Consequently, the power used by the car will be this strain times the velocity, and the scalings becomes $P \sim U^3$. This strong velocity dependance explains that to double the maximum velocity of a car, it is necessary to multiply by height the power motor !

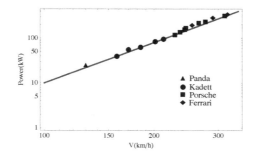

Figure 1.   Log-Log plot of the power of cars versus their maximum velocity. Data was taken from the book of H. Tennekes, The Simple Science of Flight, MIT Press, 1996.

What is the flight velocity of birds? In order to answer to that question, it is necessary to understand why birds fly. A lift force exerted at the wings counter-balances the gravity attraction. The stress at the bird wing is of order $\rho_a U^2$, so that the lift force is of order $\rho_a U^2 S$, $S$ being the surface area of the wings. This force must be of order $mg$, $m$ mass of the bird and $g$ acceleration of the gravity. This balance gives:

$$U \sim \sqrt{\frac{mg}{\rho_a S}}.$$

The mass of a living animal of size $L$ can be estimated to $\rho_w L^3$, since animals are composed almost exclusively of water, we then deduce:

$$U \sim \sqrt{\frac{\rho_w}{\rho_a} g L},$$

so that bigger birds flies faster than smallest. So that typically, a 10 cm bird flies at almost 30 m.s$^{-1}$. In order to check this trivial scaling law, I picked measures from an experiment,[2] where flying velocity of different birds have been evaluated in a wind tunnel. This very crude model permits to give a

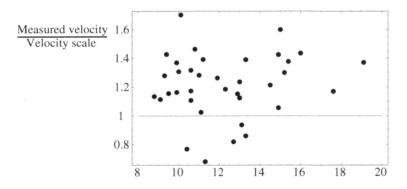

Figure 2. Scaling laws of the velocity of birds vs their size.

description, at a qualitative model. In order provide a better prediction, it is necessary to introduce the shape of the bird, or other peculiarities as feathers for example.

The last simple scaling I wanted to present is the locomotion velocity of snails. These animals can move because they exude a liquid that separate their foot from the substrate. Some recent works proposed to explain the ability of locomotion of the snails via the non-newtonian behavior of the mucus.[3] Here, assuming that this fluid is Newtonian, we propose a mechanism for locomotion of the snail over its mucus. This thin liquid film of thickness $h$ creates a viscous force that permits the movement of the snail. In lubrication theory, that will be described in more details, there

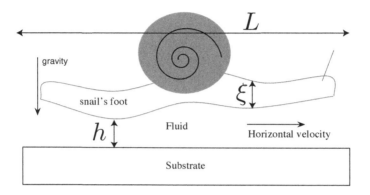

Figure 3.    Schematic representations of a moving snail.

is a relation between the pressure $P$ in the fluid and the velocity $u$ of the fluid: $\partial_x P = \eta \partial_{zz} u$. The viscosity of the fluid is $\eta$. Then it appears that $P \sim \frac{\eta u L}{h^2}$, $h$ being the thickness of the fluid and $L$ the typical size of the snail as pictured in the Fig. 3. The order of magnitude of the pressure is related to the Archimede law with $P \sim \Delta \rho g \xi$, $\Delta \rho$ is the difference of density of the snail body with the fluid in the film, and $\xi$ is the thickness of the snail's body. The snail body will follow the fluid in the film, and we then deduce a very crude scaling law for snails:

$$u \sim \frac{\Delta \rho g \xi h^2}{L \eta}$$

With $h = 10^{-4}$ m, $L = 5$ cm, $\xi = 1$ cm, $\Delta \rho = 10^3$ kg.m$^{-3}$, et $\eta = 10^{-3}$ Pa.s, this rough model gives a locomotion velocity of $u = 3$ mm.s$^{-1}$.

In the previous examples, fluid mechanics play a central role in the determination of the locomotion velocity. What make the locomotion problem so interesting is that it is necessary to compute the flow in the body environment in order to assess the forces necessary for the displacement. The first question I want to address is the drag induced by the fluid during the locomotion. The fluid dynamics are described with the well celebrated incompressible Navier-Stokes equation[4] to which the velocity $\vec{u}$ obeys:

$$\rho\left(\partial_t \vec{u} + (\vec{u}.\vec{\nabla})\vec{u}\right) = -\vec{\nabla}p + \eta\nabla^2\vec{u}$$
$$\vec{\nabla}.\vec{u} = 0.$$

The density of the fluid is $\rho$, its viscosity $\eta$, and the pressure is $P$. This quantity has to be determined such that the mass conservation equation (the second in the above set) is verified. These equations are complex but in some limits, simplifications can be done. In order to do so, it is necessary to study the importance of each component of the equations via dimensional analysis. We propose the following change of variable $t = Tt'$, $\nabla = \nabla'/L$, $p = Pp'$ $u = Uu'$, where the dotted variables do not have a dimension. We then obtain:

$$\left(\partial_t'\vec{u}' + (\vec{u}'.\vec{\nabla}')\vec{u}'\right) = -\frac{P}{\rho U^2}\vec{\nabla}p' + \frac{1}{Re}\nabla'^2\vec{u}'$$
$$\vec{\nabla}'.\vec{u}' = 0.$$

This substitution proves the existence of a non-dimensional number the Reynolds number $Re$, that plays a very important role in fluid mechanics.[4]

$$Re = \frac{\rho U L}{\eta}. \tag{1}$$

This parameter measures the ratio of the inertia of the fluid to the viscous force. The characteristic value $P$ of the pressure will depend on the Reynolds number $P = \rho U^2 f(Re)$ Two interesting limits can be studied: the high and the low Reynolds number limit. As $Re \ll 1$, clearly the viscous part becomes dominant and inertia can be dropped out:

$$0 = -\vec{\nabla}p + \eta\nabla^2\vec{u} \tag{2}$$
$$\vec{\nabla}.\vec{u} = 0. \tag{3}$$

The dimensional analysis permits us to conclude that $P \sim \frac{\eta U}{L}$. Let us compute the resistance force $F$ that fluid opposes against the movement of a sphere of radius $L$. The force $F$ will be proportional to the pressure $P$ times a surface $L^2$, and we get:

$$F = \eta U L,$$

that is known as the Stokes friction. For a cylinder, the exact computation[4] gives a pre-factor $6\pi$ : $F = 6\pi\eta U L$. It is possible to recast this drag force in terms of the Reynolds number defined in (1):

$$F = \frac{\rho U^2}{Re} L^2.$$

It is customary to use the drag coefficient $C_d = \frac{F}{\rho U^2 S}$, where $S$ is the apparent surface of the object measured in the direction of the flow. Experiments has been done to measure this coefficient, and the results are shown in Fig. 4. At low Reynolds number, it is seen experimentally that $C_d \sim Re^{-1}$. The drag coefficient does not vary so much as the Reynolds

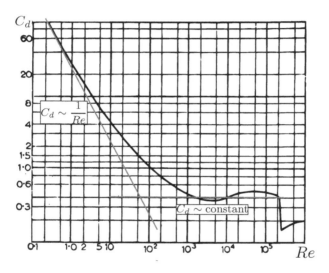

Figure 4. Drag coefficient vs the Reynolds number measured on a cylinder from the Landau's textbook. It is seen that from $Re = 10^2$ to $Re = 10^5$, $C_d$ is decreased just by a factor 2.

number becomes high enough. For example in Fig. 4, $C_d$ varies of a factor 2 when $Re$ increases from $10^2$ till $10^5$ so that it can be considered as constant, and as it was used previously in the introduction the drag force then becomes:

$$F = \rho U^2 L^2.$$

Two researchers received the Ig-Nobel prize in 2005, thanks to their work around the question "Will Humans Swim Faster or Slower in Syrup?".[5] The two scientists proposed to ten competitive swimmer to swim and six recreational swimmer to swim 50 meters in in a water pool filled with syrup. Swimming velocities have beens measured, and show no dependance at all with the change in viscosity: clearly the explanation of their results is that as for a swimming human, the Reynolds number is of order $10^5$, the drag force is that presented above: the velocity does not depend on viscosity of the fluid. It would have been necessary to change the vicosity of the fluid with almost three order of magnitude to perceive the effect on the locomotion velocity.

In this introduction, I presented very rough arguments for locomotion in fluids. There exists a huge literature on fish swimming, and I will refer to a recent survey[6] for a more general description of swimming gaits. In this lecture, I will focus on locomotion induced by the deformations of elastic plates into fluid, and I will avoid the discussions on propulsion mediated by rigid plates where recent advances have been performed.[7] Following the description of Lighthill,[8] the swimmers can be classified into two categories. At low Reynolds number, the forces in action are viscous and said to be resistive. In that case, the organism can move because of the friction with the fluid. In the other limit, the high Reynolds number, the inertia of the fluid is dominant, and the locomotion forces are said reactive: the animal can move because it settle the fluid in movement which continues moving my inertia, then by the action-reaction principle, a thrust is generated.

I will address the two limits of the locomotion at low Reynolds numbers and high Reynolds number in the two next sections.

## 2. Locomotion at low Reynolds number

When explaining the the physics of the locomotion in the Stokesian realm, it is impossible to not cite the famous article of Purcell.[9] An organism like a bacteria swims in liquids for which the Reynolds number is approximately $10^{-4}$, so that inertia is irrelevant: when the locomotion motor stops, the bacteria stops in a distance of a length of an Angstrom and in 0.3 microseconds! So the Stokesian realm is non intuitive: following the suggestion of Purcell, in order to get an idea about swimming at Low Reynolds number, human would have to try to swim in tar during a few weeks. Another non intuitive fact is that in order to move, the movement responsible of locomotion must break the temporal reversibility. There are beautiful movies made by G.I. Taylor himself in which he demonstrates that the breaking of the symmetry is responsible of the movement. In order to explain this effect, it is necessary to remark, as did Lighthill, that the propulsion is resistive. Let us decompose the movement of the animal that tries to swim in the Stokesian realm. The animal moves a part of its body, the fluid around it is pushed and stops very quickly due to the lack of inertia. By the classical law of mechanics, the action-reaction statement,[10] the animal moves forward. Nevertheless, if the animal does exactly the same movement in a reverse way, the fluid that was displaced previously is re-set at its original location: the animal moves backward. In the absence of inertia in the fluid, the organism must breaks the symmetry. The echerechia Coli has flagella that rotates forming an helix, temporal reversibility of the movement is broken, and locomotion is assured.[11]

Another possibility is to take into account the elasticity via the flexibility of the moving body parts.[12] A recent article provides nice insights about the locomotion of elastic plates at low Reynolds number.[13] One of the pioneers of this area is G.I. Taylor that wrote various articles in the earlier fifties. In this section, I will summarize nice results of Taylor.

### 2.1. *Undulating plate*

Here I present a very nice answer that G.I. Taylor[15] gave to the following question: is an undulating plate embedded in a fluid can generate a propagative force, and then move at constant velocity. In the reference frame of the plate, the question is transformed into: is a deformable obstacle embedded in fluid can settle in movement the fluid at infinity? We suppose that this plate is located at $y = h(t, x)$, and its slope is small enough such that $\partial_x h(t, x) \ll 1$. We assume that the shape of the membrane is given by

$$h = A\sin(\omega t - qx), \tag{4}$$

and there is a small non dimensional number $\varepsilon = Aq$. In order to satisfy automatically the mass conservation (3), we use the stream function defined

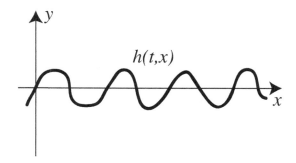

Figure 5.   Schematic representation of an undulating plate.

as

$$u = -\partial_y \psi$$
$$v = \partial_x \psi.$$

Then by taking the rotational of (2), we obtain a bi-laplacian equation for the stream function:

$$\Delta^2 \psi = 0. \tag{5}$$

It remains to impose the boundary conditions: the fluid does not slip at the membrane such that we have two more relations:

$$u(t, x, y = h(t, x)) = 0 \tag{6}$$
$$v(t, x, y = h(t, x)) = \partial_t h \tag{7}$$

Since we assume the slender slope approximation we can perform a Taylor expansion such that

$$u(t, x, 0) + h(t, x)\partial_y u(t, x, 0) + o(\varepsilon^2) = 0 \tag{8}$$
$$v(t, x, 0) + h(t, x)\partial_y v(t, x, 0) + o(\varepsilon^2) = \partial_t h, \tag{9}$$

where the second term of the each equation l.h.s is of order $\varepsilon$. The bulk equation for $\psi$ is linear but the boundary condition is not. The complete set of equations must be solved pertubatively, and we set:

$$\psi = \psi_0(t, x, y) + \varepsilon \psi_1(t, x, y) + o(\varepsilon^2) \tag{10}$$

At the lowest order the bi-laplacian equation together with the boundary condition at infinity give a simple expression for $\psi_0$

$$\psi_0 = B \sin(\omega t - qx)(1 + |y|q)e^{-q|y|}. \tag{11}$$

The boundary condition (6) is also satisfied, and the only constant we have to determine is $B$. By inserting the above expression into (7), we deduce that $B = -Aw/q$. Consequently we finally get:

$$\psi_0 = -Aw/q \sin(\omega t - qx)(1 + |y|q)e^{-q|y|}. \tag{12}$$

Clearly at the lowest order, the fluid is at rest at $y \to \pm\infty$, and it becomes necessary to process the next order. $\psi_1$ is also solution of the bi-laplacian bulk equation, but its boundary condition depends on $\psi_0$:

$$\partial_y \psi_1(t, x, 0) + h(t, x)\partial_{yy}\psi_0(t, x, 0) \tag{13}$$
$$\partial_x \psi_1(t, x, 0) + h(t, x)\partial_{yx}\psi_0(t, x, 0). \tag{14}$$

We are not going to present $\psi_1$ in a closed form, but derive a simple argument which gives the velocity at infinity. We analyse the first relation (13):

$$\partial_y\psi_1(t,x,0) = -h(t,x)\partial_{yy}\psi_0(t,x,0) = -A\omega q^2\sin^2(\omega t - qx),$$

which generates a term $-y\frac{A}{2}\omega q^2\sin^2$ in $\psi_1$, and the fluid is set in movement at infinity with a velocity $-\frac{A}{2}\omega q^2$. In his article, Taylor developed the solution till the fourth order.

I presented the elegant computation of Taylor. But it appears we could have guessed this result on grounds of symmetry. The locomotion velocity should be invariant by changing $A \to -A$, yielding $U \sim A^2$. Since we want to compute a velocity, it is necessary to introduce another length: $1/q$, then $U \sim A^2 q$. To conclude we peek the only frequency $\omega$ to construct a velocity: $U \sim A^2 wq$.

## 2.2. *Snail locomotion, and flying carpet*

In the next sub-section I would like to discuss on the locomotion of an elastic plate submitted to gravity. When we let fall down a transparency, or a piece of paper, it levitates sometimes some instant before to touch the ground or the table. This levitation is due to the formation of a thin film of air between the ground and the plate. In order to get closer, the plate must expel the air in the film, and this takes some time.

Is it possible to take advantage of this effect for locomotion. For sure there exists the over-craft, which pushes air against the ground and permits levitation. Here, we would like to know if an undulating plate can move autonomously. Here we will assume that the thickness of the film is so small such that the lubrication approximation can used. In such a case, the

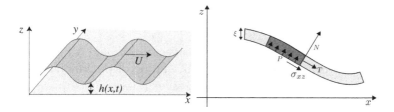

Figure 6.    Setup for the falling of a flexible plate with the notation definition.

momentum equations and the mass conservation write:

$$0 = -\partial_x P + \eta \nabla^2 u \tag{15}$$

$$0 = -\partial_z P - \rho g \tag{16}$$

$$\partial_x u + \partial_z v = 0. \tag{17}$$

The horizontal and vertical components of the velocity are named $u$ and $v$. From the mass conservation equation, we deduce that $v \ll u$, since the thickness of the fluid $h$ is much smaller than the typical horizontal distance. The Eq. (16) can be solved, yielding the pressure: $P = -\rho g z + p(x)$, where $p(x)$ is the deviation from hydrostatic pressure. Then the horizontal momentum equation can be solved:

$$u = \frac{\partial_x p}{12\eta}(z^2 - zh) + U\frac{z}{h}, \tag{18}$$

were the no slip boundary conditions have been used. The deformable plate is moving at a velocity $U$. Since the fluid flow is known, an equation for the position $h(t, x)$ of the interface is derived by integrating the mass conservation equation along the $z$ direction between 0 and $h(t, x)$. We obtain the Reynolds Equation (1883)[4]

$$\partial_t h + \frac{1}{2}U\partial_x h + \partial_x\left(\frac{1}{12\eta}h^3\partial_x p\right) = 0. \tag{19}$$

We can compute the fluid stresses exerted on the plate. Following the schematic Fig. 6. We write

$$\rho_s \xi \partial_t^2 X = -\sigma_{xz} - p\partial_s h + \partial_s(T) - \partial_s(N\partial_s h) \tag{20}$$

$$\rho_s \xi \partial_t^2 h = \sigma_{xz}\partial_s h + p + \partial_s(T\partial_s h) + \partial_s(N) - \Delta\rho g\xi, \tag{21}$$

where $s$ is the curvilinear distance from the edge of the plate. The viscous shear is $\sigma_{xz}$, the internal tension of the plate is $T$, and the internal normal forces is $N$. With the slender slope approximation, we can neglect term like the tension $T$, and $\partial_s h\sigma_{xz}$. Consequently, we can consider that the elastic plate is translating horizontally and homogeneously. By integrating the Eq. (20) over the horizontal length we write:

$$\rho_s \xi \partial_t U = -\frac{1}{L}\int_0^L \sigma_{xz} + p\partial_s h \, ds \tag{22}$$

$$0 = p + \partial_s(N), \tag{23}$$

where the vertical inertia has been dropped as it is dominated by the horizontal one. The thickness of the membrane is measured with the variable $\xi$.

The plate is supposed to be autonomous, and this is why the normal force must be equal to zero at the edges of the plate. The buoyancy force has been included with the term $\Delta\rho g\xi$, where $\Delta\rho$ is density difference of the membrane to the fluid and $g$ is the gravitational acceleration. It remains to evaluate the normal force $N$. It is associated with the internal torque $M$ with $\partial_s N + M = 0$. There are two components which constitute $M$: the elastic resistance to deflexion, proportional to the bending stiffness $B$ of the plate times its curvature and a distribution of torque along the plate F, responsible of the locomotion, and then write:

$$M = B\partial_s^2 h + F.$$

Consequently the equations of motion of the active plate are:

$$\rho_s \xi \partial_t U = -\frac{1}{L}\int_0^L \sigma_{xz} + \partial_s hpds \tag{24}$$

$$0 = p - B\partial_s^4 h - \partial_s^2 F - \Delta\rho g\xi \tag{25}$$

$$\partial_t h + \frac{1}{2}U\partial_x h + \partial_x\left(\frac{1}{12\eta}h^3\partial_x p\right) = 0. \tag{26}$$

To go further, it is necessary to solve these equations and to choose a form for the active torque. Since there are a lot of possibilities, we reversed the problem, by imposing the shape of the plate and from the geometry deduce the induced torque. Boundary conditions must be imposed to the above set of equations,: in the lubrication approximation, the pressure is so high in the thin fluid layer, that it can be assumed to be zero at the borders. The normal forces must be zero at the borders of the sheet, as the plate is set free. If we choose a general shape for $h$, it there are four boundary conditions for a system that has a second order derivative in space. Consequently, we must add two free parameters, to provide flexibility in order to solve the system. Since the functional space spanned by $f$ is infinite, here we study the problem using an inverse method, imposing the shape of the sheet, and using this to deduce the form of $f$:

$$h(t,x) = h_0(t) + \gamma(t)x + A\sin(\omega t - qx). \tag{27}$$

where the third term represents the actively generated oscillatory part of the motion with $A$ the amplitude of oscillations, $\omega$ the temporal frequency of the oscillations, and $q$ the wave number, and we have used a single mode approximation. The equations (25), (26) together with the boundary conditions constitutes a fourth order system for the functions $f, p$ and the parameters $h_0(t), \gamma(t)$. For a given ansatz (27) we use a Newton-Raphson method to determine $h_0, \gamma$ by solving (26), (25) subject to the boundary conditions, and determine $U$ by using (24). The plate will levitate, if it moves with $U$ different from 0, and a non zero time averaged $h_0$. The balance between the first twwo terms in the Eq. (26) gives the relation

$$A\omega \sim U\gamma.$$

The tilt angle can be approximated in average by $\gamma \sim \frac{h_0}{L}$. It then remains as unknown the average thickness of the film $h_0$. By balancing the second and third term of the Eq. (26), we get

$$P \sim \frac{\eta U L}{h^2}.$$

The order of magnitude of the pressure is obtained by investigating the vertical Eq. (25), and we get $p = \Delta \rho g \xi$. From these balances, we get the scaling laws for the locomotion:

$$h_0 \sim \quad \left( \frac{A\omega}{q} \frac{\eta L}{\Delta \rho g} \right)^{\frac{1}{3}} \tag{28}$$

$$U \sim \left( \frac{A\omega}{q} \right)^{\frac{2}{3}} \left( \frac{\Delta \rho g \xi}{\eta L} \right)^{\frac{1}{3}}. \tag{29}$$

Let's apply this scaling to snail locomotion: if the snail has a length of 5 cm, and a thickness $\xi = 1$ cm. We take as example the viscosity of water $\eta = 10^{-3}$ Pa.s, and $\Delta \rho = 10^3$ kg.m$^{-3}$. This simple laws give the locomotion velocity to be 1 cm.s$^{-1}$ and a fluid thickness of 15 $10^{-6}$ m, when the amplitude of the undulation of the body is $10^{-5}$ m.

## 2.3. *Anguilliform gait and C. Elegans movements*

In the previous sub section, I presented a nice computation done by Taylor for describing the locomotion of an undulating plate in a viscous fluid. In this sub-section, I would like to propose a locomotion model[16] for the *Caenorhabditis* Elegans. It is a small nematod with a typical length of 1mm, a radius of a tenth of a mm. It has been heavily studied by the biology community, because it is used as a model organism: at the adult stage, it is composed of 959 cells, which constitutes a simple but not minimal organism. It is the first pluri-cellular organism to be sequenced in 1998,[14] and now biologists have constructed very significant knowledge, which permits to easily create mutants. For example, they can modify the nematod, such that some neurons at work can express fluorescent proteins.

These worms live in humid soil, and microscope observations demonstrates the existence of a meniscus along the elongated body. What amazed

Figure 7. (left) An electron Photography done by JM DiMéglio of the C. Elegans. Clearly a meniscus is located along the body (right) The C. Elegans moving over an agar gel: note the sinusoidal.

us[16] and has been previously interested some scientists, is the perfect sinusoidal deformation that exhibits the nematod as it moves. It would be very interesting to predict to locomotion velocity. Again G.I. Taylor is one of the first person to give some insights to that question.[17] He considered a beam moving in a Low Reynolds, number fluid, and predicted a robust phenomenon. The deformation wave along the body responsible of the locomotion propagates against the direction of displacement of the organism; in other words, deformation waves propagates from the head to the tail of the beam.

### 2.3.1. *Undulating in viscous fluid*

This the head to tail wave propagation wave can be demonstrated from a physical point of view: for example swimming snakes just exhibits this fact, and following the Taylor article, I will describe how to demonstrate this striking effect. As in the previous sub-section, we model the snake-

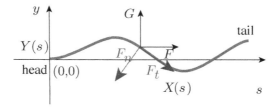

Figure 8. Setup of the undulating beam (in red) for describing the Taylor computation.

worm with a deformable beam, located at point $(X(t, s), Y(t, s))$, then the dynamics write:

$$mX_{tt} = \partial_s(T\partial_s X) + \partial_s(N\partial_s Y) + F \tag{30}$$

$$mY_{tt} = \partial_s(T\partial_s Y) - \partial_s(N\partial_s X) + G, \tag{31}$$

where the masss per unit length is $m$. The fluid exerts a horizonal force $F$ and a vertical force $G$. As in the previous computation, the internal tension is defined as $T$, and the normal constrain applied to the section of the body, both are drawn if the Fig. 5. At that point it is necessary to define what represent $F$ and $G$. Since the organism is small, the inertia can be neglected in the two above equations. We can integrate the equation (30) over the length of the body. Since the beam is free to move, $T$ and $N$ must be equal to zero at the boundaries. As consequence, the locomotion velocity will be given by the relation

$$\int_0^L F ds = 0.$$

We need now to evaluate this integral. As seen in the introduction, if the body of the organism was rigid and dragged into the fluid, we should measure a force proportional to $U$, with a prefactor that depends on the shape of the body. Since the body is slightly deformed we introduce the two prefectors $C_n$ and $C_t$ corresponding to the normal drag and tangential drag of the body:

$$F_t = c_t U_t \tag{32}$$

$$F_n = c_n U_n, \tag{33}$$

where $F_t$ and $F_n$ are forces per unit length that resist to the displacement of the body. The two friction coefficients:

The normal and tangential velocity are:

$$U_t = \frac{1}{\sqrt{(\partial_s X)^2 + (\partial_s Y)^2}} (\partial_s X \partial_t X - \partial_s Y \partial_t Y) \tag{34}$$

$$U_n = \frac{1}{\sqrt{(\partial_s X)^2 + (\partial_s Y)^2}} (\partial_s Y \partial_t Y + \partial_s X \partial_t X). \tag{35}$$

With the slender slope approximation, we have the relations $\partial_s X = 1$, $\partial_t X = U$, and $\partial_s Y \ll 1$. Using the previous relations together with (34-35) into (32) and (33) we get

$$F_t = c_t(U + Y_tY_s) \tag{36}$$
$$F_n = c_n(-UY_s + Y_t). \tag{37}$$

We can project the forces onto the horizontal direction to get

$$F = F_nY_s - F_t = -U(c_nY_s^2 + c_t) + Y_tY_s(c_n - c_t)$$

If we now prescribe the shape of the beam, the locomotion velocity can be computed. Since the beam is slightly deformed, as a first guess, we choose the following shape:

$$Y = A\sin(\omega t - qs), \tag{38}$$

where the $\omega$ is the temporal fequency oscillation, and $q$ is the wave number of the deformation. Since $Aq \ll 1$ the horizontal force is $F = -Uc_t + Y_tY_s(c_n - c_t)$, and we finally get the locomotion velocity at first order:

$$U = -\pi\frac{A\omega}{c_t}(c_n - c_t)Aq + o((Aq)^2). \tag{39}$$

The locomotion velocity is conform to the previous symmetry arguments I have presented before. The interesting aspect of this simple computation is the appearance of the prefactor $c_n - c_t$. For a perfect and rigid cylinder, $c_n \sim c_t$.[18] Such that it is almost the same to drag a cylinder tangentially or normally ! What we learn from this relation is that the locomotion velocity is proportional to this prefactor: when $\omega$, $A$ and $q$ are positive, the deformation wave propagates from left to right, whereas the organism moves from right to left. In fact this analysis captures the physics of the locomotion. A human walking on an iced soil have difficulties in moving, because, his feet are slipping. The solution to move over such material is to use ice skates. The strong anisotropy of these objects creates strong difference between the normal and tangential friction coefficient. As a consequence the high normal resistance of the ice skates, permits to transfer momentum in the human moving direction. And, the Eq. (39) shows that the displacement velocity will be enhanced as the ratio $\frac{c_n}{c_t}$ becomes high enough. For example, it becomes possible to compute the velocity of crawling snakes using such approach.[19] Why the wave propagates in the opposite direction of the body displacement? If the organism can not slide normally, it must move at the same velocity as the deflection wave. If the organism slips, then in the reference frame of the organism, the deformation sine goes in the opposite direction of movement.

### 2.3.2. *A model for the C. Elegans locomotion*

Maybe one of the first scientists to present interest in nematode locomotion is sir Wallace.[20] He studied the influence of a water film in their ability to move. If the nematodes are completely immersed in water, they cannot move in a given direction: they just wiggle. But if they are partially immersed in a film, they can move. The description we gave in the previous sub section may then not be appropriate for the understanding of the C. Elegans locomotion. The presence of a meniscus seems then highly important for propelling the nematod: the order of magnitude of the radius of the meniscus is of order of the diameter of the nematod: $10^{-4}$ m. We can then estimate how much gravity forces play a role in the mechanics. The Pressure induced by capillarity effects is equal to $\gamma/r$, $\gamma$ being the surface tension and $r$ being the radius of curvature of the interface fluid-air along the nematod body. The gravity stress is of order $\rho g r$, $\rho$ being the density of water, and $r$, the typical height over which the gravity acceleration $g$ works. In order to capture the order of magnitude of gravity vs capillarity, we take $r$ to the radius of the worm. This assumption appears to be compatible with the observation (see for example the Fig. 7), and we get a ratio of 700; gravity plays no role in the locomotion of the C. Elegans.

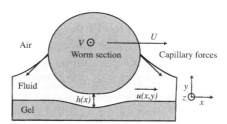

Figure 9.   Cross section of the nematod.

Capillarity forces push the nematod against the substrate. In laboratory conditions, the worm are grown over an agar gel. When observing the nematod moving, the gel appears to be clearly deformed. It is then necessary to introduce the elastic deformation of the gel and the worm. We assume the worm body to be a perfectly elastic cylinder of radius $R$. The thickness of the film between the worm and the gels is:

$$h(x) = h_0 + \frac{x^2}{2R} - \frac{1}{\pi E} \int_{-\infty}^{\infty} p(x') \log|x - x'| dx', \qquad (40)$$

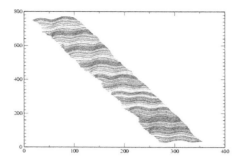

Figure 10. Spatio-temporal diagram of the worm. Each line represents the skeleton numerically computed from a digital movie of worm in movement. For each time-step, skeleton is translated vertically, so the vertical axis represents time, whereas abscissa measures the distance.

where $\frac{1}{E} = \frac{3}{8}\left(\frac{1}{E_w} + \frac{1}{E_s}\right)$ is a composite Young modulus. $E_w$ and $E_s$ are respectively the Young modulus of the worm and the gel. When no pressure is applied, the film thickness is parabolic, as the integral term vanishes. The above relation can be explained as follows: a pressure $p(x)$ that takes the form of a dirac function induces a deformation that is proportional to $\log|x - x'|$. If we add all the possible contributions of the pressure, we get that the deformation is the the integral term of (40). The composite Young Modulus corresponds the deformation of the gel *and* the deformation the worm, this is why the composite Young modulus has such a form.

The pressure can be evaluated in the film, that is considered as a low Reynolds number, and we can use the stationary version of the Stokes Equation (26):

$$\frac{1}{2}U\partial_x h + \partial_x\left(\frac{1}{12\eta}h^3\partial_x p\right) = 0. \tag{41}$$

The worm is pushed against the substrate along the wetted line with the surface tension: we then write

$$\gamma = \int_l p\,dx, \tag{42}$$

where $l$ is the length where pressure applies.

The numerical solutions solutions to this set of equations can be found in[16] and we will focus just on a scaling approach. First, we evaluate the typical horizontal length $l$. By balancing the two first terms of the right hand side of Eq. (40), we get that $l \sim \sqrt{Rh}$. Then we deduce the order of magnitude of the pressure in film:

$$P \sim \frac{\gamma}{\sqrt{Rh}}. \tag{43}$$

Then, by equilibrating the integral term of the deformation relation (40), we deduce $P \sim E \frac{h}{\sqrt{Rh}}$

$$h \sim \frac{\gamma}{E} \tag{44}$$

$$P \sim \sqrt{\frac{E\gamma}{R}}. \tag{45}$$

An agar gel has a Young Modulus $E \sim 10^7$ Pa, and surface tension of water is $\gamma \sim 0.01$ N.m$^{-1}$, such that we get $h \sim 0.1$ $\mu$m. From these two evaluations, we get the typical horizontal velocity with the Stokes Equation (41): $U \sim P\frac{h^2}{\eta l}$:

$$U = \alpha \frac{\gamma^2}{\eta R E}. \tag{46}$$

The pre-factor $\alpha$ is a non dimensional number. Assuming the fluid viscosity to be that of water, we obtain $U \sim 10^{-2}$ m.s$^{-1}$, that is over-evaluated, since the pre-factor[16] $\alpha$ of this scaling is of order to $10^{-3}$. Taking into account this remark, we then get $U \sim 10^{-5}$ m.s$^{-1}$. The inclusion of surfactant during the gel preparation will affect the locomotion properties of the nematod: its sliding velocity should decrease, and its displacement velocity should increase. Experiments,[16] confirm this prediction: the measured velocity of the worms is increased as the soap concentration is increased.

At that point, is would be desirable to predict the locomotion velocity. We now include kinematic ingredients. Suppose that the body of the worm is defined as in the Fig. 8

$$X = s - Vt \tag{47}$$

$$Y = A \sin(\omega t - qs), \tag{48}$$

where $s$ is the curvlinear distance from the animal's head. In order to compute the sliding velocity, it is necessary to compute the trajectory of the worm by injecting $s$ of the Eq. (47) into the Eq. (48) to get the trajectory:

$$Y = A \sin((\omega - qV)t - qX).$$

The sliding velocity is just the time derivative of this relation. If $\omega = qV$, no sliding occurs as the nematod follows strictly the sinusoidal form, whereas

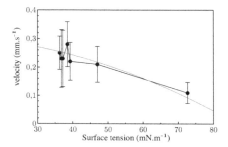

Figure 11. Velocity of the nematods vs surface tension. The measures have been done over a high number of worms. The error bars represent the standard deviation in the velocity distribution probability. The red line is the prediction of the model (50).

if $\omega \neq qV$, the trajectory depends on time. Then the sliding velocity can be taken to be the amplitude of the oscillation of $\partial_t Y$. Then following the sliding velocity $U$ as defined in the Fig. 9 is

$$U = A(\omega - qV). \tag{49}$$

Using the elasto-hydrodynamics, we have evaluated the sliding velocity $U$ as function of the physical parameters. Consequently we can deduce the forward locomotion velocity $V$ with the Eq. (49) and we finally get:

$$V = \frac{\omega}{q}\left(1 - \frac{\gamma^2}{\alpha \eta R E A \omega}\right). \tag{50}$$

This relation demonstrates that the reduction of the surface tension induces an increase of the worm locomotion velocity. The result of this computation is compared with experiments in the Fig. (11): the agreement is good.

In this section, we have only scratched some aspects of locomotion in fluid at low Reynolds number. We have only described locomotion of a moving sheet and proposed a mechanical model to predict the velocity of displacement of nematods.

## 3.  Locomotion at high Reynolds number

The swimmers with high Reynolds number are in general big organisms swimming with a high velocity. Typically, an adult human crawling in water generates a $10^5$ Reynolds number flow. A 10 cm long fish is associated with a $10^4$ Reynolds number. These swimmers interact with the fluid via reactive forces[21]: the organism is set into dsiplacement by setting the fluid in movement; this exchange of momentum is the mechanism for the locomotion. This remark is the basis of the beautiful slender body approximation.[21] In 1961, Wu introduced the first model for fish locomotion[22] assuming the fish to be a plate. By imposing the shape of the plate, the hydrodynamic forces are calculated using a mode decomposition. In the reference,[23] with a heavy numerical computation, a tridimensional fish body is set into water. Experimentally, the movements of a swimming giant danio,[24] that has been previously recorded in experiments is used to move the digitalized fish body in the dynamical fluid simulation. The fish body has been later on modeled as an elastic beam,[25] this first order approximation is shown to be crude as elasticity of living tissues is quite non-linear. Since the fish tissue is hydrated, the viscosity of the fish body has also been introduced[27] Later on, the fish body has been modeled as a viscoelastic media by Wu in.[22,25,27] The last improvement for describing fish locomotion is to consider the muscular forcing. The fish body is assumed to be an active bending beam, where the muscles are modeled as a distribution of torque along the body.[25,27,29]

In these lecture, I focus on elongated animals, and I present the nice approach proposed by Lighthill,[21] that we complement on new aproach based on results obtained in the first part of the century by Theordorssen[30]

### 3.1. *Lighthill's small amplitude theory*

Lighthill assumes in his model that the body of the swimmer is elongated, and the swimming gait is obtained by movements localized near the caudal fin of the fish. The study is restricted to a body that is only slightly

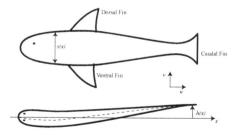

Figure 12. Setup and definition of the notation for the slender body approximation.

deformed from its horizontal equilibrium state, such that $\partial_x h \ll 1$. The fish is also assumed to move dominantlty in the direction of its body, and lateral rate displacement $\partial_t h$ is assumed to be small compared to the horizontal velocity $U$ of the fish: $\partial_t h \ll U$. The length of the fish body is $L$. The fluid surrounding the fish has a flow that is a perturbation around the equilibrium state $u = U$, and in the reference frame of the fish, the velocity of the fish $u$ is assumed to be almost $U$. We introduce the notation of the total derivative:

$$D = \partial_t + U\partial_x.$$

The vertical velocity of the fluid along the fish body is $W(t,x) = v(t,h(t,x))$:

$$W = \partial_t h + U\partial_x h.$$

The important contribution of this Lighthill's work is to have underlined the importance of the added mass $m(x)$. The added mass is the apparent mass of water necessary to displace when the fish body is in movement.[18] The lateral force exerted by the fluid over the fish $F_\perp$ is then $F_\perp = D(mW)$. From this relation, the power of the lateral motion can be evaluated:

$$P = \int_0^L F_\perp \partial_t h\, dx = \int_0^L D(mW)\partial_t h\, dx. \tag{51}$$

An integration by part of the integral gives:

$$P = \int_0^L \left( D(mW\partial_t h) - mW\partial_t Dh \right) dx. \tag{52}$$

By using the relation between $W$ and $h$, the power is written in the form

$$P = \partial_t \int_0^L m\left( W\partial_t h - \frac{1}{2}mW^2 \right) dx + [UmW\partial_t h]_0^L. \tag{53}$$

Lighthill assumes that the added mass at the head of the fish is almost zero, i.e. $m(0) = 0$. Consequently the temporal mean rate power just depends on the ending edge of the body:

$$<P> = <UmW\partial_t h|_L>, \tag{54}$$

where $<.> = \frac{1}{\tau}\int_0^\tau .dt$. The second argument of the elongated body theory is to balance the above avaraged power by the swimmer to that of the rate of power dissipated in the fluid. There are two contributions to this element: one due to the horizontal thrust $T$ of the moving body, and another one due the vertical velocity:

$$<P> = UT + \int_0^L <D\left(\frac{1}{2}mW^2\right)> dx. \tag{55}$$

By using the definition of the operator $D$, we then obtain the following relation for the rate of power injected in the fluid:

$$< P >= UT + U < \frac{1}{2}mW^2 \bigg|_L >$$ (56)

The averaged thrust $T$ can therefore be computed by comparison between the relations (54) and (56). With the assumption that the added mass is zero at the trailing edge of the fish the thrust is deduced:

$$T = m(L) < \left(W\partial_t h - \frac{1}{2}W^2\right)\bigg|_L >$$ (57)

With such a simple expression for the average thrust, it is tempting to impose for the deformation $h$ of the body with a sinusoidal propagative wave:

$$h = A\sin(\omega t - qx).$$ (58)

The averaged thrust then takes the following form:

$$T = \frac{1}{4}A^2\omega^2 m(L)\left(1 - \left(\frac{U}{V}\right)^2\right).$$ (59)

The deformation wave with velocity $V = \frac{\omega}{q}$ must propagate faster than the fish. The faster the wave propagates, higher is the thrust. This simple approach permits also to define a swimming efficiency: the ratio between the power used for the locomotion over the power injected by the swimmer body:

$$\eta = \frac{UT}{< P >}.$$ (60)

The efficiency form can be expressed in a better form using the two relations for the averaged power (54) and (56).

$$\eta = \frac{UT}{< P >} = \frac{< P > - U < \frac{1}{2}mW^2|_L >}{< P >} = 1 - \frac{< W^2 >}{2 < W\partial_t h >}.$$ (61)

In the case of a simple propagative linear wave as described in the Eq. (58), this efficiency becomes:

$$\eta = \frac{1}{2}\left(1 + \frac{U}{V}\right).$$ (62)

Hence, when, the the deformation wave $V$ is high, the thrust is maximized, but the efficiency tends to one half. On the contrary, as $V$ tends to $U$, the thrust $T$ tends to zero, but the movements becomes efficient as $\eta \rightarrow 1$.

### 3.2. *The flag fluttering instability and its associated model for fish swimming*

The Lighthill approach depicted in the previous section gives nice insights on the locomotion properties for elongated swimmers. Here, we would like to follow the Lighthill point of view, by formulating a theory able to predict the swimmer velocity, the forces.... The main assumption is that the swimmer is elongated, and just slightly deformed. From these hypotheses, it is possible in principle to link the flow surrounding the fish with that of a rigid wing. By perturbing this problem, we compute the forces applied by the fish body onto the fluid. The fish is assumed to be an almost flat elastic and thin surface with thickness $h$. The lengths along the fish body

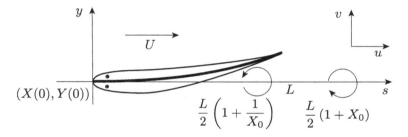

Figure 13.   Setup for the fish body, and definition of the notations.

are measured with the arc-length $s$. The head of the body is defined to be at $s = 0$ and the caudal fin located at $s = L$. The transversal sections of the fish body are labeled in the horizontal direction with $X(s)$ and in the vertical direction with $Y(s)$. In the reference frame of the fish, the flow is seen at infinity to move with the velocity $U$, and it is possible to write a balance equation for the conservation of the momentum:

$$\rho_s h(\mathbf{R}_{tt} - U_t \mathbf{x}) = \partial_s(T\tau) + \partial_s(N\eta) + \Delta P\eta + \sigma_{xy}\tau \qquad (63)$$

where $\mathbf{R}$ is the vector defining the location of the transverse section located at $= (X(S), Y(S))$. $\rho_s$ is the mass density of the fish body, $h$ is the thickness of the fish. Consequently the right hand side of the equation (63) defines the acceleration of the material of the fish in the reference frame of its center of mass. The two first terms of the right hand side of the equations corresponds to the forces applied to the internal section of the fish body. The tension is measured with $T$ whereas the normal forces to the section are labeled with $N$. The two vectors $\tau$ and $\eta$ are respectively the tangential and

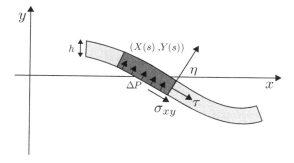

Figure 14. Definition of the tranverse sections, and the differents stresses applied to the body of the swimmer.

normal vectors to the transverse section. The pressure difference applied across the swimmer body is $\Delta P$. The last term of the r.h.s of the Eq.(63) is the stress due to viscous friction. In the next subsections, I will evaluate the different unknown ($T$, $N$,$\Delta P$ and $\sigma_{xy}$) introduced in the above equation.

### 3.2.1. *Approximation of the viscous stress*

Since the swimmer is assumed to move with high velocity, such that the Reynold's number is assumed to be high, the viscous effects are confined in the boundary layer close to the moving body. The assumption that the swimmer is just slightly deformed from the horizontal direction, parallel to the locomotion velocity, indicates that the boundary layer theory could be useful for determining the viscous shear stress. Since the deformations are assumed small, we can approximate the shear stresses by those a perfectly plane object. In this case, the corresponding flow is known as the Blasius flow.[4] In such a case, the boundary layer thickness $\delta$ takes the following form:

$$\delta = \sqrt{\frac{\mu s}{\rho_f U}},$$

where $s$ measures the distance from the head of the plate, $\mu$ is the viscosity of the fluid and $\rho_f$ its density. The shear tress induced by viscous effects takes the form $\sigma_{xy} = 2\mu\partial_y u$, with $u$ the horizontal velocity of the fluid located along the fish body. There is a factor two in the shear, because both sides of the fish are wetted. Since the horizontal velocity varies along in the boundary layer, the shear stress can be approximated with $\sigma_{xy} \sim 2\mu U/\delta$. It is by numerical integration of the Blasius flow that the numerical prefactor can be obtained[4]:

$$\sigma_{xy} = 2\alpha\rho_f \sqrt{\frac{\nu U^3}{s}}, \tag{64}$$

with $\alpha = 0.332$.

### 3.2.2. *Computation of the tension T*

The internal tension of the fish can be computed if a state equation links the extension of the fish body with $T$. Here we will assume that the fish is inextensible. Mathematically, it corresponds to impose

$$\mathbf{R}_s^2 = 1. \tag{65}$$

This constrain is the extra equation which permits to compute the unknown $T$. By differentiate it twice with respect to time, one gets:

$$\mathbf{R}_{tts}\mathbf{R}_s = 0. \tag{66}$$

As we investigate just small deformations, the nonlinear term $(\mathbf{R}_{st})^2$ has been neglected. The equation for the tension becomes explicit using the relation (63):

$$\partial_s \left( \partial_s T + \sigma_{xy} \right) = 0, \tag{67}$$

where again all nonlinear terms have been neglected. This relation can be integrated once:

$$\partial_s T + \sigma_{xy} = \beta. \tag{68}$$

The constant of integration $\beta$ will be determined by imposing the boundary values of the tension at the edges of the plate. By imposing that the tension is zero at the edges of the swimmer body $T(s = 0) = 0$ and $T(s = L) = 0$, the integration constant $\beta$ can be evaluated:

$$\beta = 4\alpha\rho_f U^2 \sqrt{\frac{\nu}{UL}}. \tag{69}$$

Finally the tension is obtained:

$$T = 4\alpha\sqrt{\frac{\nu U^3}{L}} \left( s - \sqrt{\frac{s}{L}} \right). \tag{70}$$

Clearly, the fish body is under compression.

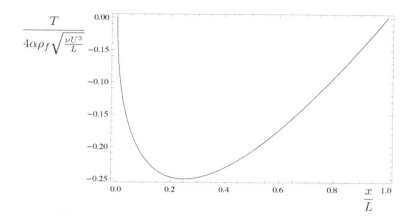

Figure 15.   Non dimensional internal tension induced by the viscous shear stress.

### 3.2.3. *Computation of the normal forces N*

In order to compute the internal normal forces, it is necessary to evaluate the torque $M$ applied to a small section of the plate, and this gives:

$$\partial_s M + N = 0. \tag{71}$$

Since no normal forces are applied to the extremities of the fish, the boundary condition is that $N|_{s=0,L} = 0$, such that

$$\partial_s M|_{s=0,L} = 0. \tag{72}$$

The torque $M$ is decomposed into two parts,[26] an elastic one $M_e$, that resists to the deformation of the foil, and an active one $M_f$. This last element takes into account the effects of the muscles inside the swimmer, this is why we name it as an active torque. This is a crucial part for the locomotion, because without the muscular forcing, there is no energy injection, and the fish can not move autonomously. We then write the internal torque as:

$$M = M_e + M_m. \tag{73}$$

In order to determine the elastic response to deflection, we need to take further assumptions: the swimmer body behaves as a linear elastic media, an hypothesis that is not verified in general. Viscosity of the flesh of the fish[27] could be in principle taken into account, by adding an extra torque response proportional to the rate of change of the curvature of the fish. In this work, we will assume that the muscle activity $M_f$ compensate exactly this effect. This rather simple approach permits to go further in the computations. The fish body is also assumed to be elongated, such that its thickness $h$ is considered small compared to its length $L$, this small aspect ratio justifies the utilisation of the plate elastic plate theory.[31] Consequently, the elastic torque response is written $M_e = B\kappa$, where $\kappa$ is the curvature of the plate, and $B$ the flexural rigidity defined as $B = Eh^3 I$, where $E$ is the Young Modulus, and $I$ is non dimensional the second moment of inertia. We choose

$$I = I_0 + W(s/L)^3, \tag{74}$$

with $W(x) = \sqrt{x(1 - x^3)}$[32] Such that $I_0$ measures the second moment of inertia of the skeleton of the swimmer and the rhs of the definition (74) is just the second moment of inertia of the fish body (flesh). For small deformations the curvature can be approximated with the usual form $\kappa = \partial_s^2 Y$. Nevertheless, mimicking animal physiology, where no muscle can act at the extremities of the body, the torque induced by the fish must be zero at both ends of the body. This imposes that

$$M_m|_{s=0,L} = 0. \tag{75}$$

The relation (75) together with (72) implies that the elastic torque must be zero at boundaries:

$$M_e|_{s=0,L} = 0. \tag{76}$$

No normal forces are applied to the leading and ending edge of the body, such that another set of boundary conditions appears:

$$\partial_s M_e|_{s=0,L} = 0, \ \partial_s M_f|_{s=0,L} = 0. \tag{77}$$

It remains to determine the muscular forcing, and this should be correlated with datas obtained through measures in vivo.[27] Here for sake of simplicity, and to keep our approach as simple as possible, we suppose that $M_m$ is a linear propagative wave. As the boundary conditions imposes that the torque and its spatial derivative to be zero at extremities, we propose the following form for the muscular forcing torque:

$$M_m = M_0(s)\cos(\omega t - qs), \tag{78}$$

where $\omega$ is a frequency forcing, $q$ is the wave number of the wave, and $M_0(s)$ is a function whose value and derivative must be zero at boundaries. We choose the following simple form:

$$M_0(s) = A\left(\tanh(\alpha s)\tanh(\alpha(L - s))\right)^2, \tag{79}$$

where $\alpha$ is assumed to be high.

### 3.2.4. *Hydrodynamics and evaluation of* $\Delta P$

Most of the mechanical strains have been evaluated in the previous sub-sections, with the help of the small aspect ratio, and small deflections approximations. It remains to compute the difference in pressure $\Delta P$ applied across the swimmer body, and this happens to be a difficult task. The main reason is the difficulty to evaluate the flow around the fish body because the Navier-Stokes equations have to be solved for each deformation of the plate. Previous approaches have been proposed, as described in the introduction of the section 3.2. Here, using the fish body that plate is almost flat, we use results obtained for rigid airfoils introduced by Theodorsen in a 1935 NACA article.[30] The perturbation of the former approach to the case to of slightly flexible plates permits the evaluation of $\Delta P$. For the flag flutter problem,[33] we derived:

$$\Delta P = -\rho_f f\left(\frac{s}{L}\right)\left((UY_t + U^2Y_s)C[\gamma] + sY_{tt}\right), \tag{80}$$

where we introduced the dimensionless functions $f(s) = 2\sqrt{\frac{1-s}{s}}$. The fluid density is defined with $\rho_f$. There is a divergence in the leading edge of the plate for the function $f(s)$. It is related to the assumption of a zero thickness plate. As the plate is assumed to be finite, this divergence disappears, we assume $f(s) = 0$ when $s < \mu$, $\mu$ being the ratio of the thickness over the length of the plate. A derivation of the computation of the cut-off is presented in another reference.[33] In the following, we will assume that $\mu$ is small compared to one, and that there is a thin boundary layer at the leading edge of the waterfoil, where the function $f(s)$ goes to zero. We now discuss the three different terms introduced in the pressure difference. As depicted in the introduction, there is a lift force that is proportional to the square of the locomotion velocity. Its prefactor contains the slope of the plate, analogous to angle of attack for wing theory.[18] Let us discuss the term proportional to $\partial_{tt}Y$. The comparison of the mechanical relation (63) with (80), explains its physical relevance: it is the added mass effect.[18] When the fish body has to move, it also need to displace volumes of fluid, and this makes the mass of the fish higher. It is here interesting to compare our approach to the elegant theory of Lighthill depicted in the previous section. The perturbation of the rigid airfoil theory of Theodorsen, permits to compute precisely the distribution of the added mass term $m$. The last term we would like to address in the Eq. (80) is the one proportional to $\partial_t Y$. Again, with the introspection of the mechanical balance (63), it appears that this terms acts as a damping force which is not surprising because the fish body exchanges momentum with the fluid. For incompressible fluid, the pressure obeys to a Poisson equation, and it would be natural to have an integral form of the pressure, in our approach, the slightly deformed shape of the waterfoil permits to get a local equation. It remains to describe the function $C[\gamma]$, following Theodorsen's work, it takes into account the emission of vertices at the ending edge of the plate. It measures the effects of the latency of change of the lift force when geometry is changed. $\gamma$ represents the distribution sheet of vorticity at the ending edge of the fish. It permits to capture subtle hydrodynamical features like the Wagner effect.[34] In non viscous fluids, the vorticity is conserved through the Kelvin's Therorem.[18] Nevertheless, lift forces are consequence of the presence of vorticity.[4] When the angle of attack is varied, the lift force changes, yielding a change in the vorticity. In order to maintain the vorticity conserved, a vertex is emitted at the instant at which the angle of attack changes. Later on, it is advected by the flow. For example, in such a case, the asymptotic temporal dependence of the function is

$$C = 1 - \frac{L}{2tU}, \tag{81}$$

such that the lift forces becomes stationary as the emitted vortex becomes far away from the plate. The vorticity function may also take closed form, if the movement of the tail of the fish becomes periodic, then it can be expressed in term of Hankel functions.[30]

### 3.2.5. ElastoHydrodynamics of the elongated swimmer

In the previous section, we assumed that the variation of the longitudinal velocity to be small. We project the Eq. (63) into the horizontal and vertical directions:

$$\rho_s h(\mathbf{X}_{tt} - U_t) = \partial_s T - \Delta P \partial_s Y + \sigma_{xy} \tag{82}$$

$$\rho_s h \mathbf{Y}_{tt} = \partial_s (T \partial_s Y) - \partial_{ss} (B \partial_{ss} Y) + \Delta P + \sigma_{xy} \partial_s Y - \partial_{ss} M_m, \tag{83}$$

where we took the lowest order in space derivative of the deflection $Y$. We then average the horizontal momentum balance such that we get:

$$\rho_s h \partial_t U = \frac{1}{L} \int_0^L \Delta P \partial_s Y ds - \frac{4\alpha L}{\sqrt{Re}} \rho_f U^{\frac{3}{2}}$$

$$\rho_s h \left(1 + \frac{\rho_f s}{\rho_s h} f\left(\frac{x}{L}\right)\right) Y_{tt} = T \partial_{ss} Y + \partial_{ss} (B \partial_{ss} Y) \tag{84}$$

$$- \rho_f U C[\gamma] f\left(\frac{x}{L}\right) (Y_t + U Y_x) - \partial_{ss} M_m.$$

We use non dimensional variables and parameters by introducing the following scalings: $Y = L\eta$, $s = Ls'$, $t = U_b/Lt'$ and $U = U_b u$, where $U_b$ is the velocity of the bending wave of deformation of the fish, defined as $U_b = \frac{h}{L}\sqrt{\frac{E}{\rho_s}}$. The equations of motion then become

$$\dot{u} = -\rho u^2 I_1 - \rho u I_2 - \rho I_1 - \frac{4\alpha}{\sqrt{Re}}\rho u^{\frac{3}{2}} \tag{85}$$

$$I_1 = \int_0^1 f(s)\left(\partial_s\eta\right)^2 ds \tag{86}$$

$$I_2 = \int_0^1 f(s)\partial_s\eta\dot{\eta}ds \tag{87}$$

$$I_3 = \int_0^1 sf(s)\partial_s\eta\dot{\eta}ds \tag{88}$$

$$(1 + \rho s f(s))\ddot{\eta} = -\partial_{ss}\left(b(s)\partial_{ss}\eta\right) - \rho u f(s)(\dot{\eta} + u\partial_s\eta) \tag{89}$$

$$-\frac{4\alpha}{\sqrt{Re}}\rho u^{\frac{3}{2}}\left(\sqrt{s} - s\right)\partial_{ss}\eta - \partial_s^2 m_m \tag{90}$$

where the three integrals $I_1$, $I_2$ and $I_3$ have been introduced from the integral in the equation (85). We introduced the parameter $\rho$ that measures the added mass effect $\rho = \frac{\rho_f L}{\rho_s h}$. The density of the flesh of the fishes are of order of those of the water, but the the small aspect ratio $\frac{h}{L}$ infers that $\rho$ is in principle higher than one.

The integral $I_1$ being always negative (see 86), we deduce that the first term of the right hand side of the horizontal momentum equation is a damping. Since this term is proportional to $\rho u^2$ we understand it as pressure drag, since its physical origin is the pressure; it disappears as the fish body becomes aligned with the flow. The last term of the r.h.s is the viscous drag induced by the boundary layer confined near the fish body. The sign of the integrals $I_2$ and/or $I_3$ should be negative, in order to have stationary solution of (85).

We have an integro-partial differential equation. If we assume that the muscular torque $m_m$ just have one temporal frequency, the above system can be reduced a set of coupled ordinary differential equations, that can be easily solved using continuation software like AUTO.[36] By using the active torque from (78), we managed to get the velocity as function of the various parameters of the model.

In the Fig. 16, we plotted the behavior of the locomotion velocity $u$ and the power $P$, for a constant bending stiffness $b(s) = 1$. It appears that the velocity of locomotion is reached for small frequency forcing $\omega$. The

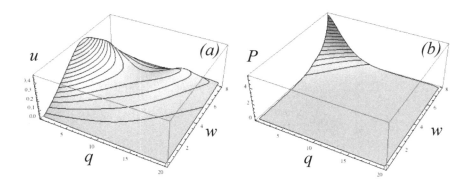

Figure 16.    Dependance of $u$ (a), $P$ (b) as function of the parameters $\omega$ and $q$. For the numerical solution $\rho = 10$, $Re = 10^4$, and $A = 0.71$.

used power in displacement increases significantly as $\omega$ is of order 10. The efficiency dependance as function of the parameters $(\omega, q)$ is nontrivial and shows that there are various peaks in efficiency. The first one is obtained for small $q$ and $\omega$. For $q$ small, we plot in Fig. 16, the ratio of the pressure and viscous drag over the thrust. Clearly, for small frequency $\omega$, the velocity is small, and the drag is dominated by the friction induced by the boundary layer. In this limit $\sigma_{xy} \sim \rho_f L Y_{tt} Y_s$, we then deduce the scaling for the velocity:

$$U \sim \left(\frac{A^4 w^4 q^2}{4\alpha^2 \nu}\right)^{1/3} \sim \left(\frac{L^2 \pi^2 \omega^2}{100\alpha^2}\right)^{1/3} \tag{91}$$

where we have assume as suggested in experiments,[35] that the amplitude of oscillation is one-tenth of the length of the fish. $f$ is the tail beating frequency, and that the viscosity of water is $10^{-2}$. This simple scaling have good prediction when compared to measures of swimming fishes like daces. In the Fig. 17, we compare our prediction with measures made on living swimming fishes.

Note that in the reference,[35] there is an evidence that the velocity per unit length is linear with respect to $f$. Here we predict a small deviation from this statement, since $U/L \sim f^{\frac{4}{3}}$.

To conclude, we also show in Fig. 18 the different gaits or simple model can predict.

In these lectures notes, I wanted to present some example of computations and scalings for predicting locomotion. I have only scratched all the

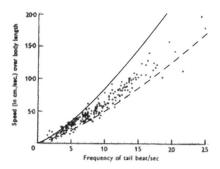

Figure 17. The dots represents the velocity of daces per unit of their body length, vs the frequency of tail beat per second. The measures have been taken from Bainbridge. The black (dashed) curve represents the velocity of a 3.6 cm (5cm)long dace. All the points have been obtained with daces of length from 3.6 to 5 cm.

possible kinds of gaits of limbless animals. Nevertheless, the great variety of swimming and flying gaits suggests that there is a lot of experimental, numerical theoretical work to accomplish before we have a unified understanding of animal locomotion.

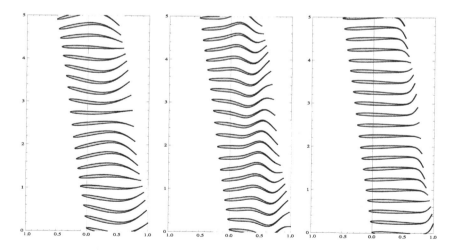

Figure 18. Spatio-temporal diagram of the swimming gait for various values of parameters. The fish body is translated vertically each 1/25th of period. (a) Constant thickness and bending stiffness, $q = 1$, $P = 1.4\ 10^{-3}$. (b) Constant thickness and bending stiffness, $q = 10$, $P = 1.5\ 10^{-3}$. (c) Constant thickness and variable bending stiffness, $q = 0$, $P = 1.2\ 10^{-5}$. Parameters are $w = 1$. $Re = 10^4$, $\rho = 10$.

# References

1. R. McNeill Alexander , *Principles of Animal Locomotion*, Princeton University Press, 2003.
2. Rayner, J. Exp. Biol. **202**, 3449, 1999.
3. N. Balmforth, B. Chan, and A. Hosoi, Phys. Fluids **17**, 113101,2005.
4. L. D. Landau and E. M. Lifshitz, *Fluid mechanics*, Pergamon Press New York (1987).
5. B. Gettelfinger & E. L. Cussler, AIChE Journal **50 11**, 2646, 2004.
6. Sfakiotakis, Lane, Davies, ieee J. of oceanic enginering **24**, 2, 1999.
7. N. Vandenberghe, J. Zhang and S. Childress, Symmetry breaking leads to forward flapping flight J. Fluid Mech. **506**, 147, 2004.
8. J. Lighthill, Mathematical Biofluid-dynamics (Soc. Indust. Appl. Math., Philadelphia), 1975.
9. E.M. Purcell, Am. J. Physics **45**, 3, 1977.
10. L. D. Landau and E. M. Lifshitz, *Mecanics*, Pergamon Press New York, 1987.
11. H. C. Berg. E. coli in motion, Springer, 2003.
12. C. H. Wiggins, D. X. Riveline, A. Ott & R. Goldstein, Phys. Rev. E **56**, R1330, 1997.
13. E. Lauga, Phys. Rev. E75, 041916 (2007),
14. The C. elegans Sequencing Consortium, *Genome Sequence of the Nematode Caenorhabditis elegans. A Platform for Investigating Biology*, Science **282**, 2012, (1998).
15. G.I. Taylor, Proc. Roy. Soc A, **209**, 4447, 1951.
16. P. Sauvage, M. Argentina, J. Drappier, JM diMÕglio, ., submitted to J.E.B.
17. G. I. Taylor, Proc. R. Soc. London A **211**, 225,1952.
18. Batchelor, G. K. An Introduction to Fluid Dynamics. Cambridge, New York: Cambridge University Press, 1973.
19. Z. V. Guo* and L. Mahadevan, Limbless undulatory propulsion on land, PNAS **105**, 2008.
20. H. R. Wallace, *The movements of eelworms in water films*, Annals of Applied Biology, **46**, 86, 1958.
21. J. L. Lighthill, *Mathematical biofluiddynamics*, SIAM, **17**, Society for Industrial and Applied Mathematics, Philadelphia, 1975.
22. Wu, T.Y-T.: Swimming of a waving plate. J. Fluid Mech. **10**, 321, 1961.
23. M.J. Wolfang, J.M. Anderson, M.A. Grosenbaugh, D.K.P. Yue & M. S. Triantafyllou, *near Body Flow dynamics in swimming fish*, J.E.B, **202**, 2303, 1999.
24. M.S., Triantafyllou, G.S., ÒAn EfÞcient Swimming Machine,Ó ScientiÞc American, 64-70, 1995.
25. Wu, T.Y-T.: Hydrodynamics of swimming propulsion. Part 1. Swimming of a two-dimensional flexible plate at variable forward speeds in an inviscid fluid. J. Fluid Mech. **46**, 337, 1971.
26. J. Y. Cheng & R. Blickhan, Bending Distribution along swimming Fish, JEB, **168**, 337, 1994.
27. Cheng, J.-Y., Pedley, T. J. and Altringham, J. D. A continuous dynamic

beam model for swimming fish. Phil. Trans. R. Soc. Lond. B **353**, 981, 1998.

28. YCB Fung - American Journal of Physiology, 1967.
29. Videler, J. and Hess, F. Fast continuous swimming of two pelagic predators, saithe (Pollachius virens) and mackerel (Scomber scombrus): a kinematic analysis. J. Exp. Biol. **109**, 209, 1984.
30. T. Theodorsen, NACA Report 496, 1935;
    http://naca.larc.nasa.gov/reports/1935/naca-report-496.
31. L. D. Landau and E. M. Lifshitz, *Elasticity*, Pergamon Press New York, 1987.
32. M. J. McHenri, C.A. Pell & J. H. Long Jr, Mechanical control of swimming speed: stiffness and axial wave form in undulating fish model. JEB **198**, 2293, 1995.
33. M. Argentina & L. Mahadevan, Fish swimming, to be submitted in PNAS.
34. S. Sane *The aerodynamics of insect SSight*, JEB **206**, 4191, 2003.
35. R. Bainbridge *The speed of swimming of Þsh as related to size and to the frequency*, JEB **35**, 109, 1958.
36. E.J. Doedel, et al *AUTO 2000: Continuation and bifurcation software for ordinary differential equations (with HomCont)* Technical Report, Caltech (2001). http://sourceforge.net/projects/auto2000

# Chapter 2

# Applications of Equilibrium Statistical Mechanics to Atmospheres and Oceans

Freddy Bouchet* and Antoine Venaille

*Ecole Normale Supérieure de Lyon et Université de Lyon 1,
ENS-Lyon, Laboratoire de Physique, CNRS, 46 allée d'Italie,
69364 Lyon cedex 07, France.*
*Freddy.Bouchet@ens-lyon.fr*

We discuss statistical mechanics models of ocean and atmosphere flows. This theoretical approach explains the self-organization of turbulent flows described by the quasi-geostrophic equations, and predicts the output of their long time evolution. On these short lectures, emphasize has been placed on examples with available analytical treatment in order to favor better understanding of the physics and dynamics. We obtain quantitative models of the Great Red Spot and other Jovian vortices, ocean jets like the Gulf-Stream, ocean vortices, and we make quantitative predictions of vertical energy transfers in the ocean. A detailed comparison between these statistical equilibria and real flow observations or numerical simulations is provided.

## Contents

## 1. Introduction

A striking and beautiful property of geophysical turbulent flows is their propensity to spontaneously self-organize into robust, large scale coherent structures. The most famous example is perhaps the Jovian Great Red Spot, an anticyclonic vortex observed many centuries ago. Similarly, the Earth's oceans are filled with long lived coherent structure: $300km$-diameter vortices with a ring shape are observed everywhere at the surface of the oceans. Mid-basin eastward jets, analogous to the Gulf-Stream or the Kuroshio exist in every oceanic basin, and strong bottom trapped recirculations such as the Zapiola anticyclone in the Argentine basin have been recently discovered. Understanding the physical mechanism underlying the formation and the persistence of these coherent structures in a turbulent flow remains a major theoretical challenge.

On the one hand, the problem of the self-organization of a turbulent flow involves a huge number of degrees of freedom coupled together via complex non-linear interactions. This situation makes any deterministic approach illusory, if not impossible. On the other hand, there can be abrupt and drastic changes in the large scale flow structure when varying a single parameter such as the energy of the flow, or its circulation. It is then appealing to study this problem with a statistical mechanics approach, which reduces the problem of large-scale organization of the flow to the study of states depending on a few key parameters only. The first attempt to use equilibrium statistical mechanics ideas to explain the self-organization of 2D turbulence was performed by[1] in the framework of the point vortex model. There exists now a theory, the Robert–Sommeria–Miller (RSM hereafter) equilibrium statistical mechanics, that explains the spontaneous organization of unforced and undissipated two-dimensional and geophysical flows.[2–5] From the knowledge of the energy and the global distribution of potential vorticity levels provided by an initial condition, this theory predicts the large scale flow as the most probable outcome of turbulent mixing.

The aim of these lectures is to present applications of this equilibrium statistical mechanics theory specifically to the description of geophysical flows. There already exist several presentations of the equilibrium statistical mechanics of two-dimensional and geostrophic turbulent flows,[6-8] some emphasizing kinetic approaches of the point-vortex model,[9] other focusing on the legacy of Onsager.[10] Interest of applications of statistical mechanics to climate problems is also discussed in a recent letter.[11] A more detailed and precise explanation of the statistical mechanics basis of the RSM theory, actual computations of a large class of equilibrium states and further references can be found in a recent review.[12]

We present in the first section the quasi-geostrophic dynamics, a simple model for geophysical turbulent flows, and introduce the RSM statistical mechanics for this system. This theory is used to interpret the formation of Jovian vortices and oceanic rings in the second section, where a formal analogy between bubble formation and the self-organization phenomenon in geophysical flows is presented. The third section is devoted to the interpretation of eastward jets as marginal equilibrium states in oceanic basins. The last section deals with vertical energy transfers in geophysical flows, with application to the formation of bottom-trapped recirculations such as the Zapiola anticylone. The equilibrium statistical mechanics has a limited range of applicability since it neglects the effect of forcing and dissipation. The validity of this hypothesis will be carefully addressed for each geophysical application in the related chapter, and will be discussed at a more general level in the conclusion.

## 2. Statistical mechanics of quasi-geostrophic flows

### 2.1. *The 1.5 layer quasi-geostrophic model*

We introduce here the simplest possible model for the dynamics of the Jovian atmosphere or the Earth's oceans: the quasi-geostrophic equations. There are several books discussing this model in more details, among which.[13-16]

We assume that the dynamics takes place in an upper active layer, and there is a lower denser layer either at rest or characterized by a prescribed stationary current.

The flow is assumed to be strongly rotating with a Coriolis parameter $f = 2\Omega \sin \theta$, where $\theta$ is the latitude and $\Omega$ the rotation rate of the planet. Strongly rotating means that we consider the limit of small Rossby numbers

$\epsilon = U/fL$, where $L$ is a typical length scale of the domain where the flow takes place, and $U$ is a typical velocity.

Another key quantity of this system is the Burger number $R/L$ where $R = (Hg\Delta\rho/\rho)^{1/2}/f$ is called the Rossby radius of deformation. It depends on the relative density difference $\Delta\rho/\rho$ between both layers, on the gravity $g$, on the Coriolis parameter $f$, and on the mean depth $H$ of the upper layer.

The fluids we consider are stably stratified, at hydrostatic balance on the vertical and at geostrophic balance on the horizontal. Geostrophic balance means that horizontal pressure gradients compensate the Coriolis force. These flows are called *quasi two-dimensional fluids* because one can show that the velocity field in the upper layer is horizontal and depth-independant (Taylor-Proudman theorem). In order to capture the dynamics, the quasi-geostrophic model is obtained through an asymptotic expansion of the Euler equations in the limit of small Rossby number $\epsilon$.[14] The full dynamical system reads

$$\frac{\partial q}{\partial t} + \mathbf{v} \cdot \nabla q = 0, \quad \text{with } \mathbf{v} = \mathbf{e}_z \times \nabla\psi , \tag{1}$$

$$\text{and } q = \Delta\psi - \frac{\psi}{R^2} + \eta_d . \tag{2}$$

The complete derivation of the quasi-geostrophic equations shows that the streamfunction gradient $\nabla\psi$ is proportional to the pressure gradient along the interface between the two layers. The dynamics (1) is a non-linear transport equation for a scalar quantity, the potential vorticity $q$ given by (2). The potential vorticity is a central quantity for geostrophic flows. The term $\Delta\psi = \omega$ is the relative vorticity[a].

The term $\psi/R^2$ is related to the interface pressure gradient and thus to the interface height variations through the hydrostatic balance. We see that $R$ appears as a characteristic length of the system. Although the rigorous derivation of quasi-geostrophic equations requires $R \sim L$, this model is commonly used in the regime $R \ll L$, either in numerical or theoretical works. This is usually a first step before considering the (more complex) shallow water equations which is the relevant model in this limit, see e.g.[14]

The term $\eta_d = \beta y + \psi_d/R^2$ represents the combined effects of the planetary vorticity gradient $\beta y$ and of a given stationary flow $\psi_d$ in the deep layer. We assume that this deep flow is known and unaffected by the dynamics of the upper layer. The streamfunction $\psi_d$ induces a permanent

---

[a]The term "relative" refers to the vorticity $\omega$ in the rotating frame.

deformation of the interface with respect to its horizontal position at rest[b]. This is why the deep flow acts as a topography on the active layer. The term $\beta y$ accounts at lowest order for the Earth's sphericity: the projection of the planet rotation vector on the local vertical axis is $f = 2\Omega \cdot \mathbf{e}_z = f_0 + \beta y$, with $f_0 = 2\Omega \sin \theta_0$, where $\theta_0$ is the mean latitude where the flow takes place, and $\beta = 2\Omega \cos \theta_0 / r_e$ with $r_e$ the planet's radius. This is called the beta-plane approximation because the term $\beta y$ appears as an effective topography in the quasi-geostrophic dynamics.

For the boundary conditions, two cases will be distinguished, depending on the domain geometry $\mathcal{D}$. In the case of a closed domain, there is an impermeability constraint (no flow across the boundary), which amounts to a constant streamfunction along the boundary. To simplify the presentation, the condition $\psi = 0$ at boundaries will be considered[c]. In the case of a zonal channel, the streamfunction $\psi$ is periodic in the $x$ direction, and the impermeability constraint applies on northern and southern boundaries. In the remaining two sections, length scales are nondimensionalized such that the domain area $|\mathcal{D}|$ is equal to one.

Because we consider here a model with one active layer above another layer, it is called a 1.5 layer quasi-geostrophic model, which is also sometimes referred to as the "equivalent barotropic model".

## 2.2. *Dynamical invariants and their consequences*

According to Noether's Theorem, each symmetry of the system is associated with the existence of a dynamical invariant, see e.g.[15] These invariants are crucial quantities, because they provide strong constraints for the flow evolution. Starting from (1), (2) and the aforementioned boundary conditions one can prove that quasi-geostrophic flows conserve the energy:

$$E = \frac{1}{2} \int_{\mathcal{D}} d\mathbf{r} \left[ (\nabla \psi)^2 + \frac{\psi^2}{R^2} \right] = -\frac{1}{2} \int_{\mathcal{D}} d\mathbf{r} \, (q - \eta_d) \, \psi, \qquad (3)$$

---

[b]A real topography $h(y)$ would correspond to $h = -f_0 \eta_d / H$ where $f_0$ is the reference planetary vorticity at the latitude under consideration and $H$ is the mean upper layer thickness. Due to the sign of $f_0$, the signs of $h$ and $\eta_d$ would be the same in the south hemisphere and opposite in the north hemisphere. As we will discuss extensively the Jovian south hemisphere vortices, we have chosen this sign convention for $\eta_d$. We will consider the effect of a real topography when discussing vertical energy transfers, in the last section.

[c]The physically relevant boundary condition should be $\psi = \psi_{fr}$ where $\psi_{fr}$ is determined by using the mass conservation constraint $\int d\mathbf{r} \, \psi = 0$ ($\psi$ is proportional to interface variations). Taking $\psi = 0$ does not change quantitatively the solutions in the domain bulk, but only the strength of boundary jets.

with $\mathbf{r} = (x, y)$. Additionally, the quasi-geostrophic dynamics (1) is a transport by an incompressible flow, so that the area $\gamma(\sigma)\, d\sigma$ occupied by a given vorticity level $\sigma$ is a dynamical invariant. The quantity $\gamma(\sigma)$ will be referred to as the global distribution of potential vorticity. The conservation of the distribution $\gamma(\sigma)$ is equivalent to the conservation of the Casimir's functionals $\int_{\mathcal{D}} d\mathbf{r}\, f(q)$, where $f$ is any sufficiently smooth function.

Depending on the properties of the domain geometry where the flow takes place, the dynamics might be characterized by additional symmetries. Each of these symmetry would imply the existence of an additional invariant.

A striking consequence of the previous conservation laws is the fact that energy remains at large scale in freely evolving quasi-geostrophic dynamics. This property makes *a priori* possible the formation and the persistence of long lived coherent structures. This contrast with three dimensional turbulence, where the direct energy cascade (toward small scale) would rapidly dissipate such structures. These long lived coherent structures are by definition steady (or quasi-steady) states of the dynamics. The stationary points of the quasi-geostrophic equations (1), referred to as *dynamical equilibrium states*, satisfy $\mathbf{v} \cdot \nabla q = \nabla \psi \times \nabla q = 0$. It means that dynamical equilibria are flows for which streamlines are isolines of potential vorticity. Then, any state characterized by a $q - \psi$ functional relationship is a dynamical equilibrium. More details on the consequence of the conservation laws in two-dimensional and geophysical turbulent flows can be found in.[12]

At this point, we need a theory i) to support the idea that the freely evolving flow dynamics will effectively self-organize into a dynamical equilibrium state ii) to determine the $q - \psi$ relationship associated with this dynamical equilibrium iii) to select the dynamical equilibria that are likely to be observed. This is the goal and the achievement of equilibrium statistical mechanics theory, presented in the next subsection.

### 2.3. *The equilibrium statistical mechanics of Robert–Sommeria–Miller (RSM)*

The RSM statistical theory initialy developed by[2–5] is introduced on a heuristic level in the following. There exist rigorous justifications of the theory see[12] for detailed discussions and further references.

A microscopic state is defined by its potential vorticity field $q(\mathbf{r})$. If taken as an initial condition, such a fine grained field would evolve toward a state with filamentation at smaller and smaller scales, while keeping in general a well defined large scale organization. Then, among all the possible fine grained states, an overwhelming number are characterized by these complicated small scale filamentary structures. This phenomenology gives a strong incentive for a mean-field approach, in which the flow is described at a coarse-grained level. For that purpose, the probability $\rho(\sigma, \mathbf{r})\mathrm{d}\sigma$ is introduced to measure a potential vorticity level $\sigma$ at a point $\mathbf{r} = (x, y)$. The probability density field $\rho$ defines a macroscopic state of the system. The corresponding averaged potential vorticity field, also referred to as *coarse-grained*, or *mean-field*, is $\overline{q}(\mathbf{r}) = \int_\Sigma \mathrm{d}\sigma\ \sigma\rho(\sigma, \mathbf{r})$, which is related to the streamfunction through $\overline{q} = \Delta\psi - \psi/R^2 + \eta_d$, and where $\Sigma = ]-\infty, +\infty[$.

Many microscopic states can be associated with a given macroscopic state. The cornerstone of the RSM statistical theory is the computation of the most probable state $\rho_{eq}$, that maximizes the Boltzmann-Gibbs (or mixing) entropy

$$S[\rho] \equiv - \int_\mathcal{D} \mathrm{d}\mathbf{r} \int_\Sigma \mathrm{d}\sigma\ \rho \log \rho\ , \tag{4}$$

while satisfying the constraints associated with each dynamical invariant. The mixing entropy (4) is a quantification of the number of microscopic states $q$ corresponding to a given macroscopic state $\rho$. The state $\rho_{eq}$ is not only the most probable one: an overwhelming number of microstates are effectively concentrated close to it.[17] This gives the physical explanation and the prediction of the large scale organization of the flow.

To compute statistical equilibria, the constraints must be expressed in term of the macroscopic state $\rho$:

- The local normalization $N[\rho](\mathbf{r}) \equiv \int_\Sigma \mathrm{d}\sigma\ \rho(\sigma, \mathbf{r}) = 1$,
- The global potential vorticity distribution $D_\sigma[\rho] \equiv \int_\mathcal{D} \mathrm{d}\mathbf{r}\ \rho(\sigma, \mathbf{r}) = \gamma(\sigma)$,
- The energy $\mathcal{E}[\rho] \equiv -\frac{1}{2} \int_\mathcal{D} \mathrm{d}\mathbf{r} \int_\Sigma \mathrm{d}\sigma\ \rho(\sigma - \eta_d)\psi = E$.

Because of the overwhelming number of states with only small scale fluctuations around the mean field potential vorticity, and because energy is a large scale quantity, contributions of these fluctuations to the total energy are negligible with respect to the mean-field energy.

The first step toward computations of RSM equilibria is to find critical points $\rho$ of the mixing entropy (4). In order to take into account the constraints, one needs to introduce the Lagrange multipliers $\zeta(\mathbf{r})$, $\alpha(\sigma)$, and $\lambda$ associated respectively with the local normalization, the conservation of the global vorticity distribution and of the energy. Critical points are solutions of:

$$\forall \, \delta\rho \quad \delta\mathcal{S} - \lambda\delta\mathcal{E} - \int_{\Sigma} \mathrm{d}\sigma \, \alpha\delta D_{\sigma} - \int_{\mathcal{D}} \mathrm{d}\mathbf{r} \, \zeta\delta N = 0 \, , \tag{5}$$

where first variations are taken with respect to $\rho$. This leads to $\rho = N \exp\left(\lambda\sigma\psi\left(\mathbf{r}\right) - \alpha(\sigma)\right)$ where N is determined by the normalization constraint $\left(\int \mathrm{d}\sigma \, \rho = 1\right)$. Statistical equilibria are dynamical equilibria characterized by a functional relation between potential vorticity and streamfunction:

$$\bar{q} = \frac{\int_{\Sigma} \mathrm{d}\sigma \, \sigma e^{\lambda\sigma\psi(\mathbf{r}) - \alpha(\sigma)}}{\int_{\Sigma} \mathrm{d}\sigma \, e^{\sigma\lambda\psi(\mathbf{r}) - \alpha(\sigma)}} = g\left(\psi\right). \tag{6}$$

It can be shown that $g$ is a monotonic and bounded function of $\psi$ for any global distribution $\gamma(\sigma)$ and energy $E$. These critical points can either be entropy minima, saddle or maxima. To find statistical equilibria, one needs then to select the entropy maxima.

At this point, two different approaches could be followed. The first one would be to consider a given small scale distribution $\gamma(\sigma)$ and energy $E$, and then to compute the statistical equilibria corresponding to these parameters. In practice, especially in the geophysical context, one does not have empirically access to the microscopic vorticity distribution, but rather to the $q - \psi$ relation (6) of the large scale flow. The second approach, followed in the remaining of these lectures, is to study statistical equilibria corresponding to a given class of $q - \psi$ relations.

## 3. Statistical equilibria and jet solutions, application to ocean rings and to the Great Red Spot of Jupiter

Most analytical results on the equilibrium states are obtained in cases where the $q - \psi$ relation is close to be linear, or equivalently in the limit of a quadratic Energy–Casimir functional, see e.g.[12] and references therein. More general solutions are very difficult to find analytically, and may require numerical computations, for instance using continuation algorithms.

There are however other limits where an analytical description becomes possible. This is for instance the case in the limit of large energies.[18] The second interesting limit is the limit of Rossby deformation radius $R$ much smaller than the size of the domain ($R \ll L$), where the nonlinearity of the vorticity-stream function relation becomes essential. This limit case and its applications to the description of coherent structures in geostrophic turbulence is the subject of this section.

In the limit $R \ll L$, the variational problems of the statistical theory are analogous to the Van Der Waals–Cahn Hilliard model that describes phase separation and phase coexistence in usual thermodynamics. The Van Der Waals–Cahn Hilliard model describes for instance the equilibrium of a bubble of a gas phase in a liquid phase, or the equilibria of soap films in air. For these classical problems, the essential concepts are the free energy per unit area, the related spherical shape of the bubbles, the Laplace equation relating the radius of curvature of the bubble with the difference in pressure inside and outside the bubble (see section 3.1), or properties of minimal surfaces (the Plateau problem). We will present an analogy between those concepts and the structures of quasi-geostrophic statistical equilibrium flows.

For these flows, the limit $R \ll L$ leads to interfaces separating phases of different free energies. In our case, each phase is characterized by a different value of average potential vorticity, and corresponds to sub-domains in which the potential vorticity is homogenized. The interfaces correspond to strong localized jets of typical width $R$. This limit is relevant for applications showing such strong jet structures.

From a geophysical point of view, this limit $R \ll L$ is relevant for describing some of Jupiter's features, like for instance the Great Red Spot of Jupiter (a giant anticyclone) (here $R \simeq 500 - 2000\,km$ and the length of the spot is $L \simeq 20,000\,km$) (see section 3.3).

This limit is also relevant to ocean applications, where $R$ is the internal Rossby deformation radius ($R \simeq 50\,km$ at mid-latitude). We will apply the results of statistical mechanics to the description of robust (over months or years) vortices such as ocean rings, which are observed around mid-latitude jets such as the Kuroshio or the Gulf Stream, and more generally in any eddying regions (mostly localized near western boundary currents) as the Aghulas current, the confluence region in the Argentinian basin or the Antarctic Circumpolar circulation (see section 3.4). The length $L$ can be considered as the diameters of those rings ($L \simeq 200\,km$).

We will also apply statistical mechanics ideas in the limit $R \ll L$ to the description of the large scale organization of oceanic currents (in inertial region, dominated by turbulence), such as the eastward jets like the Gulf Stream or the Kuroshio extension (the analogue of the Gulf Stream in the Pacific ocean). In that case the length $L$ could be thought as the ocean basin scale $L \simeq 5,000\,km$ (see section 4).

## 3.1. The Van der Waals–Cahn Hilliard model of first order phase transitions

We first describe the Van der Waals–Cahn Hilliard model. We give in the following subsections a heuristic description based on physical arguments. Some comments and references on the mathematics of the problem are provided in section 3.1.4.

This classical model of thermodynamics and statistical physics describes the coexistence of phase in usual thermodynamics. It involves the minimization of a free energy with a linear constraint:

$$\begin{cases} F = \min\left\{ \mathcal{F}\left[\phi\right] \mid \mathcal{A}\left[\phi\right] = -B \right\} \\ \text{with } \mathcal{F} = \int_{\mathcal{D}} \mathrm{d}\mathbf{r} \left[ \frac{R^2 (\nabla \phi)^2}{2} + f(\phi) \right] \text{ and } \mathcal{A}\left[\phi\right] = \int_{\mathcal{D}} \mathrm{d}\mathbf{r}\, \phi \end{cases} \quad (7)$$

where $\phi$ is the non-dimensional order parameter (for instance the non-dimensionality local density), and $f(\phi)$ is the non-dimensional free energy per unit volume. We consider the limit $R \ll L$ where $L$ is a typical size of the domain. We assume that the specific free energy $f$ has a double well shape (see Fig. 1), characteristic of a phase coexistence related to a first order phase transition. For a simpler discussion, we also assume $f$ to be even; this does not affect the properties of the solutions discussed bellow.

### 3.1.1. First order phase transition and phase separation

At equilibrium, in the limit of small $R$, the function $f(\phi)$ plays the dominant role. In order to minimize the free energy, the system will tend to reach one of its two minima (see Fig. 1). These two minima correspond to the value of the order parameters for the two coexisting phases, the two phases have thus the same free energy.

The constraint $\mathcal{A}$ (see Equation 7) is related to the total mass (due to the translation on $\phi$ to make $f$ even, it can take both positive and negative values). Without the constraint $\mathcal{A}$, the two uniform solutions $\phi = u$ or $\phi = -u$ would clearly minimize $\mathcal{F}$: the system would have only one phase. Because of the constraint $\mathcal{A}$, the system has to split into sub-domains: part of it with phase $\phi = u$ and part of it with phase $\phi = -u$. In a two dimensional space, the area occupied by each of the phases are denoted $A_+$ and $A_-$ respectively. They are fixed by the constraint $\mathcal{A}$ by the relations $uA_+ - uA_- = -B$ and by $A_+ + A_- = 1$ (where 1 is the total area). A sketch of a situation with two sub-domains each occupied by one of the two phases is provided in Fig. 2.

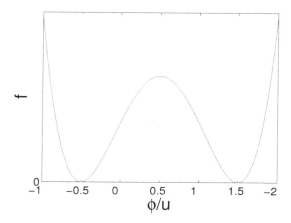

Figure 1.   The double well shape of the specific free energy $f(\phi)$ (see equation (7)). The function $f(\phi)$ is even and possesses two minima at $\phi = \pm u$. At equilibrium, at zeroth order in $R$, the physical system will be described by two phases corresponding to each of these minima.

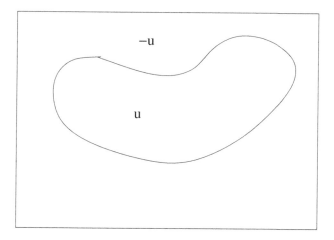

Figure 2.   At zeroth order, $\phi$ takes the two values $\pm u$ on two sub-domains $A_\pm$. These sub-domains are separated by strong jets. The actual shape of the structure, or equivalently the position of the jets, is given by the first order analysis.

Up to now, we have neglected the term $R^2 (\nabla\phi)^2$ in the functional (7). In classical thermodynamics, this term is related to non-local contributions to the free energy (proportional to the gradient rather than to only point-wise contributions). Moreover the microscopic interactions fix a length scale $R$ above which such non-local interactions become negligible. Usually for a macroscopic system such non-local interactions become negligible in the thermodynamic limit. Indeed as will soon become clear, this term gives finite volume or interface effects.

We know from observations of the associated physical phenomena (coarsening, phase separations, and so on) that the system has a tendency to form larger and larger sub-domains. We thus assume that such sub-domains are delimited by interfaces, with typical radius of curvature $r$ much larger than $R^{\mathrm{d}}$. Actually the term $R^2 \left(\nabla\phi\right)^2$ is negligible except on an interface of width $R$ separating the sub-domains. The scale separation $r \gg R$ allows to consider independently what happens in the transverse direction to the interface on the one hand and in the along interface direction on the other hand. As described in next sections, this explains the interface structure and interface shape respectively.

### 3.1.2. *The interface structure*

At the interface, the value of $\phi$ changes rapidly, on a scale of order $R$, with $R \ll r$. What happens in the direction along the interface can thus be neglected at leading order. To minimize the free energy (7), the interface structure $\phi(\zeta)$ needs thus to minimize a one dimensional variational problem along the normal to the interface coordinate $\zeta$

$$F_{int} = \min \left\{ \int \mathrm{d}\zeta \left[ \frac{R^2}{2} \left(\frac{d\phi}{d\zeta}\right)^2 + f(\phi) \right] \right\}. \tag{8}$$

Dimensionally, $F_{int}$ is a free energy $F$ divided by a length. It is the free energy per unit length of the interface.

We see that the two terms in (8) are of the same order only if the interface has a typical width of order $R$. We rescale the length by $R$: $\zeta = R\tau$. The Euler-Lagrange equation of (8) gives

$$\frac{d^2\phi}{d\tau^2} = \frac{df}{d\phi}. \tag{9}$$

This equation is a very classical one. For instance making an analogy with mechanics, if $\phi$ would be a particle position, $\tau$ would be the time, Equation (9) would describe the conservative motion of the particle in a potential

---

$^{\mathrm{d}}$This can indeed be proved mathematically, see section 3.1.4

$V = -f$. From the shape of $f$ (see Fig. 1) we see that the potential has two bumps (two unstable fixed points) and decays to $-\infty$ for large distances. In order to connect the two different phases in the bulk, on each side of the interface, we are looking for solutions with boundary conditions $\phi \to \pm u$ for $\tau \to \pm\infty$. It exists a unique trajectory with such limit conditions: in the particle analogy, it is the trajectory connecting the two unstable fixed points (homoclinic orbit).

This analysis shows that the interface width scales like $R$. Moreover, after rescaling the length, one clearly sees that the free energy per length unit (8) is proportional to $R$: $F_{int} = eR$, where $e > 0$ could be computed as a function of $f$.[19,20]

### 3.1.3. *The interface shape: an isoperimetrical problem*

In order to determine the interface shape, we come back to the free energy variational problem (7). In the previous section, we have determined the transverse structure of the interface, by maximizing the one dimensional variational problem (8). We have discussed the quantity $F_{int} = Re$, a free energy per unit length, which is the unit length contribution of the interface to the free energy. The total free energy is thus

$$\mathcal{F} = eRL, \tag{10}$$

where we have implicitly neglected contributions of relative order $R/r$, where $r$ is the curvature radius of the interface.

In order to minimize the free energy (10), we thus have to minimize the length $L$. We must also take into account that the areas occupied by the two phases, $A_+$ and $A_-$ are fixed, as discussed in section 3.1.1. We thus look for the curve with the minimal length, that bounds a surface with area $A_+$

$$\min\{eRL \,|\, \text{Area} = A_+\}. \tag{11}$$

This type of problem is called an isoperimetrical problem. In three dimensions, the minimization of the area for a fixed volume leads to spherical bubbles or plane surface if the boundaries does not come into play. When boundaries are involved, the interface shape is more complex (it is a minimal surface- or Plateau- problem). This can be illustrated by nice soap films experiments, as may be seen in very simple experiments or in many science museums. Here, for our two dimensional problem, it leads to circle or straight lines, as we now prove.

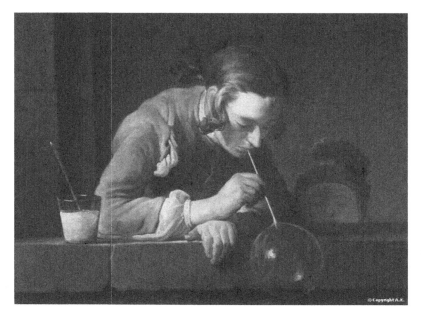

Figure 3.   Illustration of the Plateau problem (or minimal area problem) with soap films: the spherical bubble minimizes its area for a given volume (Jean Simeon Chardin, *Les bulles de savon*, 1734).

It is a classical exercise of variational calculus to prove that the first variations of the length of a curve is proportional to the inverse of its curvature radius $r$. The solution of the problem (11) then leads to

$$\frac{eR}{r} = \alpha, \tag{12}$$

where $\alpha$ is a Lagrange parameter associated with the conservation of the area. This proves that $r$ is constant along the interface: solutions are either circles or straight lines. The law (12) is the equivalent of the Laplace law in classical thermodynamics, relating the radius of curvature of the interface to the difference of pressure inside and outside of the bubble[e].

We have thus shown that the minimization of the Van Der Waals–Cahn Hilliard functional, aimed at describing statistical equilibria for first order

---

[e]Indeed, at next order, the Lagrange parameter $\alpha$ leads to a slight imbalance between the two phase free energy, which is related to a pressure difference for the two phases. This thus gives the relation between pressure imbalance, radius of curvature and free energy per unit length (or unit surface in the 3D case).

phase transitions, predicts phase separation (formation of sub-domains with each of the two phases corresponding to the two minima of the free energy). It predicts the interface structure and that its shape is described by an isoperimetrical problem: the minimization of the length for a fixed enclosed area. Thus equilibrium structures are either bubbles (circles) or straight lines. In the following sections, we see how this applies to the description of statistical equilibria for quasi-geostrophic flows, describing vortices and jets.

### 3.1.4. *The mathematics of the Van Der Waals–Cahn Hilliard problem*

The study of the Van Der Waals–Cahn Hilliard functional (7) was a mathematical challenge during the 1980s. It's solution has followed from the analysis in the framework of spaces of functions with bounded variations, and on results from semi-local analysis. One of the main contributions to this problem was achieved by.[21] This functional analysis study proves the assumptions of the heuristic presentation given in the previous subsections: $\phi$ takes the two values $\pm u$ in sub-domains separated by transition area of width scaling with $R$.

As a complement to these mathematical works, a more precise asymptotic expansion based on the heuristic description above, generalizable at all order in $R$, with mathematical justification of the existence of the solutions for the interface equation at all order in $R$, is provided in.[22] Higher order effects are also discussed in this work.

## 3.2. *Quasi-geostrophic statistical equilibria and first order phase transitions*

The first discussion of the analogy between statistical equilibria in the limit $R \ll L$ and phase coexistence in usual thermodynamics, in relation with the Van Der Waals–Cahn Hilliard model is given in.[19,22] This analogy has been recently put on a more precise mathematical ground, by proving that the variational problems of the RSM statistical mechanics and the Van Der Waals–Cahn Hilliard variational problem are indeed related.[23] More precisely, any solution to the variational problem:

$$\begin{cases} F = \min \{ \mathcal{F}[\phi] \mid \mathcal{A}[\phi] = -B \} \\ \text{with} \quad \mathcal{F} = \int_{\mathcal{D}} d\mathbf{r} \left[ \frac{R^2 (\nabla \phi)^2}{2} + f(\phi) - R\phi h \right] \quad \text{and} \quad \mathcal{A}[\phi] = \int_{\mathcal{D}} d\mathbf{r}\, \phi \end{cases} \quad (13)$$

where $\psi = R^2 \phi$ is a RSM equilibria of the quasi-geostrophic equations (1).

Considering the problem (13), using a part integration and the relation $q = R^2 \Delta \phi - \phi + Rh$ yields

$$\delta \mathcal{F} = \int d\mathbf{r} \; (f'(\phi) - \phi - q) \, \delta \phi \quad \text{and} \quad \delta \mathcal{A} = \int d\mathbf{r} \; \delta \phi. \qquad (14)$$

Critical points of (13) are therefore solutions of $\delta \mathcal{F} - \alpha \delta \mathcal{A} = 0$, for all $\delta \phi$, where $\alpha$ is the Lagrange multiplier associated with the constraint $\mathcal{A}$. These critical points satisfy

$$q = f' \left( \frac{\psi}{R^2} \right) - \frac{\psi}{R^2} - \alpha.$$

We conclude that this equation is the same as (6), provided that $f' \left( \frac{\psi}{R^2} \right) = g(\beta \psi) + \frac{\psi}{R^2} - \alpha$.

In the case of an initial distribution $\gamma$ with only two values of the potential vorticity: $\gamma(\sigma) = |\mathcal{D}| \, (a\delta(\sigma_1) + (1 - a)\delta(\sigma_2))$, only two Lagrange multipliers $\alpha_1$ and $\alpha_2$ are needed, associated with $\sigma_1$ and $\sigma_2$ respectively, in order to compute $g$, equation (6). In that case, the function $g$ is exactly tanh function. There exists in practice a much larger class of initial conditions for which the function $g$ would be an increasing function with a single inflexion point, similar to a tanh function, especially when one considers the limit of small Rossby radius of deformation.[19,20] give physical arguments to explain why it is the case for Jupiter's troposphere or oceanic rings and jets.

When $g$ is a tanh-like function, the specific free energy $f$ has a double well shape, provided that the inverse temperature $\beta$ is negative, with sufficiently large values.

### 3.2.1. Topography and anisotropy

The topography term $\eta_d = Rh\,(y)$ in (13) is the main difference between the Van Der Waals–Cahn Hilliard functional (7) and the quasi-geostrophic variational problem (13). We recall that this term is due to the beta plane approximation and a prescribed motion in a lower layer of fluid (see section 2.1). This topographic term provides an anisotropy in the free energy. Its effect will be the subject of most of the theoretical discussion in the following sections.

Since we suppose that this term scales with $R$, the topography term will not change the overall structure at leading order: there will still be phase separations in sub-domains, separated by an interface of typical width $R$, as discussed in section 3.1. We now discuss the dynamical meaning of this overall structure for the quasi-geostrophic model.

### 3.2.2. *Potential vorticity mixing and phase separation*

In the case of the quasi-geostrophic equations, the order parameter $\phi$ is proportional to the stream function $\psi$: $\psi = R^2\phi$. At equilibrium, there is a functional relation between the stream function $\psi$ and the macroscopic potential vorticity $q$, given by Eq. (6). Then the sub-domains of constant $\phi$ are domains where the (macroscopic) potential vorticity $q$ is also constant. It means that the level of mixing of the different microscopic potential vorticity levels are constant in those sub-domains. We thus conclude that the macroscopic potential vorticity is homogenized in sub-domains that corresponds to different phases (with different values of potential vorticity), the equilibrium being controlled by an equality for the associated mixing free energy.

### 3.2.3. *Strong jets and interfaces*

In section 3.1.3, we have described the interface structure. The order parameter $\phi$ varies on a scale of order $R$ mostly in the normal to the interface direction, reaching constant values far from the interface. Recalling that $\phi$ is proportional to $\psi$, and that $\mathbf{v} = \mathbf{e}_z \wedge \nabla\psi$, we conclude that:

(1) The velocity field is nearly zero far from the interface (at distances much larger than the Rossby deformation radius $R$). Non zero velocities are limited to the interface areas.
(2) The velocity is mainly directed along the interface.

These two properties characterize strong jets. In the limit $R \ll L$, the velocity field is thus mainly composed of strong jets of width $R$, whose path is determined from an isoperimetrical variational problem.

## 3.3. *Application to Jupiter's Great Red Spot and other Jovian features*

Most of Jupiter's volume is gas. The visible features on this atmosphere, cyclones, anticyclones and jets, are concentrated on a thin outer shell, the troposphere, where the dynamics is described by similar equations to the ones describing the Earth's atmosphere.[24,25] The inner part of the atmo-

Figure 4. Observation of the Jovian atmosphere from Cassini (Courtesy of NASA/JPL-Caltech). One of the most striking feature of the Jovian atmosphere is the self organization of the flow into alternating eastward and westward jets, producing the visible banded structure and the existence of a huge anticyclonic vortex $\sim 20,000\ km$ wide, located around 20 South: the Great Red Spot (GRS). The GRS has a ring structure: it is a hollow vortex surrounded by a jet of typical velocity $\sim 100\ m.s^{\{-1\}}$ and width $\sim 1,000\ km$. Remarkably, the GRS has been observed to be stable and quasi-steady for many centuries despite the surrounding turbulent dynamics. The explanation of the detailed structure of the GRS velocity field and of its stability is one of the main achievement of the equilibrium statistical mechanics of two dimensional and geophysical flows (see Fig. 3.3 and section 3).

sphere is a conducting fluid, and the dynamics is described by Magneto-hydrodynamics (MHD) equations.

The most simple model describing the troposphere is the 1-1/2 quasi-geostrophic model, described in section 2.1. This simple model is a good one for localized mid latitude dynamics. Many classical work have used it to model Jupiter's features, taking into account the effect of a prescribed steady flow in a deep layer acting like an equivalent topography $h(y)$. We emphasize that there is no real bottom topography on Jupiter.

Some works based on soliton theory aimed at explaining the structure and stability of the Great Red Spot. However, none of these obtained a velocity field qualitatively similar to the observed one, which is actually a strongly non-linear structure. Structures similar to the Great Red Spot have been observed in a number of numerical simulations, but without reproducing in a convincing way both the characteristic annular jet structure of the velocity field and the shape of the spot. Detailed observations and fluid mechanics analysis described convincingly the potential vorticity structure and the dynamical aspects of the Great Red Spot (see[24–26] and references therein). The potential vorticity structure is a constant vorticity inside the spot surrounded by a gentle shear outside, which gives a good fluid mechanics theory.[26] In this section we explain this potential vorticity structure thanks to statistical mechanics. Statistical mechanics provides also more detailed, and analytical theory of the shape of Jupiter vortices.

The explanation of the stability of the Great Red Spot of Jupiter using the statistical mechanics of the quasi-geostrophic model is cited by nearly all the papers from the beginning of the Robert–Sommeria–Miller theory. Some equilibria having qualitative similarities with the observed velocity field have been computed in.[27] The theoretical study in the limit of small Rossby deformation radius, especially the analogy with first order phase transitions[19,28] gave the theory presented below: an explanation of the detailed shape and structure and a quantitative model. These results have been extended to the shallow-water model.[29][30] argued on the explanation of the position of the Great Red Spot based on statistical mechanics equilibria.

We describe in the following the prediction of equilibrium statistical mechanics for the quasi-geostrophic model with topography. We start from the Van Der Waals–Cahn Hilliard variational problem in presence of small topography (13), recalling that its minima are statistical equilibria of the quasi-geostrophic model (see section 3.2).

The Rossby deformation radius at the Great Red Spot latitude is evaluated to be of order of $500 - 2000$ km, which has to be compared with the size of the spot: $10,000 \, X \, 20,000$ km. This is thus consistent with the limit

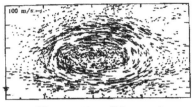

Observation (Voyager)            Statistical Equilibrium

Figure 5. Left: the observed velocity field is from Voyager spacecraft data, from Dowling and Ingersoll (1988) ; the length of each line is proportional to the velocity at that point. Note the strong jet structure of width of order $R$, the Rossby deformation radius. Right: the velocity field for the statistical equilibrium model of the Great Red Spot. The actual values of the jet maximum velocity, jet width, vortex width and length fit with the observed ones. The jet is interpreted as the interface between two phases; each of them corresponds to a different mixing level of the potential vorticity. The jet shape obeys a minimal length variational problem (an isoperimetrical problem) balanced by the effect of the deep layer shear.

$R \ll L$ considered in the description of phase coexistence within the Van Der Waals–Cahn Hilliard model (section 3.1), even if the criteria $r \ll R$ is only marginally verified where the curvature radius $r$ of the jet is the larger.

In the limit of small Rossby deformation radius, the entropy maxima for a given potential vorticity distribution and energy, are formed by strong jets, bounding areas where the velocity is much smaller. Figure 3.3 shows the observation of the Great Red Spot velocity field, analyzed from cloud tracking on spacecraft pictures. The strong jet structure (the interface) and phase separation (much smaller velocity inside and outside the interface) is readily visible. The main difference with the structure described in the previous section is the shape of the vortex: it is not circular as was predicted in the case without topography or with a linear topography. We consider the effect of a more general topography in the next section.

### 3.3.1. Determination of the vortex shape: the typical elongated shape of Jupiter's features

In order to determine the effect of topography on the jet shape, we consider again the variational problem (13). We note that the topography $\eta_d = Rh$ has been rescaled by $R$ in the term $Rh(y)\phi$ appearing in the variational problem. This corresponds to a regime where the effect of the topography is of the same order as the effect of the jet free energy. Two other

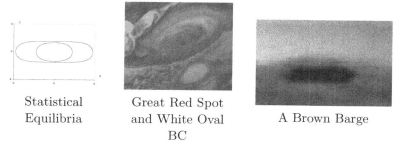

|  |  |  |
|:---:|:---:|:---:|
| Statistical Equilibria | Great Red Spot and White Oval BC | A Brown Barge |

Figure 6.   Left panel: typical vortex shapes obtained from the isoperimetrical problem (curvature radius equation (12)), for two different values of the parameters (arbitrary units). The characteristic properties of Jupiter's vortex shapes (very elongated, reaching extremal latitude $y_m$ where the curvature radius is extremely large) are well reproduced by these results. Central panel: the Great Red Spot and one of the White Ovals. Right panel: one of the Brown Barge cyclones of Jupiter's north atmosphere. Note the very peculiar cigar shape of this vortex, in agreement with statistical mechanics predictions (left panel).

regimes exist: one for which topography would have a negligible impact (this would lead to circular vortices, as treated in section 3.2) and another regime where topography would play the dominant role. This last regime may be interesting in some cases, but we do not treat it in this review.

Due to the scaling $Rh\phi$, the topography does not play any role at zeroth order. We thus still conclude that phase separation occurs, with subdomains of areas $A_+$ and $A_-$ fixed by the potential vorticity constraint (see section 3.1.1), separated by jets whose transverse structure is described in section 3.1.3. The jet shape is however given by minimization of the free energy contributions of order $R$. Let us thus compute the first order contribution of the topography term $RH = \int_{\mathcal{D}} d\mathbf{r} \, (-R\phi h(y))$. For this we use the zeroth order result $\phi = \pm u$. We then obtain $H = -u \int_{A_+} d\mathbf{r} \, h + u \int_{A_-} d\mathbf{r} \, h = H_0 - 2u \int_{A_+} d\mathbf{r} \, h$, where $H_0 \equiv u \int_{\mathcal{D}} d\mathbf{r} \, h$. We note that $H_0$ does not depend on the jet shape.

Adding the contribution of the topography to the jet free energy (10), we obtain the first order expression for the modified free energy functional

$$\mathcal{F} = RH_0 + R \left( eL - 2u \int_{A_+} d\mathbf{r} \, h(y) \right), \qquad (15)$$

which is valid up to correction of order $e\,(R/r)$ and of order $R^2 H$. We recall that the total area $A_+$ is fixed. We see that, in order to minimize the free energy, the new term tends to favor as much as possible the phase

$A_+$ with positive values of stream function $\phi = u$ (and then negative values of potential vorticity $q = -u$) to be placed on topography maxima. This effect is balanced by the length minimization.

In order to study in more details the shape of the jet, we look at the critical points of the minimization of (15), with fixed area $A_+$. Recalling that first variations of the length are proportional to the inverse of the curvature radius, we obtain

$$2uRh(y) + \alpha = \frac{eR}{r}, \tag{16}$$

where $\alpha$ is a Lagrange parameter associated with the conservation of the area $A_+$. This relates the vortex shape to the topography and parameters $u$ and $e$. From this equation, one can write the equations for $X$ and $Y$, the coordinates of the jet curve. These equations derive from a Hamiltonian, and a detailed analysis allows to specify the initial conditions leading to closed curves and thus to numerically compute the vortex shape.[19]

Figure 6 compares the numerically obtained vortex shapes, with the Jovian ones. This shows that the solution to equation (16) has the typical elongated shape of Jovian vortices, as clearly illustrated by the peculiar cigar shape of Brown Barges, which are cyclones of Jupiter's north troposphere. We thus conclude that statistical mechanics and the associated Van Der Waals–Cahn Hilliard functional with topography explain well the shape of Jovian vortices.

Figure 7 shows a phase diagram for the statistical equilibria, with Jupiter like topography and Rossby deformation radius. This illustrates the power of statistical mechanics: with only few parameters characterizing statistical equilibria (here the energy $E$ and a parameter related to the asymmetry between positive and negative potential vorticity $B$), we are able to reproduce all the features of Jupiter's troposphere, from circular white ovals, to the GRS and cigar shaped Brown Barges. The reduction of the complexity of turbulent flow to only a few order parameters is the main interest and achievement of a statistical mechanics theory.

Moreover, as seen on Fig. 7, statistical mechanics predicts a phase transition from vortices towards straight jets. The concept of phase transition is an essential one in complex systems, as the qualitative physical properties of the system drastically change at a given value of the control parameters.

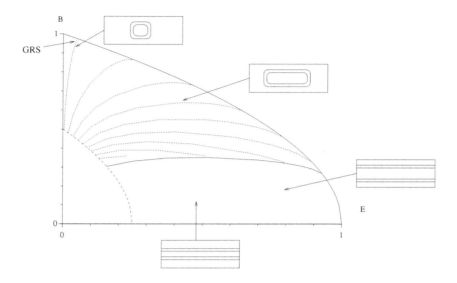

Figure 7. Phase diagram of the statistical equilibrium states versus the energy $E$ and a parameter related to the asymmetry between positive and negative potential vorticity $B$, with a quadratic topography. The inner solid line corresponds to a phase transition, between vortex and straight jet solutions. The dash line corresponds to the limit of validity of the small deformation radius hypothesis. The dot lines are constant vortex aspect ratio lines with values 2,10,20,30,40,50,70,80 respectively. We have represented only solutions for which anticyclonic potential vorticity dominate ($B > 0$). The opposite situation may be recovered by symmetry. For a more detailed discussion of this figure, the precise relation between $E$, $B$ and the results presented in this lecture, please see Bouchet Dumont.

### 3.3.2. *Quantitative comparisons with Jupiter's Great Red Spot*

In the preceding section, we have made a rapid description of the effect of a topography to first order phase transitions. We have obtained and compared the vortex shape with Jupiter's vortices. A much more detailed treatment of the applications to Jupiter and to the Great Red Spot can be found in.[19,28] The theory can be extended in order to describe the small shear outside of the spot (first order effect on $\phi$ outside of the interface), on the Great Red Spot zonal velocity with respect to the ambient shear, on the typical latitudinal extension of these vortices. A more detailed description of physical considerations on the relations between potential vorticity distribution and forcing is also provided in.[19,28]

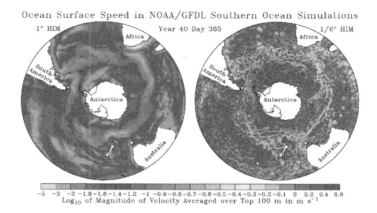

Ocean Surface Speed in NOAA/GFDL Southern Ocean Simulations

Year 40 Day 365

Log$_{10}$ of Magnitude of Velocity Averaged over Top 100 m in m s$^{-1}$

Figure 8. Snapshot of surface velocity field from a comprehensive numerical simulation of the southern Oceans (Hallberg et all 2006). Left: coarse resolution, the effect of mesoscale eddies ($\sim 100 km$) is parameterized. Right: higher resolution, without parameterization of mesoscale eddies. Note the formation of large scale coherent structure in the high resolution simulation: there is either strong and thin eastward jets or rings of diameter $\sim 200 \ km$. Typical velocity and width of jets (be it eastward or around the rings) are respectively $\sim 1 \ m.s^{-1}$ and $\sim 20 \ km$. They give a statistical mechanics explanation and model for these rings.

### 3.4. Application to ocean rings

Application of equilibrium statistical mechanics to the description of oceanic flows is a long-standing problem, starting with the work of Salmon–Holloway–Hendershott in the framework of energy-enstrophy theory.[31]

Another attempt to apply equilibrium statistical mechanics to oceanic flows had been performed by[32,33] in the framework of the Heton model of[34] for the self-organization phenomena following deep convection events, by numerically computing statistical equilibrium states of a two-layer quasi-geostrophic model.

None of these previous approaches have explained the ubiquity of oceanic rings. We show in the following that such rings can actually be understood as statistical equilibria by similar arguments that explain the formation of Jovian vortices (see[20] for more details).

#### 3.4.1. Rings in the oceans

The ocean has long been recognized as a sea of eddies. This has been first inferred from *in situ* data by Gill, Green and Simmons in the early

1970s.[35] During the last two decades, the concomitant development of altimetry and realistic ocean modeling has made possible a quantitative description of those eddies. The most striking observation is probably their organization into westward propagating rings of diameters ($L_e \sim 200\ km$), as for instance seen in Fig. 8. In that respect, they look like small Jovian Great Red Spots.

Those eddies plays a crucial role for the general ocean circulation and its energy cycle, since their total energy is one order of magnitude above the kinetic energy of the mean flow.

Those rings are mostly located around western boundary currents, which are regions characterized by strong baroclinic instabilities[f], such as the Gulf Stream, the Kuroshio, the Aghulas currents below South Africa or the confluence region of the Argentinian basin, as seen on Figs. 8 and 11. The rings can also propagate far away from the regions where they are created.

Most of those rings have a baroclinic structure, i.e. a velocity field intensified in the upper layer ($H \sim 1\ km$) of the oceans. This baroclinic structure suggest that the 1.5 layer quasi-geostrophic model introduced in the previous sections is relevant to this problem. The horizontal scale of the rings ($L_e \sim 200\ km$) are larger than the width $R \sim 50\ km$ of the surrounding jet, of typical velocities $U = 1\ m.s^{-1}$.

The organization of those eddies into coherent rings can be understood by the same statistical mechanics arguments that have just been presented in the case of Jupiter's Great Red Spot. The rings correspond to one phase containing most of the potential vorticity extracted from the mean flow by baroclinic instability, while the surrounding quiescent flow corresponds to the other phase. This statistical mechanics approach, the only one to our knowledge to describe the formation of large scale coherent structures, might then be extremely fruitful to account for the formation of such rings. It remains an important open question concerning the criteria that select the size of such coherent structures. This is an ongoing subject of investigation.

---

[f]When the mean flow present a sufficiently strong vertical shear, baroclinic instabilities release part of the available potential energy associated with this mean flow, which is generally assumed to be maintained by a large scale, low frequency forcing mechanism such as surface wind stress or heating.[16]

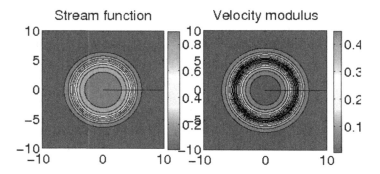

Figure 9.    Vortex statistical equilibria in the quasi-geostrophic model. It is a circular patch of (homogenized) potential vorticity in a background of homogenized potential vorticity, with two different mixing values. The velocity field (right panel) has a very clear ring structure, similarly to the Gulf-Stream rings and to many other ocean vortices. The width of the jet surrounding the ring has the order of magnitude of the Rossby radius of deformation $R$.

### 3.4.2.  *The westward drift of the rings*

In this section, we consider the consequences of the beta effect (see section 2.1), which corresponds to linear topography $\eta_d = \beta y$ in (2). We prove that this term can be easily handled and that it actually explains the westward drift of oceanic rings with respect to the mean surrounding flow.

We consider the quasi-geostrophic equations on a domain which is invariant upon a translation along the $x$ direction (either an infinite or a periodic channel, for instance). Then the quasi-geostrophic equations are invariant over a Galilean transformation in the $x$ direction. We consider the transformation $\mathbf{v}' = \mathbf{v} + V\mathbf{e}_x$, where $\mathbf{v}$ is the velocity in the original frame of reference and $\mathbf{v}'$ is the velocity in the new Galilean frame of reference.

From the relation $\mathbf{v} = \mathbf{e}_z \wedge \nabla\psi$, we obtain the transformation law for $\psi$: $\psi' = \psi - Vy$ and from the expression $q = \Delta\psi - \psi/R^2 + \beta y$ (2) we obtain the transformation law for $q$: $q' = q + Vy/R^2$. Thus the expression for the potential vorticity in the new reference frame is

$$q = \Delta\psi - \frac{\psi}{R^2} + \left(\beta + \frac{V}{R^2}\right) y.$$

From this last expression, we see that a change of Galilean reference frame translates as a beta effect in the potential vorticity. Moreover, in a reference frame moving at velocity $-\beta R^2 \mathbf{e}_x$, the $\beta$ effect is exactly canceled out.

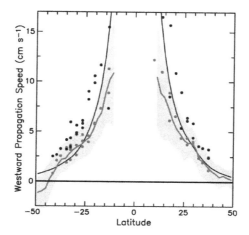

Figure 10.   Altimetry observation of the westward drift of oceanic eddies (including rings) (Chelton et al, 2007), Fig. 4. The red line is the zonal average (along a latitude circle) of the propagation speeds of all eddies with life time greater than 12 weeks. The black line represents the velocity $\beta R^2$ where $\beta$ is the meridional gradient of the Coriolis parameter and $R$ the first baroclinic Rossby radius of deformation. This eddy propagation speed is a prediction of statistical mechanics.

From this remark, we conclude that taking into account the beta effect, the equilibrium structures should be the one described by the minimization of the Van Der Waals–Cahn Hilliard variational problem, but moving at a constant westward speed $V = \beta R^2$. A more rigorous treatment of the statistical mechanics for the quasi-geostrophic model with translational invariance would require to take into account an additional conserved quantity, the linear momentum, which would lead to the same conclusion: statistical equilibria are rings with a constant westward speed $V = \beta R^2$. See also[20] for more details and discussions on the physical consequences of this additional constraint.

This drift is actually observed for the oceanic rings, see for instance Fig. 10.

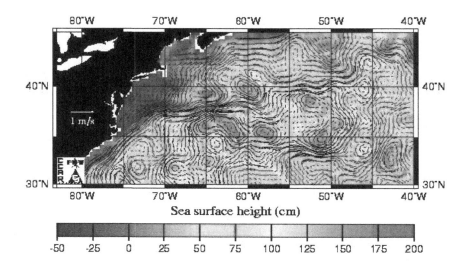

Figure 11.  Observation of the sea surface height of the north Atlantic ocean (Gulf
Stream area) from altimetry (Topex-Poseidon). For geophysical flows, the surface velocity field can be inferred from the see surface height (SSH): strong gradient of SSH are
related to strong jets. The Gulf stream appears as a robust eastward jet (in presence of
meanders), flowing along the east coast of north America and then detaching the coast
to enter the Atlantic ocean, with an extension $L \sim 2000\ km$. The jet is surrounded by
numerous westward propagating rings of typical diameters $L \sim 200\ km$. Typical velocities and widths of both the Gulf Stream and its rings jets are respectively $1\ m.s^{-1}$ and
$50\ km$, corresponding to a Reynolds number $Re \sim 10^{11}$. Such rings can be understood
as local statistical equilibria, and strong eastward jets like the Gulf Stream and obtained
as marginally unstable statistical equilibria in simple academic models (see subsections
3.4-4).

## 4.  Are the Gulf-Stream and the Kuroshio currents close to statistical equilibria?

In section 3.4, we have discussed applications of statistical mechanics ideas
to the description of ocean vortices, like the Gulf-Stream rings. We have
also mentioned that statistical equilibria, starting from the Van Der Waals–
Cahn Hilliard functional (13), may model physical situations where strong
jets, with a width of order $R$, bound domains of nearly constant potential
vorticity.

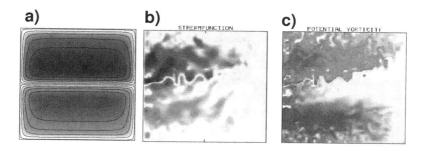

Figure 12.  b) and c) represent respectively a snapshot of the streamfunction and potential vorticity (red: positive values; blue: negative values) in the upper layer of a three layers quasi-geostrophic model in a closed domain, representing a mid-latitude oceanic basin, in presence of wind forcing. Both figures are taken from numerical simulations by P. Berloff. a) Streamfunction predicted by statistical mechanics, see section 4 for further details. Even in an out-equilibrium situation like this one, the equilibrium statistical mechanics predicts correctly the overall qualitative structure of the flow.

This is actually the case of the Gulf Stream in the North Atlantic ocean or of the Kuroshio extension in the North Pacific ocean. This can be inferred from observations, or this is observed in high resolution numerical simulations of idealized wind driven mid-latitude ocean, see for instance Fig. 12 (see[36] for more details).

It is thus very tempting to interpret the Gulf Stream and the Kuroshio as interfaces between two phases corresponding to different levels of potential vorticity mixing, just like the Great Red Spot and ocean rings in the previous section. The aim of this chapter is to answer this natural question: are the Gulf-Stream and Kuroshio currents close to statistical equilibria?

More precisely, we address the following problem: is it possible to find a class of statistical equilibria with a strong mid-basin eastward jet similar to the Gulf Stream of the Kuroshio, in a closed domain? The 1-1/2 layer quasi-geostrophic model (see section 2.1) is the simplest model taking into account density stratification for mid-latitude ocean circulation in the upper first $1000\,m$.[16,37] We analyze therefore the class of statistical equilibria which are minima of the Van Der Waals–Cahn Hilliard variational problem (13), as explained in section 3.2. We ask whether it exists solutions to

$$
\begin{cases}
F = \min\left\{ \mathcal{F}\left[\phi\right] \mid A\left[\phi\right] = -B \right\} \\
\text{with } \mathcal{F} = \int_{\mathcal{D}} d\mathbf{r} \left[ \frac{R^2 (\nabla\phi)^2}{2} + f(\phi) - R\tilde{\beta}y\phi \right] \text{ and } \mathcal{A}\left[\phi\right] = \int_{\mathcal{D}} d\mathbf{r}\,\phi
\end{cases}
\quad (17)
$$

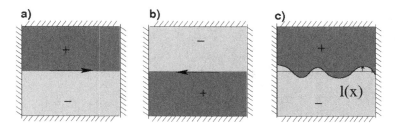

Figure 13. a) Eastward jet: the interface is zonal, with positive potential vorticity $q = u$ on the northern part of the domain. b) Westward jet: the interface is zonal, with negative potential vorticity $q = -u$ in the northern part of the domain. c) Perturbation of the interface for the eastward jet configuration, to determine when this solution is a local equilibrium (see subsection 4.2). Without topography, both (a) and (b) are entropy maxima. With positive beta effect (b) is the global entropy maximum; with negative beta effect (a) is the global entropy maximum.

in a bounded domain (let say a rectangular basin) with strong mid-basin eastward jets. At the domain boundary, we fix $\phi = 0$ (which using $\phi = R^2 \psi$ turns out to be an impermeability condition). We note that the understanding of the following discussion requires the reading of sections 4.1 to 4.3.

The term $R\tilde{\beta}y$ is an effective topography including the beta effect and the effect of a deep zonal flow (see section 2.1). Its significance and effects will be discussed in section 5.2. As in the previous section, we consider the limit $R \ll L$ and assume $f$ be a double well function.

As discussed in chapter 3.1, with these hypothesis, there is phase separation in two subdomains with two different levels of potential vorticity mixing. These domains are bounded by interfaces (jets) of width $R$. In view of the applications to mid-basin ocean jets, we assume that the area $A_+$ occupied by the value $\phi = u$ is half of the total area of the domain (this amounts to fix the total potential vorticity $\Gamma_1$). The question is to determine the position and shape of this interface. The main difference with the cases treated in subsection 3.1 is due to the effect of boundaries and of the linear effective topography $R\tilde{\beta}y$.

## 4.1. Eastward jets are statistical equilibria of the quasi-geostrophic model without topography

The value $\phi = \pm u$ for the two coexisting phases is not compatible with the boundary condition $\phi = 0$. As a consequence, there exists a boundary jet (or boundary layer) in order to match a uniform phase $\phi = \pm u$ to the boundary conditions. Just like inner jets, treated in section 3, these jets contribute to the first order free energy, which gives the jet position and shape. We now treat the effect of boundary layer for the case $h = 0$ ($\widetilde{\beta} = 0$ in this case). As explained in section 3.1.3, the jet free energy is the only contribution to the total free energy.

We first quantify the unit length free energy, $F_b$, for the boundary jets. Following the reasoning of section 3.1.3, we have

$$F_b = \min \left\{ \int d\zeta \left[ \frac{R^2}{2} \frac{d^2\phi}{d\zeta^2} + f(\phi) \right] \right\}.$$

This expression is the same as (8), the only difference is the different boundary conditions: it was $\phi \to_{\zeta \to +\infty} u$ and $\phi \to_{\zeta \to -\infty} -u$, it is now $\phi \to_{\zeta \to +\infty} u$ and $\phi(0) = 0$. Because $f$ is even, one easily see that a boundary jet is nothing else than half of a interior domain jet. Then

$$F_b = \frac{1}{2} F_{int} = \frac{e}{2} R,$$

where $F_{int}$ and $e$ are the unit length free energies for the interior jets, as defined in section 3.1.3. By symmetry, a boundary jet matching the value $\phi = -u$ to $\phi = 0$ gives the same contribution[g]. Finally, the first order free energy is given by

$$\mathcal{F} = eR \left( L + \frac{L_b}{2} \right),$$

where $L_b$ is the boundary length. Because the boundary length $L_b$ is a fixed quantity, the free energy minimization amounts to the minimization of the interior jet length. The interior jet position and shape is thus given by the minimization of the interior jet length with fixed area $A_+$. We recall that the solutions to this variational problem are interior jets which are either straight lines or circles (see section 3.1.3).

In order to simplify the discussion, we consider the case of a rectangular domain of aspect ratio $\tau = L_x/L_y$. Generalization to an arbitrary closed domain could also be discussed. We recall that the two phases occupy the same area $A_+ = A_- = \frac{1}{2} L_x L_y$. We consider three possible interface configurations with straight or circular jets:

(1) the zonal jet configuration (jet along the $x$ axis) with $L = L_x$,
(2) the meridional jet configuration (jet along the $y$ axis with $L = L_y$,
(3) and an interior circular vortex, with $L = 2\sqrt{\pi A_+} = \sqrt{2\pi L_x L_y}$ .

---

[g]We have treated the symmetric case when $f$ is even. The asymmetric case could be also easily treated.

The minimization of $L$ for these three configurations shows that the zonal jet is a global minimum if and only if $\tau < 1$. The criterion for the zonal jet to be a global RSM equilibrium state is then $L_x < L_y$. We have thus found zonal jet as statistical equilibria in the case $h = 0$.

An essential point is that both the Kuroshio and the Gulf Stream are flowing eastward (from west to east). From the relation $\mathbf{v} = \mathbf{e}_z \times \nabla \psi$, we see that the jet flows eastward ($v_x > 0$) when $\partial_y \psi < 0$. Recalling that $\phi = R^2 \psi$, the previous condition means that the negative phase $\phi = -u$ has to be on the northern part of the domain, and the phase $\phi = u$ on the southern part. From (2), we see that this corresponds to a phase with positive potential vorticity $q = u$ on the northern sub-domains and negative potential vorticity $q = -u$ on the southern sub-domain, as illustrated in the panel (a) of Fig. (13).

Looking at the variational problems (17), it is clear that in the case $\widetilde{\beta} = 0$, the minimization of $\phi$ is invariant over the symmetry $\phi \to -\phi$. Then solutions with eastward or westward jets are completely equivalent. Actually there are two equivalent solutions for each of the case 1, 2 and 3 above. However, adding a beta effect $h = R\widetilde{\beta}y$ will break this symmetry. This is the subject of next section.

We conclude that in a closed domain with aspect ratio $L_x/L_y < 1$, without topography, equilibrium states exist with an eastward jet at the center of the domain, recirculating jets along the domain boundary and a quiescent interior. For $L_x/L_y > 1$, these solutions become metastable states (local entropy maximum). This equilibrium is degenerated, since the symmetric solution with a westward jet is always possible.

### 4.2. *Addition of a topography*

For ocean dynamics, the beta effect plays a crucial role. Let us now consider the case where the topography is $\eta_d = \beta y + \frac{\psi_d}{R^2}$. The first contribution comes from the beta-effect (the variation of the Coriolis parameter with latitude). The second contribution is a permanent deviation of the interface between the upper layer and the lower layer. For simplicity, we consider the case where this permanent interface elevation is driven by a constant zonal flow in the lower layer: $\psi_d = -U_d y$, which gives $\eta_d = \left(\beta - \frac{U_d}{R^2}\right) y = R\widetilde{\beta}y$. Then the combined effect of a deep constant zonal flow and of the variation of the Coriolis parameter with latitude is an effective linear beta effect.

In the definition of $\widetilde{\beta}$ above, we use a rescaling with $R$. This choice is considered in order to treat the case where the contribution of the effective beta effect appears at the same order as the jet length contribution. This allows to easily study how the beta effect breaks the symmetry $\phi \to -\phi$ between eastward and westward jets. Following the arguments of section 3.3.1, we minimize

$$\mathcal{F} = RH_0 + R\left(eL - 2u\int_{A_+}\mathrm{d}\mathbf{r}\,\tilde{\beta}y\right), \tag{18}$$

(see equation (15)), with a fixed area $A_+$. The jet position is a critical point of this functional: $e/r - 2u\tilde{\beta}y_{jet} = \alpha$ (see equation (16)), where $\alpha$ is a Lagrange parameter and $y_{jet}$ the latitude of the jet. We conclude that zonal jets (curves with constant $y_{jet}$ and $r = +\infty$) are solutions to this equation for $\alpha = -2uR\tilde{\beta}y_{jet}$. Eastward and westward jets described in the previous section are still critical points of entropy maximization.

### 4.2.1. *With a negative effective beta effect, eastward jets are statistical equilibria*

We first consider the case $\tilde{\beta} < 0$. This occurs when the zonal flow in the lower layer is eastward and sufficiently strong ($U_d > R^2\beta$). If we compute the first order free energy (18) for both the eastward and the westward mid-latitude jet, it is easy to see that in order to minimize $\mathcal{F}$, the domain $A_+$ has to be located at the lower latitudes: taking $y = 0$ at the interface, the term $-2u\int_{A_+}d^2\mathbf{r}\,\tilde{\beta}y = u\tilde{\beta}L_xL_y/4$ gives a negative contribution when the phase with $\phi = u$ (and $q = -u$) is on the southern part of the domain ($A_+ = (0, L_x) \times (-\frac{L_y}{2}, 0)$). This term would give the opposite contribution if the phase $\phi = u$ would occupy the northern part of the domain. Thus the statistical equilibria is the one with negative streamfunction $\phi$ (corresponding to positive potential vorticity $q$) on the northern part of the domain. As discussed in the end of section 4.1 and illustrated on Fig. 13, panel (b), this is the case of an eastward jet.

Thus, we conclude that taking into account an effective negative beta-effect term at first order breaks the westward-eastward jet symmetry. When $\tilde{\beta} < 0$, statistical equilibria are flows with mid-basin eastward jets.

### 4.2.2. *With a positive effective beta effect, westward jets are statistical equilibria*

Let us now assume that the effective beta coefficient is positive. This is the case when $U_d < R^2\beta$, i.e. when the lower layer is either flowing westward, or eastward with a sufficiently low velocity. The argument of the previous

paragraph can then be used to show that the statistical equilibrium is the
solution presenting a westward jet.

### 4.2.3. With a sufficiently small effective beta coefficient, eastward jets are local statistical equilibria

We have just proved that mid-basin eastward jets are not global equilibria in
the case of positive effective beta effect. They are however critical points of
entropy maximization. They still could be local entropy maxima. We now
consider this question: are mid-basin strong eastward jets local equilibria
for a positive effective beta coefficient? In order to answer, we perturb
the interface between the two phases, while keeping constant the area they
occupy, and compute the free energy perturbation.

The unperturbed interface equation is $y = 0$, the perturbed one is $y = l(x)$, see Fig. 13. Qualitatively, the contributions to the free energy $\mathcal{F}$
(18), of the jet on one hand and of the topography on the other hand,
are competing with each other. Any perturbation increases the jet length
$L = \int dx \sqrt{1 + \left(\frac{dl}{dx}\right)^2}$ and then increases the second term in equation (18)
by $\delta \mathcal{F}_1 = Re \int dx \, (dl/dx)^2$. Any perturbation decreases the third term in
equation (18) by $\delta \mathcal{F}_2 = -2Ru\tilde{\beta} \int dx \, l^2$.

We suppose that $l = l_k \sin \frac{k\pi}{L_x} x$ where $k \geq 1$ is an integer. Then

$$\delta \mathcal{F} = \delta \mathcal{F}_1 + \delta \mathcal{F}_2 = -2u\tilde{\beta} + e \left(\frac{k\pi}{L_x}\right)^2 .$$

Because we minimize $\mathcal{F}$, we want to know if any perturbation leads to
positive variations of the free energy. The most unfavorable case is for the
smallest value of $k^2$, i.e. $k^2 = 1$. Then we conclude that eastward jets are
local entropy maxima when

$$\tilde{\beta} < \tilde{\beta}_{cr} = \frac{1}{2} \frac{e}{u} \frac{\pi^2}{L_x^2} .$$

We thus conclude that eastward zonal jets are local equilibria for sufficiently
small values of $\tilde{\beta}$.

The previous result can also be interpreted in terms of the domain
geometry, for a fixed value of $\tilde{\beta}$. Eastward jets are local entropy maxima if

$$L_x < L_{x,cr} = \pi \sqrt{\frac{e}{2u\tilde{\beta}_{cr}}}.$$

Let us evaluate an order of magnitude for $L_{x,cr}$ for the ocean case, first assuming there is no deep flow ($U_d = 0$). Then $R\tilde{\beta}$ is the real coefficient of the beta plane approximation. Remembering that a typical velocity of the jet is $U \sim uR$, and using $e \sim u^2$ (see[20] for more details). Then $L_{x,cr} \approx \pi\sqrt{\frac{U}{\beta_{cr}}}$. This length is proportional to the Rhine's' scale of geophysical fluid dynamics.[16] For jets like the Gulf Stream, typical jet velocity is 1 $m.s^{-1}$ and $\beta \sim 10^{-11}$ $m^{-1}.s^{-1}$ at mid-latitude. Then $L_{cr} \sim 300$ $km$. This length is much smaller than the typical zonal extension of the inertial part of the Kuroshio or Gulf Stream currents. We thus conclude that in a model with a quiescent lower layer and the beta plane approximation, currents like the Gulf Stream or the Kuroshio are not statistical equilibria, and they are not neither close to local statistical equilibria.

Taking the oceanic parameters ($\beta = 10^{-11}$ $m^{-1}s^{-1}$, $R \sim 50$ $km$), we can estimate the critical eastward velocity in the lower layer $U_{d,cr} = 5$ $cm$ $s^{-1}$ above which the strong eastward jet in the upper layer is a statistical equilibria. It is difficult to make further conclusions about real mid-latitude jets; we conjecture that their are marginally stable. This hypothesis of marginal stability is in agreement with the observed instabilities of the Gulf-Stream and Kuroshio current, but overall stability of the global structure of the flow. A further discussion of these points will be the object of future works.

In all of the preceding considerations, we have assumed that the term $R\tilde{\beta}$ was of order $R$ in dimensionless units. This is self-consistent to compute the unstable states. To show that a solution is effectively a statistical equilibria when $R\tilde{\beta}$ in of order one, one has to use much less straightforward considerations than in the preceding paragraphs, but the conclusions would be exactly the same.

## 4.3. *Conclusion*

We have shown that when there is a sufficiently strong eastward flow in the deep layer (i.e. when $U_d > U_{d,cr}$ with $U_{d,cr} = R^2\beta_{cr}$), ocean mid-latitude eastward jets are statistical equilibria, even in presence of a beta plane. When the flow in the deep layer is lower than the critical value $U_{d,cr}$ but still almost compensate the beta plane ($0 < \beta - \frac{U_d}{R^2} < \frac{1}{2}\frac{e}{u}\frac{\pi^2}{L_x^2}R$), the solutions with the eastward jets are local equilibria (metastable states). When $\beta - \frac{U_d}{R^2} > \frac{1}{2}\frac{e}{u}\frac{\pi^2}{L_x^2}R$ the solution with an eastward jet are unstable.

We have also concluded that the inertial part of the real Gulf-Stream or of the Kuroshio extension are likely to be marginally stable from a statistical mechanics point of view.

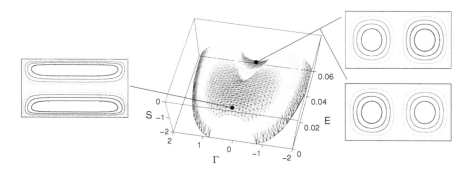

Figure 14.    Phase diagrams of RSM statistical equilibrium states of the 1.5 layer quasi-geostrophic model, characterized by a linear $q - \psi$ relationship, in a rectangular domain elongated in the $x$ direction. $S(E, \Gamma)$ is the equilibrium entropy, $E$ is the energy and $\Gamma$ the circulation. Low energy states are the celebrated Fofonoff solutions, presenting a weak westward flow in the domain bulk. High energy states have a very different structure (a dipole). Please note that at high energy the entropy is non-concave. This is related to ensemble inequivalence, which explain why such states were not computed in previous studies (Please see Venaille and Bouchet, 2009, for more details).

The statistical equilibria that we have described in this section have a flow structure that differs notably from the celebrated Fofonoff solution.[38]

The Fofonoff solution is a stationary state of the quasi-geostrophic equations (1-2) on a beta plane ($\eta_d = \beta y$) obtained by assuming a linear relationship between potential vorticity and streamfunction ($q = a\psi$), in the limit $a + R^{-2} \gg L^{-2}$, where $L$ is the domain size. In this limit, the Laplacian term in (2) is negligible in the domain bulk. Then $\psi \approx \beta/(a + R^{-2})y$, which corresponds to a weak westward flow, as illustrated in Fig. 14. Strong re-circulating eastward jets occur at northern and southern boundaries, where the Laplacian term is no more negligible.

The original work of Fofonoff was carried independently from statistical mechanics considerations. The linear $q - \psi$ relationship was chosen as a starting point to compute analytically the flow structure. Because both the Salmon–Holloway–Hendershott statistical theory[31] and the Bretherton–Haidvoguel minimum enstrophy principle[39] did predict a linear relationship between vorticity and streamfunction, it has been argued that statistical equilibrium theory predicts the emergence of the classical Fofonoff flows, which had effectively been reported in numerical simulations of freely decaying barotropic flows on a beta plane for some range of parameters.[40]

It is shown in[23] that all those theories are particular cases of the RSM statistical mechanics theory. On the one hand it has been actually proven that the classical Fofonoff solutions are indeed RSM statistical equilibria in the limit of low energies.[41] On the other hand, as illustrated by the results of this section, there exists a much richer variety of RSM equilibrium states than the sole classical Fofonoff solution. Even in the case of a linear $q - \psi$ relation, high energy statistical equilibrium states are characterized by a flow structure that differs notably from the original Fofonoff solution, as illustrated Fig. 14. These high energy states correspond actually to the RSM equilibrium states of the Euler equation, originally computed by.[42] The transition from classical Fofonoff solutions to those high energy states has been related the the occurrence of ensemble inequivalence.[43] This explains also why such high energy states have not been reported in earlier studies, where computations were always performed in the (unconstrained) canonical ensemble, see[41] for more details.

The early work of Fofonoff and the equilibrium statistical mechanics of geophysical flows presented in this review are often referred to as the inertial approach of oceanic circulation, meaning that the effect of the forcing and the dissipation are neglected.

Ocean dynamics is actually much influenced by the forcing and the dissipation. For instance the mass flux of a current like the Gulf Stream is mainly explained by the Sverdrup transport. Indeed in the bulk of the ocean, a balance between wind stress forcing and beta effect (the Sverdrup balance) lead to a meridional global mass flux (for instance toward the south on the southern part of the Atlantic ocean. This fluxes is then oriented westward and explain a large part of the Gulf Stream mass transport. This mechanism is at the base of the classical theories for ocean dynamics, see e.g.[37] Because it is not an conservative process, the inertial approach does not take this essential aspect into account. Conversely, the traditional theory explains the Sverdrup transport, the westward intensification and boundary current, but gives no clear explanation of the structure of the inertial part of the current: the strongly eastward jets.

Each of the classical ocean theory or of the equilibrium statistical mechanics point of view give an incomplete picture, and complement each other. Another interesting approach consider the dynamics from the point of view of bifurcation theory when the Reynolds number (or some other controlled parameters) are increased. These three types of approaches seem complimentary and we hope they may be combined in the future in a more comprehensive non-equilibrium theory.

## 5. The vertical structure of geostrophic turbulence

In the previous sections, we assumed that the dynamics took place in an upper active layer above a lower layer at rest. Vertical energy transfers were therefore neglected. The vertical structure of the oceanic mesoscale (from 50 to 500 km) is actually a fundamental problem in geophysical fluid dynamics, one that has has been reinvigorated by the need to interpret altimetric observation of surface velocity fields.[44,45]

It is widely accepted that the energy of oceanic mesoscale currents is mostly injected at the surface of the oceans.[46] Indeed, the primary source of geostrophic turbulence is mostly baroclinic instability, extracting turbulent energy from the potential energy reservoir set at the basin scale by large scale wind patterns,[35] and involving surface-intensified unstable modes, see e.g.[47]

It leads to the following question: what is the vertical structure of a three dimensional quasi-geostrophic flow in the presence of surface forcing? A first step to tackle this problem is to consider a simpler problem without forcing and dissipation: does an initially surface-intensified flow remain trapped at the surface or does it spread on the vertical? Here we address this issue in the framework of freely-evolving stratified quasi-geostrophic turbulence, which allows for theoretical analysis with equilibrium statistical mechanics.

We introduce in the next subsection the continuously stratified quasi-geostrophic dynamics, and explain how to compute equilibrium states in a low energy limit.

We examine in a second subsection what are the precise conditions for the oft-cited barotropization process to occur. Barotropization refers to the tendency of a quasi-geostrophic flow to reach a depth-independant flow.[48,49] We study in particular the key role played by the beta effect (the existence of planetary vorticity gradients) in such barotropization processes.

Finally, we show in a third subsection that the formation of bottom trapped-flow in the presence of bottom-topography may be accounted for by statistical mechanics arguments.

### 5.1. *Continuously stratified quasi-geostrophic flows*

Continuously stratified quasi-geostrophic flows take place in three dimensions, but their dynamics is quasi two-dimensional because the non-divergent advecting velocity field has only horizontal components, and can

be described by a streamfunction $\psi(x, y, z, t)$. Such flows are stably stratified with a prescribed buoyancy profile $N(z)$ above a topographic anomaly
$h_b(x, y)$, and are strongly rotating at a rate $f_0/2$. In the absence of forcing
and dissipation, the dynamics is expressed as the advection of potential
vorticity $q(x, y, z, t)$ (see e.g.,[16] section 5.4):

$$\partial_t q + \mathbf{v} \cdot \nabla q = 0 , \quad \mathbf{v} = \mathbf{e}_z \times \nabla \psi \tag{19}$$

$$q = \Delta \psi + \frac{\partial}{\partial z} \left( \frac{f_0^2}{N^2} \frac{\partial}{\partial z} \psi \right) + \beta y , \tag{20}$$

where $\Delta$ is the horizontal Laplacian, and where we have considered the
beta plane approximation (see section 2). The boundary condition at the
bottom ($z = -H$, where $H$ is now the averaged ocean depth) is given by

$$\left. \frac{f_0}{N^2} \partial_z \psi \right|_{z=-H} = -h_b , \tag{21}$$

The boundary condition at the surface (defined as $z = 0$, using the rigid
lid approximation), is given by the advection of buoyancy

$$\partial_t b_s + \mathbf{v}|_{z=0} \cdot \nabla b_s = 0 , \quad \left. \frac{f_0^2}{N^2} \partial_z \psi \right|_{z=0} = b_s. \tag{22}$$

The upper boundary condition (22) can be formally replaced by the condition of no buoyancy variation ($\partial_z \psi = 0$ at $z = 0$), provided that surface
buoyancy anomalies are interpreted as a thin sheet of potential vorticity
just below the rigid lid.[50] For this reason, and without loss of generality,
we will consider that $b_s = 0$ in the remainder of this course. In the following we will consider the case of a square doubly-periodic domain $\mathcal{D}$, and
we choose to adimensionalize length such that the domain length is $2\pi$, so
that the streamfunction is $2\pi$-periodic in the $x, y$ direction.

The dynamics admits similar conservation laws as the one-layer quasi-
geostrophic model considered in sections 2-3-4. Dynamical invariants include the total (kinetic plus potential) energy

$$\mathcal{E} = \frac{1}{2} \int_{-H}^{0} dz \int_{\mathcal{D}} dx dy \left[ (\nabla \psi)^2 + \frac{f_0^2}{N^2} (\partial_z \psi)^2 \right] , \tag{23}$$

and the Casimir functionals $\mathcal{C}_g(z)[q] = \int_{\mathcal{D}} dx dy \, g(q)$ where $g$ is any continuous function. These conservation laws have important physical con-
sequences. In particular, there is an inverse energy cascade that leads to
the formation of robust, large scale coherent structures filling the domain

in which the flow takes place. It is shown in[51] that the vertical structure of the large scale flow organization resulting from inviscid, freely evolving continuously stratified quasi-geostrophic dynamics can be predicted using equilibrium statistical mechanics. We sum up the main results in the next subsections.

### 5.1.1. *Equilibrium states characterized by a linear $q - \psi$ relation*

Let us call $E_0 = \mathcal{E}(q_0)$ and $Z_0(z) = (1/2) \int_{\mathcal{D}} \mathrm{d}x\mathrm{d}y \, q_0^2$ the energy and enstrophy, respectively, of the initial condition given by $q_0(x, y, z)$. Using a general result of,[23] it is shown in,[51] Appendix 1, that in the *low energy limit*, the calculation of RSM equilibrium states amounts to finding the minimizer $\bar{q}_{min}$ of the "total macroscopic enstrophy"

$$\mathcal{Z}_{cg}^{tot} [\bar{q}] = \frac{1}{2} \int_{-H}^{0} \mathrm{d}z \int_{\mathcal{D}} \mathrm{d}x\mathrm{d}y \, \frac{\bar{q}^2}{Z_0} \qquad (24)$$

among all the fields $\bar{q}$ satisfying the energy constraint

$$\mathcal{E}[\bar{q}] = \frac{1}{2} \int_{\mathcal{D}} \mathrm{d}x\mathrm{d}y \, f_0 h_b \psi|_{z=-H} - \frac{1}{2} \int_{\mathcal{D}} \mathrm{d}x\mathrm{d}y \int_{-H}^{0} \mathrm{d}z \, (\bar{q} - \beta y) \psi = E_0, \quad (25)$$

The variational problem (24-25) can be seen as a generalization to the stratified case of the phenomenological minimum enstrophy principle of.[39]

Critical states of the variational problem (24-25) are computed by introducing Lagrange multiplier $\beta_t$ associated with the energy constraint, and by solving $\delta \mathcal{Z}_{cg}^{total} + \beta_t \delta \mathcal{E} = 0$,, leading to the linear relation $\bar{q} = \beta_t Z_0 \psi$. The next step is to find which of these critical states are actual minimizers of the macroscopic enstrophy for a given energy. We perform these computations in various cases of geophysical interest in the following subsections.

## 5.2. *The presence of a beta plane favors barotropization*

### 5.2.1. *Equilibrium states without topography and without beta effect*

We consider here the case $h_b = 0$ and $\beta = 0$. Injecting $\bar{q} = \beta_t Z_0 \psi$ in (20) and projecting on Fourier modes yields

$$\frac{\partial}{\partial z} \left( \frac{f_0^2}{N^2} \frac{\partial}{\partial z} \widehat{\psi}_{k,l} \right) = (\beta_t Z_0 + K^2) \, \widehat{\psi}_{k,l} \,, \qquad (26)$$

with

$$\partial_z \widehat{\psi}_{k,l}|_{z=0,-H} = 0, \quad K^2 = k^2 + l^2, \quad \psi = \sum_{k,l} \widehat{\psi}_{k,l} \exp \left( 2i\pi \left( kx + ly \right) \right).$$

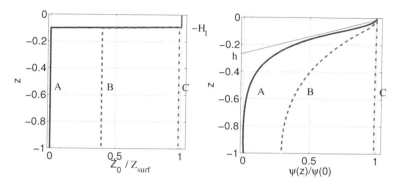

Figure 15. Left panel: three different vertical profiles of the microscopic enstrophy. Right panel: corresponding vertical structure of statistical equilibrium states ($\psi(z)/\psi(0)$ on the left panel), in the case of constant stratification ( $f_0^2/N^2 = 0.1$). The $e$-folding depth in case A is $h = f_0/NK$ (with here $K = 1$ for the statistical equilibrium state).

We see that each critical point is characterized by a given wavenumber modulus $K$. Its vertical structure and the corresponding value of $\beta_t$ must be computed numerically in the case of arbitrary profiles $Z_0(z)$. Let us consider the example shown in Fig. 15, for a two-step microscopic enstrophy profile

$$Z_0 = Z_{\text{surf}} \Theta \left( z + H_1 \right) + Z_{\text{int}} \Theta \left( -z - H_1 \right), \quad H_1 \ll H, \qquad (27)$$

where $\Theta$ is the Heaviside function, and for $Z_{\text{int}}/Z_{\text{surf}}$ varying between 0 and 1. We find that the minimum macroscopic enstrophy states are always characterized by the gravest horizontal mode on the horizontal ($K = 1$). As for the vertical structure, we observe Fig. 15 a tendency toward more barotropic flows when the ratio $Z_{\text{int}}/Z_{\text{surf}}$ tends to one. One can actually show that the equilibrium state is barotropic when $Z_{\text{int}}/Z_{\text{surf}} = 1$, and that the equilibrium state is surface intensified with $e$-folding depth $h = f_0/N(0)K$ when $Z_{\text{int}}/Z_{\text{surf}} = 0$ and $H_1 \ll H$, see.[51]

These examples show the importance of the conservation of microscopic enstrophy $Z_0(z)$ to the vertical structure of the equilibrium state. The main result is that statistical mechanics predicts a tendency for the flow to reach the gravest Laplacian mode on the horizontal ($K = 1$). The vertical structure associated with this state is fully prescribed by solving (26) with $K = 1$. Because the barotropic component of such flows are larger than solutions of (26) with $K > 1$, we can say that the inverse cascade on the horizontal is associated with a tendency to reach the gravest vertical mode compatible with the vertical microscopic enstrophy profile $Z_0$. This means a tendency toward barotropization, although in general, the fact that the profile $Z_0$ is non constant prevents complete barotropization.

### 5.2.2. *Including the $\beta$-effect*

For a given initial condition $\psi_0(x, y, z)$, increasing $\beta$ increases the contribution of the (depth independent) available potential enstrophy defined as

$$Z_p = \beta^2 \int_{\mathcal{D}} \mathrm{d}x\mathrm{d}y \; y^2, \tag{28}$$

to the total microscopic enstrophy profile $Z_0(z) = \int_{\mathcal{D}} \mathrm{d}x\mathrm{d}y \; q_0^2$, where $q_0$ is the initial potential vorticity field that can be computed by injecting $\psi_0$ in (20). For sufficiently large values of $\beta$, the potential vorticity field is dominated by the beta effect ($q_0 \approx \beta y$), $Z_0$ therefore tends to $Z_p$ and becomes depth independent. Because statistical equilibria computed in the previous subsection were fully barotropic when the microscopic enstrophy $Z_0$ was depth-independent, we expect a tendency toward barotropization by increasing $\beta$.

### 5.2.3. *Numerical experiments*

We consider in this section the final state organization of an initial surface intensified flow, varying the values of $\beta$. The initial potential vorticity field is $q_0 = q_{\mathrm{surf}}(x, y)\Theta(z + H_1) + \beta y$, such that $q_0 = \beta y$ in the interior $(-H < z < -H_1)$ and $q_0 \approx q_{\mathrm{surf}}$ in a surface layer $z > -H_1$. The surface potential vorticity $q_{\mathrm{surf}}(x, y)$ is a random field with random phases in spectral space, and a Gaussian power spectrum peaked at wavenumber $K_0 = 5$, with variance $\delta K_0 = 2$, and normalized such that the total energy is equal to one ($E_0 = 1$).

We perform simulations of the dynamics by considering a vertical discretization with 10 layers of equal depth, horizontal discretization of $512^2$, $H = 1$, and $F = (Lf_0/HN)^2 = 1$, using a pseudo-spectral quasi-geostrophic model.[52] We choose $H_1 = H/10$ for the initial condition, so that there is non zero enstrophy only in the upper layer in the absence of a beta effect.

The case with $\beta = 0$ is presented in the left panel of Fig. 16. An inverse cascade in the horizontal leads to flow structures with an horizontal wavenumber $K$ decreasing with time, associated with a tendency toward barotropization: the $e$-folding depth of the surface-intensified flow increases

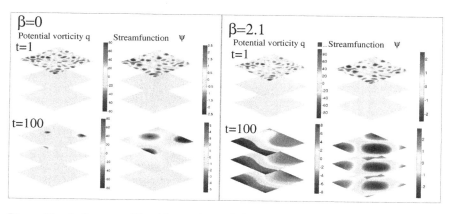

Figure 16. Left: case without beta plane. Only the fields in upper, middle and lower layer are shown. The end-state is surface-trapped. Right: case with a beta plane; the initial streamfunction is the same as on the left. The end-state is almost depth-independant. Note that the interior "beta plane" is not clearly visible in the upper panel of potential vorticity because the color scale is different than in the lower panel.

as $f_0/2NK$. The concomitant horizontal inverse cascade (most of the kinetic energy is in the gravest horizontal mode $K = 1$ at the end of the simulation) and the increase of the $e$-folding depth are observed on Fig. 16, showing good qualitative agreement between statistical mechanics and numerical simulations.

We now switch on the beta effect, with the same initial surface-intensified streamfunction $\psi_0(x, y, z)$. As a consequence, the contribution of the depth independent part of the microscopic enstrophy increases, which means a tendency toward a more barotropic equilibria, according to the previous subsection. This is what is actually observed in the final state organization of Fig. 16 in the presence of beta effect. This result reflects the fact that in physical space, the initial surface-intensified flow stirs the interior potential vorticity field (initially a beta plane), which in turn induces an interior flow, which stirs even more the interior potential vorticity field, and so on. We conclude that in this regime, the beta effect is a catalyst of barotropization, as predicted by statistical mechanics.

## 5.3. *The formation of bottom-trapped flows*

We saw in the previous section that the existence of planetary vorticity gradients (beta effect) provide a depth-independant source of microscopic

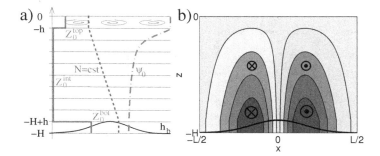

Figure 17.   a) Sketch of the flow configuration. The continuous red line represents the initial microscopic enstrophy profile (here $Z_0^{int} = 0$). The dashed blue line represents the density profile, and the dashed-dotted green line represents the streamfunction amplitude shape, which is initially surface-intensified. The thick continuous black line represents bottom topography. b) Vertical slice of the meridional velocity field $v$ of the statistical equilibrium state in the low energy limit.

enstrophy that favors barotropization. Another source of microscopic potential vorticity would be provided by the addition of bottom topography. This should in this case play against barotropization, since the topography induces potential microscopic enstrophy in the lower layer only. In fact, an initially surface intensified flow may evolve towards a bottom trapped current above the topographic anomaly, which can be explained by the statistical mechanics arguments presented above.

Bottom-intensified flows are commonly observed along topographic anomalies in the ocean. A striking example is given by the Zapiola anticyclone, a strong recirculation about 500 km wide that takes place above a sedimentary bump in the Argentine Basin, where bottom-intensified velocities of order 0.1 m.s$^{-1}$ have been reported from *in situ* measurements[53] and models.[54]

Phenomenological arguments for the formation of bottom-trapped flows were previously given by[55] in a forced-dissipated case. A complementary point of view was given by,[56] who computed critical states of equilibrium statistical mechanics for truncated dynamics. He observed that some of these states were bottom intensified in the presence of topography.[57] showed how to find the equilibrium states among these critical states, how they depend on the initial microscopic enstrophy profile, and provided numerical evidence of the spontaneous self-organization into bottom-trapped flows. We summarize in the following the main results.

We consider the configuration of Fig. 17-a: the stratification is linear in the bulk ($N$ is constant for $-H + h < z < -h$), and homogeneous in two layers of thickness $h \ll H$ at the top and at the bottom, where $N = 0^+$. In these upper and lower layers, the streamfunction is depth independent, denoted by $\psi^{top}(x, y, t)$ and $\psi^{bot}(x, y, t)$, respectively. In these layers, the dynamics is then fully described by the advection of the vertical average of the potential vorticity fields, denoted by $q^{top}(x, y, t)$ and $q^{bot}(x, y, t)$. The interior potential vorticity field is denoted by $q^{int}(x, y, z, t)$. For a given field $q^{top}, q^{int}, q^{bot}$, the streamfunction is obtained by inverting the following equations:

$$q^{top} - \beta y = \Delta \psi^{top} - \frac{f_0^2}{hN^2} \frac{\partial}{\partial z} \psi \bigg|_{z=-h} , \tag{29}$$

$$q^{bot} - \beta y - f_0 \frac{h_b}{h} = \Delta \psi^{bot} + \frac{f_0^2}{hN^2} \frac{\partial}{\partial z} \psi \bigg|_{z=h-H} , \tag{30}$$

$$q^{int} - \beta y = \Delta \psi + \frac{f_0^2}{N^2} \frac{\partial^2}{\partial z^2} \psi \quad \text{for } -H + h < z < -h , \tag{31}$$

$$\psi^{top} = \psi(x, y, -h), \ \psi^{bot} = \psi(x, y, -H + h). \tag{32}$$

Equations (29–30) are obtained by averaging Eq. (20) in the vertical direction in the upper and the lower layers, respectively, and by using the boundary condition (21). In the following, the initial condition is a surface-intensified velocity field induced by a perturbation of the potential vorticity field confined in the upper layer:

$$q_0^{top} = \beta y + q_0^{pert} , \quad q_0^{int} = \beta y , \quad q_0^{bot} = \frac{f_0}{h} h_b + \beta y. \tag{33}$$

It is assumed in the following that $\beta y \ll q_0^{pert} \ll f_0 h_b/h$. The potential vorticity fields are therefore associated with microscopic enstrophies $Z_0^{int} \ll Z_0^{top} \ll Z_0^{bot}$. The macroscopic enstrophy minimizers of this configuration are computed in the Appendix of[57] by solving the variational problem (24–25). The main result is that for a fixed topography, in the *low energy limit*, the equilibrium streamfunction is a bottom-intensified quasi-geostrophic flow such that bottom streamlines follow contours of topography with positive correlations, see Fig. 17-b.

The initial condition of Fig. 18-b is the the same surface-intensified velocity field as the one used in Fig. 16. After a few eddy turnover times, the enstrophy of the upper layer has cascaded towards small scales as shown

*F. Bouchet and A. Venaille*

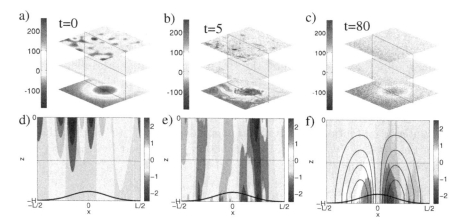

Figure 18. a),b),c) Potential vorticity field at three successive times. In each panel, only layers 1 (top), 5 (middle) and 10 (bottom) are represented. b),d),f) Vertical slices of the meridional velocity fields $v$ taken at the center of the domain ($y = 0$), and associated with the potential vorticity fields given in panels d),e),f), respectively. The bold continuous dark line represents bottom topography. The continuous black contours of panel f) give the structure of the statistical equilibrium state in the low energy (or large topography) limit, corresponding to Fig. 17-b. Contour intervals are the same as those between the different shades.

by numerous filaments in Fig. 18-c, concomitantly with an increase of the horizontal energy length scale. As in the case without topography, this inverse energy cascade on the horizontal leads therefore to a deeper penetration of the velocity field, shown in Fig. 18-d. When this velocity field reaches the bottom layer, it starts to stir the bottom potential vorticity field. This induces a bottom-intensified flow, which then stirs the surface potential vorticity field, and so on. The corresponding flow is shown in Fig. 18-e, which clearly represents a bottom-intensified anticyclonic flow above the topographic anomaly, qualitatively similar to the one predicted by statistical mechanics in the low energy limit.

These results have important consequences for ocean energetics: topographic anomalies allow transferring surface-intensified eddy kinetic energy into bottom-trapped mean kinetic energy, which would eventually be dissipated in the presence of bottom friction, as for instance in the case of the Zapiola anticyclone.[55,58] In the case of the Zapiola anticyclone, the dissipation time scale is of the order of a few eddy turnover-time. It is therefore not *a priori* obvious that the results obtained in a freely evolving configuration may apply to this case. However, one can now build upon these results to address the role of forcing and dissipation in vertical energy transfers above topographic anomalies.

## 6. Conclusion

On these lectures, we have discussed applications of equilibrium statistical mechanics of the quasi-geostrophic model to the Great Red Spot and other vortices of Jupiter, to Jupiter's jets, to ocean mesoscale eddies, ocean mid-basin jets analogous to the Gulf-Stream or the Kuroshio, and to the vertical structure of geostrophic ocean turbulence.

All these applications illustrate the power of equilibrium statistical mechanics. The theory predicts the detailed shape, relation with both external and deep shear, and the jet profile for the Great Red Spot of Jupiter, depending on only a few key control parameters. It also predicts the structure and the westward velocity of mesoscale vortices, much of the qualitative properties of mid-basin jets, and the vertical structure of quasi-geostrophic turbulence. Still more applications to ocean and atmosphere dynamics are currently under investigations.

However, the range of validity of the approach is limited. Equilibrium statistical mechanics can be valid only if the effects of forcing and dissipation can be neglected. This corresponds to two different kinds of situation. The first one, as discussed in the original papers,[2–5] is when the flow is produced by an instability, or from a prepared initial condition, and then evolves to a self-organized state during a time scale which is much smaller than the typical time scales associated to the non-inertial processes (forcing and dissipation). This framework is probably the correct one, for instance for the formation of ocean mesoscale eddies from the instability either of the Gulf Stream (Gulf Stream rings) or of the Agulhas current downstream of Cape Agulhas.

Most of geophysical and other natural flows are however in another regime. Very often they have settled down from a very long time to a statistically stationary solution, for which forces balance dissipation on average. In this case, one can still compare the typical time scale for inertial organization on one hand (usually turnover times, or typical times for wave propagation) to the forcing and dissipation time scale on the other hand (spin up or spin down time scale). If these two time scales are well separated, then we still expect equilibrium statistical mechanics to describe at leading order the flow structure, and its qualitative properties. Usefulness of equilibrium statistical mechanics in this second framework, for instance close to a phase transition, is illustrated in.[59] We nevertheless note a limitation of equilibrium statistical mechanics in this second framework. It does not predict which of the set of possible statistical equilibria (parameterized by the inertial invariants) is actually selected by the long term effect of forces and dissipation. This should be determined at next order by computing the vanishingly small fluxes of conserved quantities.

Still most of ocean and atmosphere flows, for instance large scale organization of the atmosphere or the ocean, fulfill these separation of time scale hypothesis only marginally. Then a truly non-equilibrium statistical mechanics approach has to be considered. This is the subject of a number of current approaches, using kinetic theory,[60,61] related approaches such as stochastic structural stability theory (see[62–64] and references therein), or cumulant expansions (see[65,66]and references therein), or instanton theory. Section 6 of the review[12] contains a more complete discussion of such non-equilibrium approaches; whereas the review by[11] stresses the interest of statistical mechanics for climate applications.

## Acknowledgments

This work was supported through the ANR program STATFLOW (ANR-06-JCJC-0037-01) and through the ANR program STATOCEAN (ANR-09-SYSC-014), and partly by DoE grant DE-SC0005189 and NOAA grant NA08OAR4320752.

## References

1. L. Onsager, Statistical hydrodynamics, *Nuovo Cimento.* **6 (No. 2 (Suppl.))**, 249–286 (1949).
2. R. Robert, Etats d'équilibre statistique pour l'écoulement bidimensionnel d'un fluide parfait, *C. R. Acad. Sci.* 1, 311:575–578 (1990).
3. J. Miller, Statistical mechanics of euler equations in two dimensions, *Phys. Rev. Lett.* **65**(17), 2137–2140 (1990). doi: 10.1103/PhysRevLett.65.2137.
4. R. Robert, A maximum-entropy principle for two-dimensional perfect fluid dynamics, *J. Stat. Phys.* **65**, 531–553 (1991).
5. R. Robert and J. Sommeria, Statistical equilibrium states for two-dimensional flows, *J. Fluid Mech.* **229**, 291–310 (1991).
6. J. Sommeria. Two-Dimensional Turbulence. In ed. S. Berlin, *New trends in turbulence*, vol. 74, *Les Houches*, pp. 385–447 (2001).
7. C. Lim and J. Nebus, *Vorticity, Statistical Mechanics, and Monte Carlo Simulation.* Springer Monographs in Mathematics (ISSN 1439-7382), New York, NY : Springer Science+Business Media (2007).

8.  A. J. Majda and X. Wang, *Nonlinear Dynamics and Statistical Theories for Basic Geophysical Flows*. Cambridge University Press (2006).
9.  P. H. Chavanis. Statistical mechanis of two-dimensional vortices and stellar systems. In eds. T. Dauxois, S. Ruffo, E. Arimondo, and M. Wilkens, *Dynamics and Thermodynamics of Systems With Long Range Interactions*, vol. 602, *Lecture Notes in Physics*, pp. 208–289, Springer-Verlag (2002).
10. G. L. Eyink and K. R. Sreenivasan, Onsager and the theory of hydrodynamic turbulence, *Rev. Mod. Phys.* **78**, 87–135 (2006).
11. B. Marston, Looking for new problems to solve? Consider the climate, *Physcs Online Journal.* 4:20 (Mar., 2011).
12. F. Bouchet and A. Venaille, Statistical mechanics of two-dimensional and geophysical flows, *Physics Reports.* **515**, 227–295 (2012).
13. A. E. Gill, *Atmosphere-Ocean Dynamics* (1982).
14. J. Pedlosky, *Geophysical fluid dynamics* (1982).
15. R. Salmon, *Lectures on Geophysical Fluid Dynamics.* Oxford University Press (1998).
16. G. K. Vallis, *Atmospheric and Oceanic Fluid Dynamics* (2006). doi: 10.2277/0521849691.
17. J. Michel and R. Robert, Large deviations for young measures and statistical mechanics of infinite dimensional dynamical systems with conservation law, *Communications in Mathematical Physics.* **159**, 195–215 (1994).
18. B. Turkington, On steady vortex flow in two dimensions, I, *Communications in Partial Differential Equations.* **8 (9)**, 999–1030 (1983).
19. F. Bouchet and J. Sommeria, Emergence of intense jets and Jupiter's Great Red Spot as maximum-entropy structures, *Journal of Fluid Mechanics.* **464**, 165–207 (Aug., 2002).
20. A. Venaille and F. Bouchet, Ocean rings and jets as statistical equilibrium states, *Journal of Physical Oceanography.* **10**, 1860–1873 (2011).
21. L. Modica, The gradient theory of phase transitions and the minimal interface criterion, *Archive for Rational Mechanics and Analysis.* **98**, 123–142 (1987). doi: 10.1007/BF00251230.
22. F. Bouchet, *Mécanique statistique des écoulements géophysiques.* PHD, Université Joseph Fourier-Grenoble (2001).
23. F. Bouchet, Simpler variational problems for statistical equilibria of the 2d euler equation and other systems with long range interactions, *Physica D Nonlinear Phenomena.* **237**, 1976–1981 (2008).
24. T. E. Dowling, Dynamics of jovian atmospheres, *Annual Review of Fluid Mechanics.* **27**, 293–334 (1995).
25. A. P. Ingersoll and A. R. Vasavada, Dynamics of Jupiter's atmosphere., *IAU Special Session.* **1**, 1042–1049 (1998).
26. P. S. Marcus, Jupiter's Great Red Spot and other vortices, *Ann. Rev. Astron. Astrophys.* **31**, 523–573 (1993).
27. J. Sommeria, C. Nore, T. Dumont, and R. Robert, Statistical theory of the Great Red SPOT of Jupiter, *Academie des Science Paris Comptes Rendus Serie B Sciences Physiques.* **312**, 999–1005 (1991).
28. F. Bouchet and T. Dumont, Emergence of the great red spot of jupiter from

random initial conditions, *cond-mat/0305206* (2003).

29. F. Bouchet, P. H. Chavanis, and J. Sommeria, Statistical mechanics of Jupiter's Great Red Spot in the shallow water model, *Preprint, to be submitted* (2012).

30. B. Turkington, A. Majda, K. Haven, and M. Dibattista, Statistical equilibrium predictions of jets and spots on Jupiter, *PNAS.* **98**, 12346–12350 (2001).

31. R. Salmon, G. Holloway, and M. C. Hendershott, The equilibrium statistical mechanics of simple quasi-geostrophic models, *Journal of Fluid Mechanics.* **75**, 691–703 (1976).

32. M. T. Dibattista and A. J. Majda, An Equilibrium Statistical Theory for Large-Scale Features of Open-Ocean Convection, *Journal of Physical Oceanography.* **30**, 1325–1353 (June, 2000).

33. M. T. Dibattista, A. J. Majda, and J. Marshall, A Statistical Theory for the "Patchiness" of Open-Ocean Deep Convection: The Effect of Preconditioning, *Journal of Physical Oceanography.* **32**, 599–626 (Feb., 2002).

34. N. G. Hogg and H. M. Stommel, The Heton, an Elementary Interaction Between Discrete Baroclinic Geostrophic Vortices, and Its Implications Concerning Eddy Heat-Flow, *Royal Society of London Proceedings Series A.* **397** (Jan., 1985).

35. A. E. Gill, J. S. A. Green, and A. Simmons, Energy partition in the large-scale ocean circulation and the production of mid-ocean eddies, *Deep-Sea Research.* **21**, 499–528 (1974).

36. P. Berloff, A. M. Hogg, and W. Dewar, The Turbulent Oscillator: A Mechanism of Low-Frequency Variability of the Wind-Driven Ocean Gyres, *Journal of Physical Oceanography.* **37**, 2363 (2007).

37. J. Pedlosky, *Ocean Circulation Theory.* New York and Berlin, Springer-Verlag (1998).

38. N. P. Fofonoff, Steady flow in a frictionless homogeneous ocean., *J. Mar. Res.* **13**, 254–262 (1954).

39. F. P. Bretherton and D. B. Haidvogel, Two-dimensional turbulence above topography, *Journal of Fluid Mechanics.* **78**, 129–154 (1976).

40. J. Wang and G. K. Vallis, Emergence of Fofonoff states in inviscid and viscous ocean circulation models, *Journal of Marine Research.* **52**, 83–127 (1994).

41. A. Venaille and F. Bouchet, Solvable Phase Diagrams and Ensemble Inequivalence for Two-Dimensional and Geophysical Turbulent Flows, *Journal of Statistical Physics.* **143**, 346–380 (Apr., 2011).

42. P. H. Chavanis and J. Sommeria, Classification of self-organized vortices in two-dimensional turbulence: the case of a bounded domain, *J. Fluid Mech.* **314**, 267–297 (1996).

43. A. Venaille and F. Bouchet, Statistical Ensemble Inequivalence and Bicritical Points for Two-Dimensional Flows and Geophysical Flows, *Physical Review Letters.* **102**(10), 104501 (Mar., 2009).

44. R. B. Scott and F. Wang, Direct Evidence of an Oceanic Inverse Kinetic Energy Cascade from Satellite Altimetry, *Journal of Physical Oceanography.* **35**, 1650 (2005). doi: 10.1175/JPO2771.1.

45. G. Lapeyre, What Vertical Mode Does the Altimeter Reflect? On the Decomposition in Baroclinic Modes and on a Surface-Trapped Mode, *Journal of Physical Oceanography*. **39**, 2857 (2009).
46. R. Ferrari and C. Wunsch, Ocean Circulation Kinetic Energy: Reservoirs, Sources, and Sinks, *Annual Review of Fluid Mechanics*. **41**, 253–282 (Jan., 2009).
47. K. S. Smith, The geography of linear baroclinic instability in Earth's oceans, *Journal of Marine Research*. **65**, 655–683 (2007).
48. J. G. Charney, Geostrophic Turbulence., *Journal of Atmospheric Sciences*. **28**, 1087–1094 (Sept., 1971).
49. P. Rhines, ed., *The dynamics of unsteady currents*, Vol. 6. Wiley and Sons (1977).
50. F. Bretherton, Critical layer instability in baroclinic flows, *Quart. J. Roy. Meteor. Soc.* **92**, 325–334 (1966).
51. A. Venaille, G. Vallis, and S. Griffies, The catalytic role of the beta effect in barotropization processes, *Journal of Fluid Mechanics* (August, 2012).
52. K. S. Smith and G. K. Vallis, The Scales and Equilibration of Midocean Eddies: Freely Evolving Flow, *Journal of Physical Oceanography*. **31**, 554–571 (2001). doi: 10.1175/1520-0485(2001)031.
53. P. M. Saunders and B. A. King, Bottom Currents Derived from a Shipborne ADCP on WOCE Cruise A11 in the South Atlantic, *Journal of Physical Oceanography*. **25**, 329–347 (Mar., 1995).
54. A. P. de Miranda, B. Barnier, and W. K. Dewar, On the dynamics of the Zapiola Anticyclone, *J. Geophys. Res.* **104**, 21137 – 21150 (Sept., 1999).
55. W. Dewar, Topography and barotropic transport control by bottom friction, *J. Mar. Res.* **56**, 295–328 (1998).
56. W. J. Merryfield, Effects of stratification on quasi-geostrophic inviscid equilibria, *Journal of Fluid Mechanics*. **354**, 345–356 (Jan., 1998).
57. A. Venaille, Bottom-trapped currents as statistical equilibrium states above topographic anomalies, *Journal of Fluid Mechanics*. **699**, 500–510 (May, 2012).
58. A. Venaille, J. Le Sommer, J. Molines, and B. Barnier, Stochastic variability of oceanic flows above topography anomalies., *Geophysical Research Letters*. **38**(16611) (2011).
59. F. Bouchet and E. Simonnet, Random Changes of Flow Topology in Two-Dimensional and Geophysical Turbulence, *Physical Review Letters*. **102**(9), 094504 (Mar., 2009).
60. C. Nardini, S. Gupta, S. Ruffo, T. Dauxois, and F. Bouchet, Kinetic theory for non-equilibrium stationary states in long-range interacting systems, *Journal of Statistical Mechanics: Theory and Experiment*. **1**, L01002 (Jan., 2012).
61. F. Bouchet and H. Morita, Large time behavior and asymptotic stability of the 2D Euler and linearized Euler equations, *Physica D Nonlinear Phenomena*. **239**, 948–966 (June, 2010).
62. B. F. Farrell and P. J. Ioannou, Structural Stability of Turbulent Jets., *Journal of Atmospheric Sciences*. **60**, 2101–2118 (Sept., 2003).

63. B. Farrell and P. J. Ioannou, A Theory of Baroclinic Turbulence, *Journal of the atmospheric sciences.* **66**(8), 2444–2454 (2009).
64. K. Srinivasan and W. R. Young, Zonostrophic Instability, *Journal of the atmospheric sciences.* **69**(5), 1633–1656 (2011).
65. J. B. Marston, Statistics of the general circulation from cumulant expansions, *Chaos.* **20**(4), 041107 (Dec., 2010).
66. J. B. Marston, E. Conover, and T. Schneider, Statistics of an Unstable Barotropic Jet from a Cumulant Expansion, *Journal of Atmospheric Sciences.* **65**, 1955 (2008).

# Chapter 3

# Stochastic Perturbations of Nonlinear Dispersive Waves

Anne De Bouard

*Centre de Mathématiques Appliquées*
*CNRS and Ecole Polytechnique, Route de Saclay*
*91128 Palaiseau Cedex, FRANCE*
*E-mail: debouard@cmapx.polytechnique.fr*

Wave propagation phenomena, in particular self focusing in certain molecular systems subject to thermal fluctuations may be mathematically modeled with the use of systems of discrete equations with random perturbations, which in the continuous limit give rise to stochastic nonlinear dispersive equation, as *e.g.* the stochastic nonlinear Schrödinger equation. From the point of view of models, we study here the influence of different kinds of stochastic perturbations, which will always be white noise, *i.e.* $\delta$-correlated in time, on the dynamical behavior of the solutions, as *e.g.* wave collapse, or soliton propagation. It appears, as a result of theoretical studies as well as numerical computations, on the stochastic models, that the dynamical behavior of the waves strongly depends on the spatial correlations of the noise.

## Contents

## 1. Introduction

Nonlinear dispersive waves in general, and solitons in particular are universal objects in physics. They may as well describe the propagation of certain hydrodynamic waves as localized waves in plasma physics, signal transmission in fiber optics, or also more microscopic phenomena such as energy transfer in excitable molecular systems, or dynamical properties of biological molecules like DNA or proteins. In all those domains, the formation of stable, coherent spatial structures have been experimentally observed, and may be mathematically explained by the theory of nonlinear integrable (or soliton) equations. However, none of those systems is exactly described by soliton equations, and those equations may only be seen as asymptotic models for the description of the physical phenomena. Moreover, as soon as microscopic systems are under consideration, thermal fluctuations may not be negligible. They give rise in general to stochastic fluctuations in the corresponding model, and the interaction of those fluctuations with the waves have to be studied. In some other situations, the underlying asymptotic model is not even an integrable equation, even though it is a nonlinear dispersive equation. Solitary waves may still exist in this situation, but they may also not be stable. In particular, collapse in finite time of the waves may be possible (of course in general the asymptotic model is not physically valid up to the collapse time). This is *e.g.* the case for the two-dimensional Nonlinear Schrödinger (NLS) equation that will be considered in the next sections. The mathematical theory is then much less developed than it is in the integrable case, but the hamiltonian structure of the equation may still be used to obtain dynamical properties of the waves. The wave collapse in the two-dimensional (NLS) equation has for example been investigated from a mathematical point of view, independently by Zakharov[1] and Vlasov, Petrishev and Talanov.[2]

Our aim in these lectures is to give an idea of the phenomena that occur when those nonlinear dispersive waves interact with stochastic fluctuations, coming *e.g.* from thermal fluctuations, or from the presence of amplifiers in fiber optics. The mathematical theory is even more difficult in this case, since the equations involved are no more hamiltonian systems. Moreover, in some situations, adding some physical noise in the model will not necessarily lead to a well defined mathematical equation, as is the case when one has to deal with space-time white noise. Then, numerical simulations may be helpful to give an idea of whether the model indeed has a solution or not, and what this solutions looks like. We will be mainly interested in two kinds of questions. On the one hand, how does a the presence of a noise influence the collapse phenomenon, and on the other hand, how does it act on the propagation of stable solitons? Concerning the first question, a mathematical analysis can be performed when the noise is spatially correlated, and, rather surprisingly, it does not require that the noise amplitude is small, which is in general the case in physical situations. For the soliton propagation however, the analysis is asymptotic in terms of the noise amplitude.

These notes do not pretend to give an exhaustive overview of the existing methods and results on the subject. One may consult the book *Nonlinear Random Waves*,[3] for a review of the results related to perturbation methods of integrable equations.

## 2. A stochastic model of energy transfer

The starting point of the derivation of the model which will be our main example of a stochastic model involving both nonlinearity and dispersion was an experiment performed by D. Möbius and H. Kuhn.[4] In the experiment, a Scheibe aggregate is considered, which is a highly ordered compact molecular monolayer, here composed of oxacyanine dyes as donors, doped with thiacyanine acceptors. The oxacyanine dye is then excited with low intensity UV radiation and the rate of energy transfer from donors to acceptors is measured at different values of the temperature $T$, between 20 and $300K$. This energy transfer is then found to be particularly efficient at room temperature, and the measurements show its proportionality to the absolute temperature $T$. The authors's explanation of the phenomenon was based on a simple model of coherent exciton extending over a certain domain, inversely proportional to the temperature, so that the lifetime of the exciton is indeed proportional to $T$. This kind of behavior was also observed in other experimental studies on Scheibe aggregates, or on microcrystals (see[5] and[6]).

Latter on, Bang, Christiansen, If, Rasmussen and Gaididei[7] proposed a more detailed model, taking into account both thermal fluctuations and strong coupling between exciton and phonon, which leads to nonlinearity. Their derivation, which we partly reproduce here, was inspired by Davidov's models of energy transport in protein,[8] and was based on a postulated Hamiltonian. Note however that in the present case, and contrary to Davidov's derivation which led to a one-dimensional model, the situation is two-dimensional. The postulated Hamiltonian $\hat{H}$ has three components,

$$\hat{H} = \hat{H}_{ex} + \hat{H}_{ph} + \hat{H}_{int};$$

the first component $\hat{H}_{ex}$ corresponds to the exciton energy and may be written as

$$\hat{H}_{ex} = \sum_n E_0 \hat{B}_n^\dagger \hat{B}_n - \sum_{n \neq p} \sum_p J_{np} \hat{B}_n^\dagger \hat{B}_p$$

where the indices $n$ and $p$ stand for the molecules's sites, $\hat{B}_n$ and $\hat{B}_n^\dagger$ are creation and annihilation operators, $E_0$ is the molecular energy of each site, and $-J_{np}$ is the dipole-dipole interaction energy between molecules at sites $n$ and $p$ which allows the exciton propagation. $\hat{H}_{ph}$ is linked to the (classical) phonon energy, that is the energy of external vibrations of the molecules, which are approximated by Einstein oscillators, all oscillating with the same frequency $\omega_0$. Hence

$$\hat{H}_{ph} = \frac{1}{2} M \sum_n \left[ \left( \frac{du_n}{dt} \right)^2 + \omega_0^2 u_n^2 \right],$$

$M$ being the molecular mass, and $u_n(t)$ the elastic degree of freedom at site $n$, or the derivative of the displacement of the molecule at site $n$ from its equilibrium position. Finally,

$$\hat{H}_{int} = \chi \sum_n u_n \hat{B}_n^\dagger \hat{B}_n$$

is connected with the exciton-phonon interaction energy and $\chi$ is the coupling parameter. Introducing then the non equilibrium density matrix[9] $\hat{\rho}(t) = \sum_i p_i |\psi_i\rangle \langle\psi_i|$, where $p_i$ is the probability for the system to be in the state $|\psi_i\rangle$, together with the classical Hamiltonian $H = \text{Tr}\{\hat{\rho}(t)\hat{H}\}$ and the classical function $\rho_{nn'}(t) = \text{Tr}\{\hat{\rho}(t)\hat{B}_n^\dagger \hat{B}_{n'}\}$, which is assumed to have the form $\rho_{nn'}(t) = \phi_n^*(t)\phi_{n'}(t)$, the equation for $u_n$, given by the Hamiltonian $H$, is found to be the Newton equation

$$M \frac{d^2 u_n}{dt^2} + M\omega_0^2 u_n = \chi |\phi_n(t)|^2.$$

In order to take into account the interactions of the phonon system with a thermal reservoir at temperature $T$, Bang $et.$ $al.$ introduced a noise $\eta_n$ on each molecular site $n$, and a damping term $\lambda$ in the preceding equation, which now becomes

$$M \frac{d^2 u_n}{dt^2} + M\lambda \frac{du_n}{dt} + M\omega_0^2 u_n = \chi |\phi_n|^2 + \eta_n,$$

where $\lambda$ is the width of the infrared absorption peak. Note that the energy of $u_n$ is no more preserved in this new equation, reflecting the fact that phonons are created and destroyed by thermal effects. The noise $(\eta_n)_n$ is a

Gaussian zero mean white noise, acting independently on each site, so that its autocorrelation function has the form

$$\langle \eta_n(t)\eta_{n'}(t')\rangle = D\delta(t-t')\delta_{nn'}.$$

The strength of the noise $D$ is chosen in order that thermal equilibrium is achieved, and according to the fluctuation-dissipation theorem (see[10]), it leads to impose $D = 2M\lambda kT$, with $k$ the Boltzmann's constant. The use of the quantum mechanical Liouville equation, and of the special form assumed above for the classical function $\rho_{nn'}(t)$ leads to the coupled system

$$i\hbar\frac{d\phi_n}{dt} + \sum_{p\neq n} J_{pn}\phi_p + \chi u_n \phi_n = 0 \tag{1}$$

$$M\frac{d^2 u_n}{dt^2} + M\lambda\frac{du_n}{dt} + M\omega_0^2 u_n = \chi|\phi_n|^2 + \eta_n \tag{2}$$

in which $|\phi_n|^2$ is the probability for finding the exciton at site $n$. From Eq. (2), it is not difficult to express $u_n$ in terms of $\phi_n$, using a variation of constant formula. The above system of equations may then be simplified if one assumes that $|\phi_n|^2$ varies slowly compared with $\sin\omega_0 t$. Indeed, averaging over a period of the fast oscillating terms appearing in the expression of the phonon in terms of $|\phi_n|^2$ (see Bang et. al.[7] for details), one then gets the expression

$$\bar{u}_n(t) = \frac{\chi}{M\omega_0^2}|\phi_n(t)|^2 + \bar{s}_n(t)$$

for the slowly varying part of the phonon $u_n(t)$. Here, the averaged noise $\bar{s}_n(t)$ is still stationary, but now correlated in time, and may be expressed through its spectrum

$$\frac{S(\omega)}{S(0)} = \frac{\left(\frac{\omega_1}{\pi\omega}\right)^2 \sin^2\left(\frac{\pi\omega}{\omega_1}\right)}{(\omega_0^2 - \omega^2)^2 + \lambda^2\omega^2},$$

where $\omega_1 = \sqrt{\omega_0^2 - \lambda^2/4}$ is the perturbed frequency (close to $\omega_0$) arising in the phonon expression. We recall here that the spectrum of the noise $S(\omega)$ is such that $S(\omega)\delta(\omega - \omega')$ is the Fourier transform of the autocorrelation function $\langle \bar{s}_n(t)\bar{s}_n(t')\rangle$. Plugging the preceding expression of $\bar{u}_n$ in the equation describing the evolution of the exciton $\phi_n$, one finds

$$i\hbar\frac{d\phi_n}{dt} + \sum_{p\neq n} J_{pn}\phi_p + V|\phi_n|^2\phi_n = -\sigma_n\phi_n$$

with $V = \chi^2/M\omega_0^2$ and $\sigma_n(t) = \chi\bar{s}_n(t)$. This equation shows in particular that the total probability for finding the exciton in the system, which is given by $N = \sum_n |\phi_n(t)|^2$, is preserved as time evolves. A next simplification in the model consists in considering only interactions between nearest neighbors, i.e. considering that $J_{np} = 0$ except for $|n - p| = 1$, for which $J_{np} = J_0$, and performing the gauge transform $\phi_n \mapsto e^{4iJ_0t/\hbar}\phi_n$, one gets an equation that may be "approximated" by a nonlinear Schrödinger equation

$$i\hbar\phi_t + \ell^2 J_0 \nabla^2\phi + \ell^2 V|\phi|^2\phi = -\ell^2\sigma\phi,$$

where $\ell$ is the distance between nearest neighboring molecules in the aggregate, $|\phi(x,y,t)| = |\phi_n(t)|^2/\ell^2$ is a probability density and $\sigma(x,y,t) = \sigma_n(t)/\ell^2$ is a noise density. Introducing then the dimensionless variables $x' = x/\ell$, $y' = y/\ell$, $t' = J_0t/\hbar$, $\phi' = \sqrt{V\ell^2/J_0}\phi$, $\sigma' = \chi\ell^2/J_0\sigma$ (and dropping the primes) leads to the simplified equation

$$i\phi_t + \nabla^2\phi + |\phi|^2\phi + \sigma\phi = 0.$$

As mentioned before, $\sigma$ is no more delta-correlated in time. Let us discuss an important point here, which is the domain of validity of this model. Indeed, the averaging procedure which has allowed us to remove the fast oscillating terms is valid only if the original time is much larger than the period of the oscillations, $1/\omega_0$, which is of the order $10^{-12}$ in Moëbius and Kuhn's experiment (see[7]); hence in the new scales, the time must be much larger than 10 (still according to the values given in[7]). In other words, the preceding model is valid only for large times. On the other hand, the space continuous approximation will be valid only at scales much larger than the distance between molecules, and hence in the dimensionless variables, the scale in $x$ must be much larger than one. Let us consider a special case in this domain of validity by setting $x' = \varepsilon x$ and $t' = \varepsilon^2 t$, where $\varepsilon$ is a small parameter. Setting also $\tilde{\phi} = \varepsilon\phi$, one gets the equation for $\tilde{\phi}$

$$i\partial_{t'}\tilde{\phi} + \nabla^2_{x'}\tilde{\phi} + |\tilde{\phi}|^2\tilde{\phi} + \tilde{\sigma}\tilde{\phi} = 0 \tag{3}$$

with a new noise $\tilde{\sigma}(x',y',t') = \frac{1}{\varepsilon^2}\sigma(\frac{t'}{\varepsilon^2}, \frac{x'}{\varepsilon}, \frac{y'}{\varepsilon})$, which, as $\varepsilon$ goes to zero, approximates a space-time white noise $\xi$ with correlation function

$$\langle\xi(x,y,t)\xi(x',y',t')\rangle = D\delta(x - x')\delta(y - y')\delta(t - t').$$

Hence, considering a large time scale which, in dimensionless form is of the order of the square of the spatial scale, the natural noise to put in the model is indeed the space-time white noise.

## 3. About space-time white noise

Now we would like to give a more mathematical definition to the equations that we have met in the preceding section. Indeed, there are several possible mathematical definitions, even though only one of them is probably physically relevant here. White noise is indeed not a physical noise : it is a mathematical idealization of a noise with very small correlation length. The denomination white noise comes from the fact that its spectrum (the Fourier transform of its autocorrelation function) is flat, as for the white light; hence its autocorrelation function is a delta-function. Let us denote by $\xi$ such an object, assuming for the moment that $\xi$ only depends on the time variable. Note that $\xi$ has zero mean, $\langle \xi(t) \rangle = 0$, and $\xi(t)$ is independent of $\xi(t')$ for $t \neq t'$. It is easily seen[10] that such an object cannot have a finite variance. Actually, assuming in addition that $\int_0^t \xi(s)ds$ is continuous with probability one leads to the fact that necessarily, $\xi = \dot{B}(t)$ is the time derivative of a Brownian motion.

Now, $B(t)$ is nowhere differentiable, and one has to make more precise the meaning of a stochastic equation like the Langevin equation, of the form

$$\frac{dx}{dt} = a(t, x(t)) + b(t, x(t))\dot{B}(t)dt.$$

Such a stochastic equation is an integral equation, when written in the form

$$x(t) - x(0) = \int_0^t a(s, x(s))ds + \int_0^t b(s, x(s))\dot{B}(s)ds$$

and it is sufficient to give a precise definition of the integral $\int_0^t b(s, x(s))\dot{B}(s)ds = \int_0^t b(s, x(x))dB(s)$. In general such an integral is defined as the limit of Riemann sums $\sum_{i=1}^n b(\tau_i, x(\tau_i))(B(t_{i+1}) - B(t_i))$. It appears that due to the irregularity of $B$, the definition depends on the choice of the point $\tau_i$ between $t_i$ and $t_{i+1}$, so that different choices of $\tau_i$ will lead to different values of the integral. The natural choice here is the so called "Stratonovich" integral, which corresponds to taking $\tau_i = (t_i + t_{i+1})/2$. Indeed, the diffusion-approximation theorem (see[10]) states that if $x(t)$ is a Markov process satisfying an equation of the form

$$\frac{dx}{dt} = a(x)dt + b(x)\alpha_0(t)$$

wit $\alpha_0$ a stochastic source with nonzero correlation length, then as the correlation length goes to zero, $x$ approaches the solution of

$$dx = a(x)dt + b(x) \circ dB(t)$$

where the circle means that the corresponding integral must be understood in the Stratonovich sense.

Now, let us come back to the space time white noise $\xi(t, x, y)$, *i.e.* a Gaussian process with correlation function

$$\langle \xi(t, x, y)\xi(t', x', y')\rangle = D\delta(t - t')\delta(x - x')\delta(y - y'). \tag{4}$$

Setting $\xi(t, x, y) = \partial_t B(t, x, y)$, it appears that $B(t, ., .)$ may then be seen as the infinite-dimensional generalization of a Brownian motion. Indeed, consider

$$B(t, x, y) = \sum_{k,l=0}^{+\infty} e_k(x)e_l(y)B_{k,l}(t)$$

where $e_{2k+1}(x) = \frac{1}{\sqrt{\pi}}\sin kx$, $e_{2k}(x) = \frac{1}{\sqrt{\pi}}\cos kx$ and where $B_{k,l}$ is a family of independent real valued centered Brownian motions with variance $Dt$. Then the autocorrelation function of $B$ may be computed thanks to the independence of the family, as

$$\langle B(t, x, y)B(t', x', y')\rangle = \sum_{k,l,k',l'} e_k(x)e_{k'}(x')e_l(y)e_{l'}(y')\langle B_{k,l}(t)B_{k',l'}(t')\rangle$$

$$= D\min(t, t') \sum_k \cos k(x - x') \sum_l \cos l(y - y')$$

and we recover the correlation given in (4) for $\xi = \partial_t B$ since $\sum_k \cos kx$ is the Fourier decomposition of $\delta_0(x)$. Note that the decomposition above is only a formal decomposition, the series do not converge and one can show that such a process $B$ is actually not a well defined function of $x$ and $y$. In order that $B$ becomes a well defined function of $x$ and $y$, one has to add spatial correlations in the noise. This may be done *e.g.* by introducing coefficients $\lambda_k$ in the definition of $B$, setting then

$$B(t, x, y) = \sum_{k,l=0}^{+\infty} \lambda_k\lambda_l e_k(x)e_l(y)B_{k,l}(t)$$

were the $\lambda_k$ are such that the series $\sum_k \lambda_k^2$ is convergent. In this case $B$ is still white in time, but becomes colored in space with the correlation function $D\min(t, t')c(x - x')c(y - y')$, where $c$ is now a square integrable function of $x$. We will see in the next section that the addition of spatial correlations in the noise may have drastic effects on the dynamical behavior of the solution of a model equation like (3).

## 4. Influence of spatial correlations on wave collapse

We come back to the model equation (3), but we first assume that the noise is spatially correlated. This is necessary in order to be able to perform analytical computations, and especially to make use of the energy (or Hamiltonian) of the equation. It will be indeed made clear hereafter that if the noise is a space time white noise, then the energy of the solution is infinite. We will also consider a model close to (3), but with additive noise, that is in which the noise term appears simply as a forcing term in the equation. More precisely, let us consider

$$id\phi + (\nabla^2\phi + |\phi|^2\phi)dt = \begin{cases} \phi \circ dB \\ dB \end{cases} \tag{5}$$

where the symbol $\circ$ denotes the fact that the definition we should give to the multiplicative equation, in which the noise appears as a potential term, is a Stratonovich definition, similar to that of the preceding section. This is indeed a very natural choice for the model equation (3), given the discussion at the end of Sec. 2 and the approximation-diffusion theorem. Again, here, $B$ may be seen as an infinite dimensional Brownian motion, written with the help of a complete orthonormal system $(e_j)_{j\in\mathbb{N}}$ of $L^2(\mathbb{R}^d;\mathbb{R})$, as

$$B(t,x) = \sum_{j\in\mathbb{N}} \lambda_j e_j(x) B_j(t) \tag{6}$$

and the family $(B_j)$ is an independent family of real valued, centered normalized Brownian motions. Note that we will allow the space dimension to be different from 2 here (by setting $x \in \mathbb{R}^d$), our motivation being to emphasize the influence of criticality which is a balance between the power in the nonlinearity and the space dimension. As we have mentioned in Sec. 3, the coefficients $\lambda_j$ are used to introduce spatial coloration in the noise. We will suppose that the $\lambda_j$ are all nonzero, which implies that the noise is in some sense non degenerate and we assume also that the series $\sum_{j\in\mathbb{N}} \lambda_j^2$ is finite, so that the spatial correlation function $c(x,x') = \sum_j \lambda_j e_j(x)e_j(x')$, with

$$\langle \dot{B}(t,x)\dot{B}(t',x')\rangle = c(x,x')\delta(t-t')$$

is indeed square integrable in $(x,x')$. Note that the situation here is slightly different from the one in the preceding section where the correlation function depended only on $x - x'$, *i.e.* the noise was stationary in space. The reason of this difference is the fact that here we deal with an extended system, that is an equation stated in the whole spatial domain $\mathbb{R}^d$, and that we need some integrals, like energy integrals to be finite, which excludes the possibility to work with periodic functions. We could however consider a stationary noise in space, but only in the multiplicative case, and the correlation function would no more be square integrable.

Since our purpose is to investigate the influence of such a noise on the collapse of solutions, let us first recall a few facts concerning collapse in the nonlinear Schrödinger equations in the absence of stochastic fluctuations. First of all, if we neglect the right hand side in equation (5), then two integrals of motions are conserved by the evolution equation, namely the mass (or number of particles)

$$M(\phi) = \int_{\mathbb{R}^d} |\phi(x)|^2 dx$$

and the energy or Hamiltonian

$$H(\phi) = \frac{1}{2} \int_{\mathbb{R}^d} |\nabla \phi(x)|^2 dx - \frac{1}{4} \int_{\mathbb{R}^d} |\phi(x)|^4 dx.$$

These conserved integrals allow in particular, when $d = 1$, and thanks to some functional inequalities, to get uniform bounds in time on quantities like $\int |\phi|^2 dx$ and $\int |\nabla \phi|^2 dx$; this in turn implies that the life-time of the solution is infinite, that is no collapse is possible with only one space dimension. This is coherent with the fact that the one-dimensional NLS equation is actually an integrable equation. Another consequence of the conservation of $M$ and $H$ is the stability of the soliton solution in $1 - D$. This question, and the influence that the noise may have on the evolution of the soliton will be investigated in more details in the next section for a different model equation, that is the Korteweg–de Vries equation.

Now, as the functional inequality mentioned above does not give anymore uniform bounds in time on the solution when the space dimension is two (critical) or more (super-critical), one may think that a different situation occurs. This is indeed the case, as may be explained thanks to the variance (or virial) identity, which describes the evolution of the quantity $V(\phi) = \int |x|^2 |\phi|^2 dx$, as was first observed by Zakharov[1] and Vlasov *et al.*[2] The identity is the following:

$$\frac{d^2 V(\phi(t))}{dt^2} = 16 H(\phi(t)) + 2(2 - d) \int_{\mathbb{R}^d} |\phi(x)|^4 dx. \tag{7}$$

The preceding equality together with the conservation of $H$ leads to the inequality

$$V(\phi(t)) \le V(\phi(0)) + 4G(\phi(0))t + 8H(\phi(0))t^2 \tag{8}$$

as long as $t$ is less than the life-time of the solution. Now, it is clear that this life time cannot be infinite if $V(\phi(0))$ is finite and if *e.g.* $H(\phi(0))$ is negative; indeed, if it is the case, then the right hand side above will necessarily take negative values after some time, while the left hand side is a positive quantity. It appears that the kinetic energy necessarily becomes infinite, that is

$$\lim_{t \to T^*} \int_{\mathbb{R}^d} |\nabla \phi(t)|^2 dx = +\infty$$

where $T^*$ is the collapse time. Note that $T^*$ is in general less than the time predicted by the above computation, that is the positive root of the right hand side of (8) (see *e.g.* Sulem and Sulem[11]).

Now, if we add noise in the equation and consider now the model equation (5), then the energy $H$ and mass $M$ will no more be preserved by the evolution. It is however possible to compute the evolution in time of those quantities. In the additive case, one finds an evolution for $M$ of the form

$$M(\phi(t)) = M(\phi(0)) - 2Im \sum_j \lambda_j \int_0^t \left( \int_{\mathbb{R}^d} \phi e_j dx \right) \circ dB_j(t).$$

We may then compute the mean value, using the definition of the Stratonovich integral, the independence of the $B_j$ and the equation satisfied by $\phi$ again. One finds

$$\langle M(t) \rangle = \langle M(0) \rangle + tc_1,$$

where $c_1 = \sum_j \lambda_j^2$; it becomes clear at that point that for the space time white noise, *i.e.* when $\lambda_j = 1$ for all $j$, then the number of particles $M$ cannot be finite for the additive equation. In the same way, one may write the evolution of the energy $H$ of the solution (see[12] for details) in the additive case and check that

$$\langle H(t) \rangle \leq \langle H(0) \rangle + \frac{t}{2} c_2 \tag{9}$$

with $c_2 = \sum_j \lambda_j^2 |\nabla e_j|_{L^2}^2$; hence the mean energy grows at most linearly with time.

In the multiplicative case, that is for the model derived in Sec. 2, it was already observed that the total number of particles $M$ should be conserved. This is indeed given by the equation, because the noise term is real valued, and the equation is in Stratonovich form. Concerning the evolution of energy, one may compute

$$H(t) = H(0) - Im \sum_j \lambda_j \int_0^t \left( \int_{\mathbb{R}^d} \bar{\phi} \nabla \phi . \nabla e_j dx \right) \circ dB_j$$

and this leads again to a linear growth in time of the mean value, due to the conservation of the number of particles $M$

$$\langle H(t)\rangle \leq \langle H(0)\rangle + \frac{t}{2}\langle M(0)\rangle c_3 \qquad (10)$$

with $c_3 = \sup_x \sum_j \lambda_j^2 |\nabla e_j(x)|^2$.

In both additive and multiplicative cases, the preceding bounds on the evolution of the number of particles and the energy, together with the functional inequalities which have been mentioned in the deterministic (no noise) case, show that the life time of the solution is infinite when the model is restricted to a one dimensional space; in this case collapse cannot occur, even with a spatially colored noise. On the opposite, if the space dimension is two or three, then the variance inequality may be generalized to the noisy case and one obtains *e.g.* for an additive noise the following inequality for the evolution of the mean of the variance $V$:

$$\langle V(t)\rangle \leq \langle V(0)\rangle + [4\langle G(0)\rangle + c_4]t + [8\langle H(0)\rangle + 2c_2]t^2 + \frac{4}{3}c_1 t^3. \qquad (11)$$

Again, this shows that whatever $t_0$ is, if $\langle H(0)\rangle$ is sufficiently negative compared to the constants $c_j$ arising in the variance inequality Eq. (11), then this inequality cannot hold true for $t = t_0$, and that means collapse necessarily has occurred before $t_0$ with a positive probability. Note that the collapse time $T^*$ is now a random time. The collapse condition *a priori* depends on the initial value of the energy. Actually, it does not, and any initial state will lead with a positive probability to a collapsing solution, as was proved in.[13] The reason is the following: if all the coefficients $\lambda_j$ are positive, which means that all the modes are excited, then the noise is non degenerate, *i.e.* with a positive probability, it will be close to any forcing term. Now the deterministic evolution equation is controllable by a forcing term and this together with the non degeneracy of the noise implies the irreducibility of the evolution equation: given any states $\phi_0$ and $\phi_1$, and any time $T$, the solution $\phi$ of Eq. (5) with the additive noise, and starting from $\phi_0$ at time 0 will be close to $\phi_1$ at time $T$ with a positive probability. Choosing for $\phi_1$ a state with sufficiently negative energy leads then to a solution which will evolve towards a collapsing state, and this whatever $\phi_0$ is.

The same is true for the multiplicative noise $u \circ dB$ in dimension three, see,[14] but not in dimension two for the reason that dimension two is critical with respect to the collapse, and in that case solutions which have sufficiently small number of particles will not present any collapsing state; this is due to the functional inequality

$$\int_{\mathbb{R}^2} |\phi|^4 dx \le C \int_{\mathbb{R}^2} |\phi|^2 dx \int_{\mathbb{R}^2} |\nabla\phi|^2 dx$$

which holds in dimension two, and which allows to get for the solution of Eq. (5) a bound on the kinetic energy

$$\frac{1}{2} \left( \int_{\mathbb{R}^2} |\nabla\phi|^2 dx \right) \left( 1 - \frac{C}{4} M(t) \right) \le H(t) = H(0)$$

where $C$ is the above constant. As the noise does not act on $M(\phi(t))$, the above shows that a solution with an initially small number of particles will not lead to collapse, although solutions with negative $H(\phi(0))$ do collapse in finite time.

The behavior of the (random) collapse time has been investigated numerically by L. Di Menza, M. Barton Smith and A. Debussche.[15,16] The authors used a Crank-Nicholson scheme for the time discretization and finite differences or finite elements in space, in both the one-dimensional and two-dimensional cases, and with different values of the power non-linearity, that is replacing the cubic term $|\phi|^2\phi$ by a more general power nonlinearity $|\phi|^{2\alpha}\phi$. This last trick allows to perform numerical simulations in super-critical cases, without going to dimension three. They also used a semi-implicit dicretization for the noise term, which allows to simulate efficiently the Stratonovich product. The Crank-Nicholson scheme is conservative for both the number of particles and the energy for the equation without noise, and the simulations showed strong numerical evidence of a linear growth of the average energy in the presence of noise, which tends to prove that the bounds obtained in (9) and (10) are sharp. The influence of the noise amplitude, the power of the nonlinearity and the sign of the initial energy on the collapse time has been numerically computed, both for the additive and multiplicative equations. The sign of the initial energy means that attention is paid in particular to the fact that the initial data will lead to collapse or not in the absence of noise. The numerical results are in complete agreement with the theoretical remarks above. Moreover, it was found that when the sign of the initial energy is negative, then the additive noise has a tendency to accelerate the collapse, while collapse appears later with a multiplicative noise. When the initial energy is positive, or more precisely when the corresponding solution does not collapse in the absence of noise, it was observed that the variance of the collapse time is much more important in the multiplicative case. Naturally, all those effects are amplified as the noise amplitude is increased. No collapse was observed in the critical case with the multiplicative noise, when the initial number of particles is small, while it occurred with an additive noise; again, this was in complete agreement with the above theoretical observations.

Another important observation made in[15] and[16] was concerned with the space time white noise. Indeed, it was mentioned previously that if the noise is completely spatially uncorrelated, *i.e.* if $\lambda_j = 1$ for all $j$, then the energy is almost surely infinite and the above theoretical computations are no more available. The numerical observations in this case were based on the remark that when one approximates the noise by independent Gaussian random variables on each space-time mesh, then the correlation length of this numerical noise is the size of the mesh, and such a noise does not contain scales smaller than this mesh size. On the other hand, the collapse mechanism is clearly a small scales mechanism which involves energy transfer from large scales to small scales. This explains why a spatially correlated noise, which does not contain small scales, cannot perturb the mechanism once it has started. However a spatially delta-correlated noise contains all scales and may then have a different effect. In order to capture such an effect, a space and time refinement procedure has been used, locally in space, as soon as the amplitude of the computed solution has increased by a given factor. It is expected that the refinement procedure gives a better approximation for the space-time white noise. This procedure has allowed very interesting observations. First, the amplitude of any computed solution with an additive noise increases drastically with the refinement; this is a strong numerical evidence that the additive equation with space-time white noise has no reasonable solution. The situation is completely different with a multiplicative noise. Indeed, it was found that provided the refinement procedure takes place sufficiently early, as the solution amplitude starts to increase, then the collapse process is prevented in both the critical and supercritical cases. Moreover, the higher is the amplitude of the noise, the more efficient seems to be the mechanism of prevention of collapse. Unfortunately, up to now, these very interesting numerical observations have remained unexplained from a theoretical point of view. The main reason is that, as explained above, the phenomenon is truly infinite dimensional, since it involves arbitrarily small scales. Hence any attempt of finding an approximate finite dimensional anzatz (that is with a finite number of degrees of freedom) for the evolution of the solution close to the time where the phenomenon takes place is expected to fail.

## 5. Soliton diffusion by noise in the Korteweg–de Vries equation

We will here consider another nonlinear dispersive equation, for which we will again investigate the influence of a (spatially correlated) noise, with delta correlations in time. However, the aim here is different than in the preceding section: since the Korteweg–de Vries equation is an integrable equation which possesses soliton solutions and has no collapse effect, we will concentrate on the dynamical behavior of the solutions initially close to solitons, in the presence of a noise of the same type as the noise described in Sec. 3.

The Korteweg–de Vries equation, which may be written as

$$\partial_t u + \partial_x u + \partial_x^3 u + \partial_x(u^2) = 0, \quad x \in \mathbb{R}, \quad t > 0 \tag{12}$$

is known to be an asymptotic model for the propagation of unidirectional, weakly nonlinear long waves on the surface of shallow water, see.[17] In that context, $u$ stands for the (rescaled) elevation of the water surface. The equation also occurs in the context of plasma physics see Washimi and Taniuti[18]; the equation is also integrable, both in the sense of the inverse scattering method, as was first observed by Gardner, Greene, Kruskal and Miura,[19] and as an infinite dimensional Hamiltonian system. Hence, the Korteweg–de Vries equation (12) may also be qualified as a "universal model" and has given birth to a huge literature.

We will not take specific advantage of the integrability properties of the KdV equation here; actually, the method that we will use to investigate the influence of the noise on the solutions could be used for any nonlinear dispersive equation possessing solitary waves. As mentioned in the introduction, one could consult the book by Konotop and Vazquez[3] or Garnier's paper[20] for results more specifically related to the integrability of the equation.

The question of the influence of a noise on the solutions of (12) is a natural question, as such random perturbations may have different origins, like a random pressure fields on the surface of the water, a random bottom topography, or simply thermal effects in the context of plasma physics. The first attempt in this direction was probably the works by Wadati,[21] and Wadati and Akutsu,[22] which were concerned with the case where the equation is simply perturbed by a delta correlated noise depending only on the time variable. In,[21] the following equation was considered.

$$du + (\partial_x u + \partial_x^3 u + \partial_x(u^2))dt = dW(t) \tag{13}$$

where $\dot{\xi}(t) = \frac{dW}{dt}$ is a one dimensional delta-correlated Gaussian noise,

$$\langle \dot{\xi}(t)\dot{\xi}(s)\rangle = \delta(t-s)$$

which means that $W(t)$ is a real valued Brownian motion (see Sec. 3). Using a Galilean transformation, the solution of (13) is then easily found to be given by

$$u(t,x) = U(t, x - \int_0^t W(s)ds) + W(t) \qquad (14)$$

where $U$ is a solution of the unperturbed KdV equation (12). An interesting consequence of this remark is that if the solution $U$ is a soliton solution of the unperturbed KdV equation, then one finds the asymptotic behavior of the maximum of the mean of the solution

$$\max_x \langle u(t,x)\rangle \sim ct^{-3/2}.$$

This "superdiffusion" phenomenon, as it was called by Wadati, is due to the random behavior of the center of mass $x(t)$ of the solution of the perturbed equation; in view of (14), indeed, the center of mass $x(t)$ is a Gaussian process with variance $t^3/3$.

A natural question that arises is: does this behavior of the perturbed soliton persists with a more complicated noise, depending also on the space variable? The question is of course much more difficult to answer, as there is no explicit form of the solution when the noise depends on space. However, some results remain, as shows the following, where we use a decomposition of the solution as a main part which is a soliton with randomly varying parameters and a remaining part, which at first order is Gaussian with zero mean.

Again, we will consider several types of Gaussian noises, which are white in time and depend on the space variable, but colored in space, for the same reason as before: a space-time white noise would not allow us to use the energy of the solution. We then perturb equation (12) as follows:

$$du + (\partial_x u + \partial_x^3 u + \partial_x(u^2))dt = \begin{cases} \varepsilon dB \\ \varepsilon u dB \end{cases} \qquad (15)$$

where here again, $B(t)$ is an infinite dimensional Brownian motion, that we will take as in (6) (see Sec. 4) if the noise is additive, while in the multiplicative case, it will be assumed that the process is stationary in space with correlation

$$\langle \dot{B}(t,x)\dot{B}(t',x')\rangle = c(x-x')\delta(t-t').$$

The notation $udB$ means that we do not consider a Stratonovich product here, but a so called Itô product, whose dicretization is defined as $\sum_k u(t_k)(B_j(t_{k+1}) - B_j(t_k))$.

Let us recall a few facts about the deterministic KdV equation. It is well known that soliton solutions exist, which form a two parameter family of localized traveling waves, given by $\varphi_c(x - (c - 1)t + x_0)$, with $c > 1$, and $x_0 \in \mathbb{R}$, where the profile $\varphi_c$ may be explicitly computed:

$$\varphi_c(x) = \frac{3c}{2} \operatorname{sech}^2(\sqrt{c}\frac{x}{2}).$$

The family is known to be particularly stable and robust, and many experiments show that it is possible to observe, and even produce solitons on the surface of water. The first observation of a soliton on the surface of a channel was reported by J. Scott Russel in.[23] It is also known, thanks to the integrability properties of the KdV equation, that the family of soliton solutions describes the asymptotic behavior of any localized solution, see.[24] However, we will not use the integrability properties of the KdV equation here, but rather the fact that any soliton solution is a local minimum of the Hamiltonian

$$H(u) = \frac{1}{2} \int_{\mathbb{R}} (\partial_x u)^2 dx - \frac{1}{3} \int_{\mathbb{R}} u^3 dx$$

constrained to constant "mass" (or charge)

$$M(u) = \frac{1}{2} \int_{\mathbb{R}} u^2 dx.$$

One may also consult the work by Garnier[20] for an analysis of the behavior of the soliton solution using integrability properties of the equation, but with particular noises, of the form *e.g.* $(\partial_x u)\dot{B}(t)$, $\partial_x(u^2)\dot{B}(t)$, or $(\partial_x^3 u)\dot{B}(t)$, where $B(t)$ is a real valued Brownian motion. Note that the momentum $H$ and the mass $M$ are here again, conserved by the evolution equation (12), and it follows that the functional

$$Q_c(u) = H(u) + (c - 1)M(u),$$

whose critical point is precisely the profile $\varphi_c$ of the soliton with velocity $c$, may be used as a Lyapunov functional. Indeed, if a solution $u$ of the KdV equation (12) starts at $t = 0$ close to the soliton $\varphi_c$, then using a Taylor expansion of the functional $Q_c(u(t, \cdot + x(t)))$ allows to bound the deviations from $\varphi_c(\cdot - ct)$ of the translated solution $u(t, x + x(t))$, provided that the "center of mass" $x(t)$ is well chosen. This leads to the stability of the family $\{\varphi_c(x + x_0), x_0 \in \mathbb{R}\}$, or in other words to orbital stability of the soliton.

The use of the parameter $x(t)$, together with the fact that the mass $M$ is conserved, allows to get rid of two "secular modes" in the equation, which are generated by $\varphi_c$ and $\partial_x \varphi_c$.

When the equation is perturbed by a noise as in (15), a natural question is that of the persistence time of the soliton which will necessarily be destroyed by the noise after some time. In order to answer this question, one should naturally take into account random modulations in the center of mass and velocity of the soliton. For that purpose, we write the solution $u$ of Eq. (15) which has for initial state $\varphi_{c_0}(x)$ in the form

$$u^\varepsilon(t, x) = \varphi_{c^\varepsilon(t)}(x - x^\varepsilon(t)) + \varepsilon \eta^\varepsilon(t, x - x^\varepsilon(t)), \tag{16}$$

where $c^\varepsilon(t)$ and $x^\varepsilon(t)$ are random modulation parameters, which are chosen in order that $\varphi_{c^\varepsilon}(x - x^\varepsilon)$ stays orthogonal to the two secular modes $\varphi_{c_0}$ and $\partial_x \varphi_{c_0}$, so that the remaining part $\varepsilon \eta^\varepsilon$ of the solution will be small for a longer time. Note that $H$ and $M$ are no more conserved by the perturbed evolution equation (15), but one may compute their evolution, and this allows in particular to use the functional $Q_{c_0}$ to get a bound on the exit time $\tau_\alpha^\varepsilon$ of an $\alpha$ neighborhood of the orbit of the soliton, that is the first time for which the difference between the solution and the modulated soliton is larger than $\alpha$. It is found that for any time $T > 0$,

$$\mathbb{P}(\tau_\alpha^\varepsilon \le T) \le e^{\frac{-C_\alpha}{\varepsilon^2 T}} \tag{17}$$

for some constant $C_\alpha > 0$ (see[25] or[26] for details). This means that the solution stays close to the randomly modulated soliton for a time of the order of $\varepsilon^{-2}$. Note that without using the modulation parameters $x^\varepsilon(t)$ and $c^\varepsilon(t)$, the exit time would be at most of the order $\varepsilon^{-2/3}$:

$$\mathbb{P}(\tilde{\tau}_\alpha^\varepsilon \le T) \ge e^{\frac{-C_\alpha}{\varepsilon^2 T^3}}$$

for small $\varepsilon$, where $\tilde{\tau}_\alpha^\varepsilon$ is the first time for which the solution leaves the ball centered at $\varphi(. - c_0 t)$. This is due to the presence of the secular modes in this case.

In the case the noise is multiplicative and homogeneous (stationary) in space, it is also possible to obtain the convergence of the order one part of the solution $\eta^\varepsilon$ (see (16)) as $\varepsilon$ goes to zero to a centered Gaussian process of Ornstein–Uhlenbeck type. Moreover, keeping only the order one in $\varepsilon$, the equations for the modulation parameters $c^\varepsilon$ and $x^\varepsilon$ are decoupled from the remaining part $\varepsilon \eta^\varepsilon$ and are found to be of the form

$$\begin{cases} dx^\varepsilon = c_0 dt + \varepsilon B_1 dt + \varepsilon dB_2 \\ dc^\varepsilon = \varepsilon dB_1; \end{cases}$$

in this expression, $B_1$ and $B_2$ are correlated Brownian motions with covariance

$$\langle W_i(t)W_j(s)\rangle = \sigma_{ij}\inf(t,s), \quad i,j = 1,2$$

which may be expressed in terms of the original noise $B$, and the soliton $\varphi_{c_0}$: $B_1$ and $B_2$ correspond to the projection of the noise $\varphi_{c_0}B$ on the two-dimensional space generated by the secular modes $\varphi_{c_0}$ and $\partial_x \varphi_{c_0}$.

Coming back to our problem of soliton diffusion, it is now possible, still keeping only the order one in $\varepsilon$, to compute explicitly the mean value of the modulated soliton, since the vector $(x^\varepsilon(t), c^\varepsilon)(t)$ is a Gaussian vector, and the asymptotic rate of diffusion in time:

$$\max_x \langle \varphi_{c^\varepsilon(t)}(x - x^\varepsilon(t))\rangle \sim C\varepsilon^{-1/2}t^{-5/4}, \quad \text{as} \quad t \to +\infty.$$

Of course this computation is valid only if $\varepsilon$ tends to zero while $t$ tend to infinity, with $t$ much less than $\varepsilon^{-2}$.

One may also compute the exit time when the solution does not start from a single soliton, but from a multi-soliton. A multi-soliton is a solution of the KdV equation(12) which asymptotically in time decomposes as a sum of single solitons with increasing velocities of the form

$$\sum_j \varphi_{c_j}(\, . - \delta_j^\pm - c_j t)$$

where $c_1 < c_2 < ... < c_n$ are the velocities and $\delta_j^+ \neq \delta_j^-$ denote deviations. An exponential bound of the same type as (17) is found for the exit time, which takes into account the fact that the center of mass of the different solitons cannot be much closer than they are initially.

The same analysis can be naturally performed for other equations possessing soliton solutions. It is the case *e.g.* for the nonlinear Schrödinger equation of Sec 4. However, it should be noted that for this equation, if the space dimension is $d$, then the family of solitary waves solutions is a $2d + 2$ parameter family given by

$$\psi_{\omega,\mathbf{v}}^{\mathbf{x_0},\theta} = \omega\psi(\sqrt{\omega}(\mathbf{x} - 2\mathbf{v}t - \mathbf{x_0}))e^{(\mathbf{v}\cdot\mathbf{x} - v^2 t + \omega t + \theta)}$$

where the profile $\psi$ is a radial function which has for $d = 1$ the same expression as the soliton of the KdV equation. In particular, the modulation equation will give a system of $2d + 2$ stochastic differential equations. However, even when the noise is multiplicative and homogeneous in space, it is not possible to obtain the diffusion of soliton as for the KdV equation by keeping only the first order terms in $\varepsilon$. This may be seen by noticing that the noise being real valued, it does not act on the modulus of the solution. Moreover, numerical simulations performed in the subcritical cases (when no collapse is possible, see Sec. 4), show that diffusion occurs not only on the mean value of the solution, but also on each trajectory. This phenomena is still completely unexplained.

## 6. Conclusion

Noise has to be taken into account in many physical equations, and we have tried to show how the presence of white in time random perturbations can influence classical phenomena arising in nonlinear dispersive models in two different situations, namely wave collapse and propagation of solitons. We have tried to emphasize the fact that when adding a noise in a model equation, one should be careful about the way this noise has to be defined in the equation, and in particular the way it as to be discretized in view of performing numerical simulations. We have shown that for those kinds of models, space-time white noise do not in general lead to well defined mathematical models, but that in some situations, one can still perform numerical simulations approximating solutions with uncorrelated noise in space.

## References

1. V. Zakharov, Collapse of langmuir waves, *Sov. Phys. JETP.* **35**, 908–914 (1972).
2. S. Vlasov, V. Petrishev, and V. Talanov, Average description of wave beams in linear and nonlinear media (the method of moments), *Radiophys. Quantum Electron.* **14**, 1062–1070 (1974).
3. V. Konotop and L. Vázquez, *Nonlinear Random Waves.* World Scientific, Singapore (1994).
4. D. Mobius and H. Kuhn, Energy transfer in monolayers with cyanine dye aggregates, *J. Appl. Phys.* **64**(10), 5138–5141 (1988).
5. S. De Boer and D. Wiersma, Dephasing-induced damping of superradiant emission in j-aggregates, *Chem. Phys. Lett.* **165**, 45–53 (1990).
6. T. Itoh, T. Ikehara, and Y. Iwabuchi, Quantum confinement of excitons and their relaxation processes in cucl microcrystals, *J. Lumin.* **45**, 29–33 (1990).
7. O. Bang, P. Christiansen, F. If, K. Rasmussen, and Y. Gaididei, Temperature effects in a nonlinear model of monolayer scheibe aggregates, *Phys. Rev. E.* **49**, 4627–4636 (1994).
8. A. Davydov, *Solitons in molecular systems.* Reidel, Dordrecht (1986).
9. A. Davydov, *Quantum Mechanics.* Pergamon, Oxford (1965).

10. C. Gardiner, *Handbook of stochastic methods for physics, chemistry and the natural sciences.* Springer (1986).

11. C. Sulem and P. Sulem, *The Nonlinear Schrödinger Equation, Self-focusing and Wave collapse.* Springer-Verlag, New York (1999).

12. A. de Bouard and A. Debussche, The stochastic nonlinear schrödinger equation in $h^1$, *Stochastic Anal. Appl.* **21**, 97–126 (2003).

13. A. de Bouard and A. Debussche, On the effect of a noise on the solutions of the focusing supercritical nonlinear schrödinger equation, *Probab. Theory Relat. Fields.* **123**, 76–96 (2002).

14. A. de Bouard and A. Debussche, Blow-up for the stochastic nonlinear schrödinger equation with multiplicative noise, *Annals of Proba.* **33**, 1078–1110 (2005).

15. A. Debussche and L. Di Menza, Numerical simulation of focusing stochastic nonlinear schrödinger equations, *Phys. D.* **262**, 131–154 (2002).

16. M. Barton-Smith, A. Debussche, and L. Di Menza, Numerical study of two-dimensional stochastic nls equations, *Numer. Methods Part. Diff. Eq.* **21**, 810–842 (2005).

17. D. Korteweg and G. de Vries, On the change of form of long waves advancing in a rectangular canal, and on a new type of long stationary waves, *Phil. Mag.* **39**, 422–443 (1895).

18. H. Washimi and T. Taniuti, Propagation of ion-acoustic solitary waves of small amplitude, *Phys. Rev. Lett.* **17**, 996–998 (1966).

19. C. Gardner, J. Greene, M. Kruskal, and R. Miura, Method for solving the korteweg-de vries equation, *Phys. Rev. Lett.* **19**, 1095–1097 (1967).

20. J. Garnier, Long-time dynamics of korteweg-de vries solitons driven by random perturbations, *Journal of Stat. Phys.* **105**, 789–833 (2001).

21. M. Wadati, Stochastic korteweg-de vries equation, *J. Phys. Soc. Japan.* **52**, 2642–2648 (1983).

22. M. Wadati and Y. Akutsu, Stochastic korteweg-de vries equation with and without damping, *J. Phys. Soc. Japan.* **53**, 3342–3350 (1984).

23. J. Russel. Report on waves. In *Report of the fourteenth meeting of the British Association for the Advancement of Science*, pp. 311–390, York, U.K. (1844).

24. W. Eckhaus and P. Schuur, The emergence of solitons of the korteweg-de vries equation from arbitrary initial conditions, *Math. Methods Appl. Sci.* **5**, 97–116 (1983).

25. A. de Bouard and A. Debussche, Random modulation of solitons for the stochastic korteweg-de vries equation, *Ann. IHP, Analyse non linéaire.* **24**, 251–278 (2007).

26. A. de Bouard and E. Gautier. Exit problems related to the persistence of solitons for the korteweg-de vries equation with small noise. Preprint (2008).

# Chapter 4

# On Different Aspects of Granular Physics

Christophe Josserand, Pierre-Yves Lagrée and Daniel Lhuillier

*Institut Jean Le Rond D'Alembert, CNRS UMR 7190- Univ. P & M Curie Paris VI, Case 162, 4 place Jussieu, 75005 Paris, France*

We present here some aspect of the granular physics, focusing on recent results on granular shear flows and memory effects obtained by the authors. Other recent advances are also discussed showing the large variety of applications and domains concerned by granular systems. However, this document does not pretend to be a review but rather to arbitrary select some results chosen among the exciting and very active recent works.

## Contents

The use of granular materials is obviously as old as human activities. It concerns a large variety of phenomena and can involve also many different time and length scales. Grains create naturally patterns under wind, water or erosion actions that form most of the landscape around us, from general morphogenesis, to dunes, river meandering or valleys profile (see sediment flow patterns observed on a beach as an illustration in Fig. 1). The granular physics is in fact at the crossroad between different domains and many recent results have taken advantages of fruitful interactions between physics, mathematics, mechanical and chemical engineering, and computer science. We present here different aspects of granular physics which concern both gas, liquid and solid mechanics. Indeed, grains have the specificity to behave very differently depending on the context: it is a gas in shaken sands, liquid in avalanches and solid in compact silos for instance. This document does not pretend to show an extensive overview of the domain, but rather to extract some recent results that shed on the diversity of properties of granular systems. Our presentation focus thus on dry granular system and we do not discuss in great details the role of the embedding fluid. The influence of the surrounding fluid (mostly air or water) is definitely of great interest and represents active research fields that would deserve a specific review.

Figure 1.    Rivulet patterns observed at low tide along the beach of Santa Barbara, California.

We thus exhibit here recent results showing the rich and complex behaviors of granular materials. Our presentation follows a review style and an interested reader should make use of the bibliography to obtain more details on specific results. In fact, we particularly develop in this document points where original results have been obtained by us. General introductions on the domain can be found in recent books by Guyon[1] or Duran.[2] But one cannot write a review on granular materials without citing the pioneering works of Lord Bagnold[3,4]: most of his results were initiated by field observations while he was working as brigadier of the British army in the Sahara desert during the 1930's and 40's.

## 1. Introduction

We experience granular systems every day in our life, by cooking with flour, walking on the beach and building sand-castles for instance, or when using an egg-timer to measure time. Industrial aspects can range from train ballast rocks to chemical micro-powders. Grains are in fact the second aspect of matter for human uses, beside liquids. The physical understanding of the grains features is still challenging since the apparent simplicity of grains interactions hides in fact high levels of complex behaviors. Granular physics holds for a large variability of scales, ranging from powders (few microns or less) to stellar and interstellar dynamics (thousand of kilometers). Rocks in mountains or iceberg dynamics on the poles seas are another examples of the diversity of these systems. Outside these extreme scales, granular materials are often considered for particles ranging from 100 $\mu$m to millimeter sizes.

Two important features are pointed out when characterizing granular systems: they are "frozen" or a-thermal and they can however mimic all the different states of the matter, gas, liquid and solid.[5] Indeed, by comparing the thermal energy for a grain($E_T \sim k_b T$) to the typical potential energy of gravity ($E_g \sim mgd \sim \rho g d^4$, $\rho$ being the grain mass density, $d$ their typical diameter and $g$ the gravity), we obtain that the radius of a grain should be at laboratory temperature much below tenth of micrometer scales or that the temperature for a 1mm diameter grain should be above $10^{15} K$ for brownian motion to become important. Thus, the system thermodynamic temperature will everywhere be neglected. In contrary, temperature-like motions are observed in granular systems when submitted to high vibrations or intensive mixing for instance. There, the external vibrations transmitted to the grains mimic thermal activity. It is then tempting to define a so-called granular temperature based on these mechanical induced fluctuations. As we will see below this temperature is defined in a similar way as in classical statistical physics, using the fluctuations of the kinetic energy of the grains for dilute systems. This temperature exhibits the same thermodynamic properties than the one issued from brownian motion, as long as it will be used in the same energy balance context as for its definition. Further on, temperature refers to a well-defined quantity deduced by a careful analysis of the system fluctuations and in no way the room temperature will be invoked.

We show also that granular systems react very differently to various constraints: they act like a gas when submitted to high vibrations in dilute context or in interstellar space. Liquid behaviors are observed in avalanches and more generally when sheared motions are considered. Finally we can still walk on the beach and the trains take advantage of the strong dissipation of the rocks ballasts, so that granular can support stress as much as a solid. Our presentation below will follow this general structure: in the next section we will show how granular gases can be modeled with a specific discussion on the classical Maxwell daemon paradox for agitated grains. Section 3 will then focus on granular flows, mostly created by shear stress situation like in avalanches for instance. We will in particular present a continuum mechanics approach developed recently.[22] A short overview on granular solids will then be presented (section 4). Before the conclusion (section 6), we will discuss the apparent analogy between granular systems and glasses, using the particular scope of an experiment showing memory effect during granular compaction (section 5).

## 2. Granular gas

When the grains density is small and agitation large enough for the granular system to act as a dilute system interacting only through binary collisions, we refer to granular gases by analogy with molecular gases (see[6] for a review). Such situations are thought to be met in interstellar regions where small material dusts are (relatively) concentrated (but still in the dilute regime of course). It is also important on human scale in strong cereal flows or rapid snow avalanches for instance. The main (and almost unique) difference between a usual gas of particle which interact through elastic collision and granular gases stands in the inherent inelastic nature of the collision processes between the grains. Therefore, the different models of these systems have used the general tools of kinetic theory developed for molecular gases with the precise correction coming from the inelasticity in the collision terms. This difference is however crucial since it invalidates the usual Maxwell-Boltzmann stationary solutions. We present here the hydrodynamic equation resulting from the kinetic theory models only, without the details of the calculations. We then discuss the density and cluster instabilities. An interesting consequence of the inelastic dynamics is finally presented: a sand version of the Maxwell's daemon.

## 2.1. *Kinetic theory model*

The extensions of kinetic theory to granular systems have been made by considering a gas of identical inelastic spheres.[15-17] The situation only differs from elastic sphere by a rate of dissipation in the balance law for kinetic energy which is obtained from the inelastic properties of the collision. This rate involves the coefficient of restitution $e$ for the normal component of the relative velocity of the collision and the Coulomb effective friction coefficient since one needs to carefully evaluate whether the two grains slide or do not slide during all the collision. The velocity and angular momentum for each grain before and after the collision have to be determined. It is a rather complicated calculation and the interested reader is encouraged first to consider the detailed analysis done for planar situation.[16] Similarly to elastic sphere a hierarchy system of equation for particle position and velocity distribution functions is obtained and a simple closure between the two particles and the single particle distribution functions is proposed. Finally, simplified equations of motion for granular fluids (dilute system) can be deduced[7,8,16]:

$$
\begin{cases}
\rho \frac{D\mathbf{v}}{Dt} = -\nabla \mathbf{P}, \\
\frac{3}{2}\rho \frac{D\mathbf{T}}{Dt} = -\nabla \cdot \mathbf{Q} - tr(\mathbf{P} \cdot \mathbf{D}) - \gamma, \\
\frac{\partial \rho}{\partial t} + \nabla \cdot (\rho \mathbf{v}) = 0,
\end{cases}
\tag{1}
$$

where $D/Dt = \partial/\partial t + \mathbf{v} \cdot \nabla$ is the material derivative using the mean velocity $\mathbf{v} = <\mathbf{u}>$ ($<\cdot>$ for the distribution average). $\rho$ is the mass density, $T = <(\mathbf{u} - \mathbf{v})^2>$ the granular temperature defined as the kinetic energy fluctuations. $\mathbf{P}$ is the stress tensor, $\mathbf{Q}$ the heat flux and $\gamma$ describes the dissipation rate due to inelastic collisions. All these constitutive functions can be determined using kinetic theory with the usual assumptions such as the expansion to the fourth order of the distribution function. These constitutive relations have been shown to be valid for dilute gas and not too small coefficient of restitution $e$ using Molecular Dynamics simulations.

This yields:

$$\begin{cases} \mathbf{P} = -p_g\mathbf{I} - 2\mu(\mathbf{D} - \tfrac{1}{3}div(\mathbf{v}))\mathbf{I}, \\ \qquad \mathbf{Q} = \kappa\nabla T. \end{cases} \tag{2}$$

One can also convince oneself from the balance law for inelastic collision that the pressure is $p_g = \rho T$ similarly to perfect gases. The viscosity follows $\mu \propto T^{1/2}$ as does the heat conductivity $\kappa \propto T^{1/2}$, while the dissipation rate is found to obey $\gamma \propto (1 - e^2)\rho^2 T^{3/2}$. The different coefficients can be deduced from kinetic theory and they are calculated precisely in.[16]

## 2.2. *Hydrodynamic instability*

A typical feature of granular gases is their ability to form clusters of dense particles surrounded by large regions with low density and relatively large velocity. It might play a crucial role in jamming dynamics where the system suddenly creates large stress and density fluctuations. It is also suspected to be one of the processes leading to the formation of concentrated dust rings around planets (Saturn rings for instance). This tendency to create cluster can be understood as a consequence of an hydrodynamic instability of the granular fluid equations described above. Indeed, one can easily obtain that an homogenous solution is linearly unstable. Such homogenous solution is in fact not a steady state since it dissipates through inelastic collision and it is described by $\mathbf{v} = \nabla T = \nabla \rho = 0$ with $T(t) = T(0)/(1 + t/t_0)$ where $t_0 = \pi^{1/2}d\phi(1 - e^2)T^{1/2}(0)$ is the characteristic collision time. Here, we used $d$ the grain diameter (sometimes noted $D$ later on in the manuscript) and $\phi$ the volume fraction of the grains. The effective temperature of an homogenous gas is thus simply decreasing algebraically with time due to energy dissipation in the collisions. The linear stability analysis shows unstable modes with non exponential growth due to the time dependence of the homogenous solution. A detailed analysis of this clustering instability can be found in.[8] Numerical simulation agree with this analysis and show the formation of denser and more dilute domains. The evolution of this instability can eventually lead to high density regions where many collisions can happens in a short time leading to a rapid decrease of the energy. Such "cold" domains are surrounded by almost particle-free regions with high velocity (warm) grains.

## 2.3. Inelastic collapse

Singular dynamics in the collisions can follow these highly heterogeneous regimes and are called inelastic collapses. The collapse consists of an infinite sequence of collisions in a finite time. This inelastic collapse can be illustrated by considering the fall of an inelastic sphere (a rigid marble used by kids for instance) on a solid plane normal to the gravity. If the marble falls initially from an height $h(0)$, it will bounce on the plane and its next maximal height will be $h(1) = e^2 h(0)$. The time for the sphere to touch the ground is $t(0) = \sqrt{2h(0)/g}$. The set of rebounds of the marble on the ground will correspond to successive height $h(n) = e^{2n} h(0)$ with falling times $t(n) = e^n t(0)$. Thus the marble will loose all its energy in an infinite number of rebounds $(n \to \infty)$ *but in a finite time* $\tau = t(0)(1+e)/(1-e)$. We can observe signatures of this high-school example of finite time singularity when considering an assembly of grains interacting through inelastic collisions. It has been first observed in one dimensional systems where the singularity is somehow reminiscent of the falling ball.[9–12] Molecular Dynamics simulations have also shown finite time singularity arising as a divergence of the number of collisions in a small time interval in two[13] and three[14] dimensions. It manifests along a line of grains embedded in a large dense region which come into contact through collapses. The collapse itself seems eventually to be concentrated in few grains aligned in one direction. Similarly to clustering, the inelastic collapse is a consequence of inelastic collisions which reduce the kinetic energy of the systems and thus can generate extra-cold region. It is important however to distinguish clearly between the clustering instability which produce cold high density regions surrounded by dilute hot grains with the inelastic collapse where the particles come into contact in finite time without any interparticle forces or cohesion.

## 2.4. Maxwell's daemon with grains

The natural tendency of the grains to form cluster can lead to amazing dynamics when considering shaking sands. Indeed highly agitated grains can be considered to mimic molecular gases at some imposed temperature. However, while for molecular gases the maximization of the entropy at constant temperature would lead to an homogenous solution with Maxwellian velocity distribution, the situation is more complicate for grains. A spectacular illustration of this strong difference has been obtained in an experiment designed for German high-school students[18]! It consists of a box separated

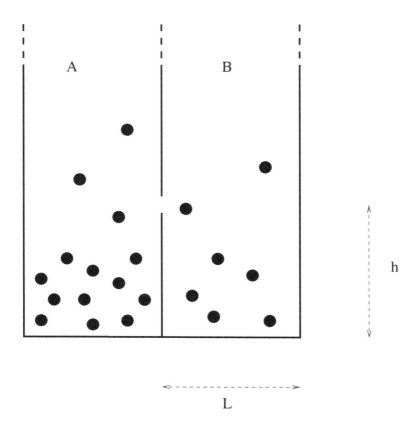

Figure 2.   Illustration of the Maxwell daemon experiments with grains. The two boxes A and B of thickness $L$ are in contact through a small hole situated at height $h$. The whole system is vibrated.

in two symmetric equal vertical parts by a wall (see Fig. 2). A small hole in the wall allows a connection between the two parts. The whole apparatus is mounted on a shaker which imposes an oscillating vibration to the system. Grains are deposited and the dynamics is observed as the amplitude of the vibrations is changed for constant frequency. For high amplitudes, the grains are highly agitated and are equally distributed between the two parts. The small hole between the two part allows for the mixing of the grains. When the amplitude is decreased, a spectacular transition occurs for finite amplitude where grains are still agitated: suddenly almost all the grains concentrate randomly on one of the two sides and the symmetry is broken (see Fig. 3 for illustration). The situation is similar to Maxwell's daemon paradox, where a daemon located at the intersection between the two box let with no effort particles preferably go from one side to the other one and thus make the system violate the second principle of thermodynamics. For grains, since collisions are inelastic and usual thermodynamics useless, the understanding of this transition comes from the rate of energy loss by collision which grows like the square of the density. Thus, if the external forcing is not strong enough, a small difference in the number of grains between the two part of the system can lead to a linear instability. The more grains, the more energy is lost through collisions and the more the particle balance deviates. A detailed description of this instability has been first proposed in.[19] The hydrodynamical model (1) described before gives in fact a good mathematical tool to describe this instability. The stability analysis of the equal partition of the grains between the two sides of the box can be interpreted qualitatively. Consider that a small disequilibrium in the number of particle is created (let call side A the one with slightly more grains than side B): on one hand, the agitation due to the bottom wall lead to a higher probability for particles to go from A to B than the opposite. This is the usual situation for molecular gases that restores the equilibrium. On the other hand, more binary collisions between grains will happen in side A than in side B, leading to an energy decreasing in side A that will lower the probability for particles in A to jump into B. The balance between these two processes leads to a second order phase transition.[19] Additional situations have been considered in more recent studies where zero-gravity experiments have been performed in particular.[20,21]

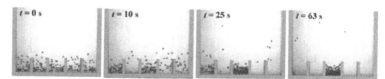

*Figure 1: Clustering beads in a row of five vertically shaken compartments.*

Figure 3. Illustration of the Maxwell daemon dynamics for grains. The grains are gathering into one single compartment of the experiment. Figure taken from reference[21] with permission from the authors.

## 3. Dense granular flows

### 3.1. *Overview*

Liquid-like dynamics are commonly observed for grains. Illustrative examples come from avalanche flows arising in many different context and in general flows occur when dense granular system are submitted to shear constraints. Liquid property is in fact crucial for grain transports. An important experimental literature on dense granular flows has been developed these last years. These flows are either due to avalanches dynamics or to some imposed shear conditions. They correspond to flows on inclined planes, dunes or rotating drums for instance while the latter cases are often illustrated by Couette flows, either cylindrical in experiments, or planar for numerics and theory. In all the situations usually considered, a stationary regime is often observed between transient dynamics due to the initiation and the end of the flow regime. These transient dynamics are more complex to understand so that most of the efforts have first concentrated on the stationary regimes. Experiments usually try to measure and determine both the velocity field and the density profiles of the flows in the steady state. The grain volume fraction is actually found to vary very little, roughly between the two natural frontiers determined as the packing fraction (noted $\phi_m$ further on) below which the grains act as a gas and the one ($\phi_M$) above which the grains cannot move without involving high deformation stress (determined by the Young modulus of the material). Somehow $\phi_m$ is close to the loose packing fraction and represents the smallest compaction compatible with the existence of a continuous network of contacts between grains. $\phi_M$ corresponds also roughly to the highest possible random packing (the so-called close packing fraction). The volume fraction is defined in an elementary cell as the ration between the volume of the grains divided by the total volume of the cell. These frontiers define a narrow range of volume fractions between which dense granular flows can be considered: in two dimension we can take $\phi_m = 0.7$ and $\phi_M \sim 0.8$; in three dimensions, we use $\phi_m = 0.5$ and $\phi_M \sim 0.65$.

A large set of recent experimental results have been gathered in a recent review paper[39] written by the "Groupe de Recherche sur les milieux divisés" (G.D.R. MiDi), a french scientific network working on granular physics. Besides a large set of experimental results on dense granular flows, this paper attempts moreover to extract a granular rheology. The main assertion of the proposed model stands in a friction coefficient depending on the shear rate (and also slightly on the compression). The various experimental results show therefore general agreement when they are presented with respect with this friction coefficient. This model commonly denoted $\mu(I)$ (where $\mu$ is the friction coefficient and $I$ the rescaled shear rate) has been applied with great success later on to many new experiments involving shear flows (see for instance a 3D generalization in[40,41]). These articles are warmly recommended to the interested reader. In the following chapter, we will describe in details a continuum mechanics models of shear flows. Instead of determining phenomenological relations such as $\mu(I)$, we try to identify the constitutive relations of dense granular shear flows. We emphasize there that although the grain volume fraction varies only a little in the flow, the transport coefficients can vary tremendously. Thus we argue that one cannot describe a general approach of granular flows without accounting for density variations. Such influence is clearly seen for jamming dynamics and in the Reynolds dilatancy. The next chapter is therefore largely inspired by the original work.[22]

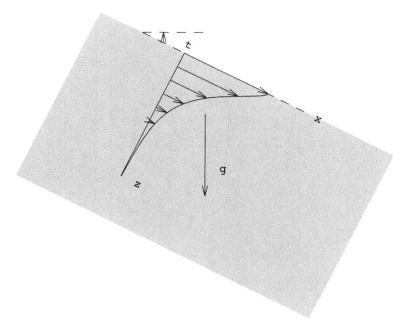

Figure 4.  Gravity-induced shear flow with free surface (over a heap or an inclined plate with an angle $\theta$ relative to the horizontal plane).

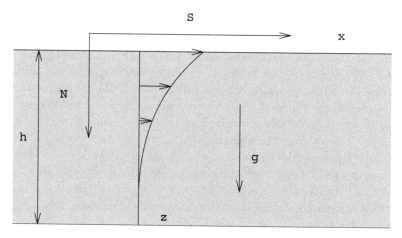

Figure 5.   Planar confined shear flow of an infinite horizontal granular layer bounded by two plates separated by a fixed distance $h$. The pressure load $P$ and the gravity are both oriented along the direction $z$, the shear stress $S$ and the flow are along direction $x$).

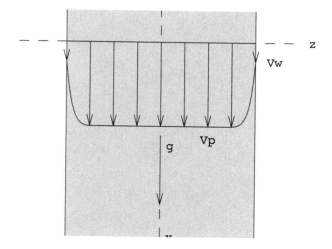

Figure 6.   Vertical chute flow between two rough plates.

## 3.2.  *A continuum mechanics approach?*

The present approach concerns the description of flows with relatively low stress levels. We are indeed not questioning the description by soil mechanics of quasi-static and highly stressed granular materials. The possibility of a general description of granular dynamics through simple relations and partial differential equations is an old problem with rather pessimistic answers (see e.g.[23-26]). For instance, for dense granular flows, experimental observations show that the flows exhibit a typical thickness of a few grain diameters only without clear links to other typical length scales. More generally, it is hard to determine the pertinent variables that would describe the grains structure and their dynamics. It is therefore difficult to deduce some kind of granular Navier–Stokes equations. However, we know from several examples in suspension mechanics that the continuum approach can cope with high velocity gradients in one direction, provided one has some statistical homogeneity in the other two directions. This situation is exactly the one met in sheared granular media, if we discard transient effects and focus on the stationary state. In this case, which actually concerns with some approximations a large class of granular flows, a rheology can be deduced and the dynamical equations can be obtained, although a precise analysis of the system needs to be performed before calculations. As discussed above, the constant density and incompressibility condition commonly used for dense granular dynamics cannot be used anymore. In fact, the main contribution of the volume fraction changes does not come from compressibility effects but mostly on the rapid dependence on the transport coefficient with small changes in the density. One can have in mind a rough analogy with the glassy transition where the viscosity can have a tremendous evolution for a small variation of the temperature while jamming occurs in granular system for small variation of the density.

Following the general properties of the flows in experiment, we will present here a stationary approach in two dimensions so that we only need two independent variables to describe the dynamics. Since we also have translational invariance in the direction of the flow, it is associated with two conservation equations, one for the mass, the other for the momentum in the orthogonal direction. As discussed above, we choose as independent variables the volume fraction and the velocity of the grains. We somehow assume that all other evolution variables (energy, angular momentum) and equations are slaved to these variables. As we will see below, the model needs further on constitutive equations to describe the stress tensor as function of the independent variables. This issue has been discussed in previous works (see for instance[27,28]), and our approach is directly inspired by these works: indeed, we introduce a stress tensor that is the sum of a frictional and a collisional contribution. Since we are concerned with dense flows, the collisional term corresponds to rebound-less impacts with zero restitution coefficient.[25] Also our model shares common features with more recent approach,[29–31] but we emphasize that our stress tensor is in fact specifically devoted to dense media instead of extending the tensor obtained through kinetic theory (not valid for the volume fractions considered here). Moreover, our model is quite simple since it denies any special role to the compaction gradient,[32] avoids the non-locality concept[33] and does not introduce an additional order parameter like in phase field models.[34]

### 3.3. *Free surface shear-flows*

The general problem can be expressed as the gravity-induced chute (over a heap or an inclined plate, see Fig. 4) with an angle $\theta$ relative to the horizontal plane as the typical shear flow. We will discuss this situation as a prototype for describing shear flows. One needs to recall that experimentally, stationary dynamics only occurs for a relatively narrow range of angle $\theta$ (roughly speaking, between 20 and 30 degrees). The velocity profiles $\mathbf{V} = V\mathbf{e}_x$ and the grain volume fraction $\phi$ are thus only dependent on the $z$ variable *distance to the free surface* and the general dynamics reads:

$$0 = -\frac{\partial \tau_{xz}}{\partial z} + \phi \rho g \sin(\theta) \,, \quad 0 = -\frac{\partial \tau_{zz}}{\partial z} + \phi \rho g \cos(\theta) \tag{3}$$

where $\rho$ is the constant mass per unit volume of the grain material, $g$ is the acceleration of gravity, and $\tau$ the stress tensor. Due to the geometry and the steadiness, only two components of the stress tensor are important for the dynamics: $\tau_{zz}$ which plays the role of a pressure and is due to the long-lived contact forces between grains and to the almost bounce-less impacts; $\tau_{xz}$ on the other hand expresses the shear stress and is due also both to solid properties (Coulomb-like relation) and viscous like effects. For further models, $\tau_{xx}$ could be determined using Janssen argument of proportionality while the symmetry property of the tensor determines $\tau_{zx}$. The contributions to the stress tensor come from both compression forces (compressible term) and shear stress. In our approach, we decompose the diagonal stress tensor component into these two contributions. The possibility of such simple decomposition is still a subject of interesting scientific debates, although it is one of the main ingredients of our approach! We introduce a typical scale $P^*$ for the compressive part of the tensor. The need for a new scaling looks rather strange since two "natural" scaling for grain compression could be already invoked: an extrinsic one calculated as the typical gravity pressure encountered by a grain $\rho g d$ and a material dependent one, Young's modulus. From experimental results,[39] $P^*$ appears to be of the order of $\rho g D$ for usual value of the diameter $D$, orders of magnitude below Young's modulus of grains. Although it has been first proposed to use $\rho g d$ as a pressure scale[22] it has been noticed that this external force could not be used to describe an intrinsic property of the grain compactions[46]! Actually, $P^*$ has to be related to the pressure needed to obtain stable dense granular system; it is in fact difficult to determine such a pressure without any well defined statistical mechanics for these systems. Such statistical physics has been proposed by S. Edwards in 1994[45] and a discussion on the pressure definition has been developed in Ref.[46] Thus, $P^*$ should be related to the weight of the different configurations of a system of grain for a given density, given a consistent definition of a "granular temperature" if any. It should in particular depend on the geometrical details of the grains (roughness for instance). One can then interpret $P^*$ as an intrinsic pressure of dense granular systems. The compressive stresses are related to the grain volume fraction as $P^*F(\phi)$, where $dF/d\phi$ is the non-dimensional rigidity of the granular medium. Since deformation is mostly due to gravity, the rigidity is in fact related to the grain configurations when sheared (see[46] for a detailed discussion). Viscous-like behavior has to enter in the tensor through rate-dependent impact stress. Such influence has been estimated by Bagnold[3,4] where the theoretical justifications are based on dilute flows. Alternatively one can determine such Bagnold-like term from purely dimensional analysis so that we propose for the full normal stress finally:

$$\tau_{zz} = \rho D^2 \mu_N(\phi) \left(\frac{dV}{dz}\right)^2 + P^*F(\phi), \qquad (4)$$

where $\mu_N(\phi)$ represents the compaction-dependent intensity to be determined later on) of the normal stress induced by the shear rate.

For the shear component of the stress tensor, since the granular medium is flowing (and thus sliding), we continue our split static-shear approach by assuming hat it is made of a Coulomb-like contribution with a friction coefficient $\mu(\phi)$ completed by a Bagnold-like contribution involving a coefficient $\mu_T(\phi)$:

$$\tau_{xz} = \rho D^2 \mu_T(\phi) \left(\frac{dV}{dz}\right)^2 + \mu(\phi)\tau_{zz}. \qquad (5)$$

One can observe from this two components stress tensor (namely $\tau_{xz}$ and $\tau_{zz}$), that our modelization had finally to introduce four unknown functions of the grain compactions (the function $F$ and the different $\mu$ coefficients). This apparent weakness of our approach should be attenuated by the fact that this quantities are due to constitutive relations and could be formally (although it is complicated) measured in experiments once forever. Moreover, simple behavior arguments can give important insights on their general dependence on the density. Firstly, these four functions are characteristic of the dense regime and have a meaning in the range $\phi_m \leq \phi \leq \phi_M$ only. Due to solid deformation, we expect $F$, $\mu_T$ and $\mu_N$ to become extremely large, of the order of Young's modulus when $\phi = \phi_M$, because no motion nor extra compaction is expected above the maximum random packing. We should then take this limit as infinity at our level of modelization. We also expect $F$ and $\mu_N$ to vanish for $\phi = \phi_m$, because the normal stresses must vanish for the most tenuous contact network (recall that the influence of the embedded fluid is neglected). Concerning the friction coefficient $\mu$, reminiscent of the Coulomb friction between the grains, it is the only coefficient which remains finite all over the considered range and it seems in fact to be a slightly decreasing function of the compaction.[35] Therefore, the three scalars $F$, $\mu_N$ and $\mu_T$ are showing very large variations for small changes of the compaction, while $\mu$ has a much smoother behaviour. A detailed discussion on the different contributions arising for the function $F$ can be found in.[46] The free-surface boundary condition allows the direct integration of the momentum relation which implies that $\tau_{xz} = \tan(\theta)\tau_{zz}$ everywhere. By sake of simplicity, we will write $P^* = \rho g D F(\phi) cos(\theta)$ from now on. This rough approximation consists in two strong simplification. First, we acknowledge that $P^*$ is of the order of $\rho g D$ although such assumptions might probably mean that the so-defined function $F(\phi)$ is now not as universal as initially proposed. The $cos(\theta)$ multiplicative term is there only for formula convenience although it should not affect the final solution much ($\theta$ varying very little in all the configurations we are interested in, so that again the angle can be absorbed by a prefactor in the function $F$ which loose its general properties however). We claim that these approximations only result in the fact that the quantitative determination of $F$ (the numerical prefactor in other words) is now problem-dependent but that we do not loose the general properties of the solutions. Finally, while solving the equations of motion (3) with the model expressions (4) and (5), one can eliminate the Bagnold contribution. The compaction profile and the velocity profile are now solution of:

$$D\frac{d\phi}{dz} = \frac{\phi}{\frac{\partial}{\partial\phi}[\frac{F}{1 - (\mu_N/\mu_T)(\tan(\theta) - \mu)}]} \qquad (6)$$

and

$$\left(\frac{D}{g}\right)^{1/2}\frac{dV}{dz} = -\left(\frac{F(\sin(\theta) - \mu\cos(\theta))}{\mu_T(1 - (\mu_N/\mu_T)(\tan(\theta) - \mu))}\right)^{1/2}. \qquad (7)$$

The boundary condition for the grain fraction at the free surface is $\phi = \phi_m$ since it delimits shear flows from dilute systems. According to (6) the solid fraction increases inside the pile towards its maximum value $\phi_M$. The typical length scale of variation is proportional to the grain diameter but depends on $\theta$ if $\mu_N/\mu_T$ is different from zero. Hence $\mu_N/\mu_T$ represents the relative magnitude of Reynold's dilatancy. The characteristic scale for velocity is $(gD)^{1/2}$ and relation (7) shows that a solution exists only for angle $\theta$ verifying the inequality $\mu(\phi) \leq \tan(\theta) \leq \mu(\phi) + \mu_T(\phi)/\mu_N(\phi)$. Indeed, for values of angle outside this range, the argument of the square-root is negative and no stationary solution of the flow can be found. More precisely, when $\tan(\theta) \leq \mu(\phi)$ the static Coulomb friction is important enough for the grains not to slide one over each other. On the other hand, if $\tan(\theta) \leq \mu(\phi) + \mu_T(\phi)/\mu_N(\phi)$ the friction terms cannot balance the gravity so that stationary flows cannot stand and accelerated flows occur. For certain values of $\theta$ the inequality might be possibly satisfied in a part only of the full range $\phi_m \leq \phi \leq \phi_M$. The four functions of the compaction introduced in the model cannot be deduced straightforwardly from the rather scarce experimental or numerical results on stationary shear flows. Moreover, the real discrete structure of granular system involves an additional difficulty to deduce the functions from the experiments. However, simple and reasonable assumptions on these functions allow for tractable solutions which present the general features found in the various experiments. *We assume henceforth that $\mu$ and $\mu_T/\mu_N$ are independent of the grain compaction.* $\mu$ is indeed found to vary only little from molecular-dynamics experiments. It comes from the intimate relation between this coefficient with the Coulomb coefficient of the grain materials. The assumption on $\mu_T/\mu_N$ cannot be justified on scientific ground. Outside of allowing simple analytical solution one can notice that it does not deny the general behavior of each of these coefficients. For the above problem, a stationary solution is then possible in a well-defined angle range $\theta_{min} \leq \theta \leq \theta_{max}$, with $\tan(\theta_{min}) = \mu$ and $\tan(\theta_{max}) = \mu + \mu_T/\mu_N$ similarly to experimental observations. As we will briefly present below, our model has been applied to all the experimental situations known by the authors where stationary dense flows are observed. It is remarkable to notice that for all these, the above model has never failed to describe the qualitative properties and regimes seen in the experiments without changing the model anymore nor the constitutive relations! As explained above, only the prefactor in the models would have to be fitted from a situation to another one in order to have quantitative agreements. The next chapters discuss some of the detailed solutions.

## 3.4. *Heap flows*

For the inclined granular flows, we distinguish in fact between heap flows and flows on inclined support. The first case describes sand dunes where the inclination angle is self-determined by the dynamics while the angle is imposed in the latter case. Notably, the velocity appears directly as an integral involving the compressibility function $F$:

$$\frac{V_{heap}(0)}{\sqrt{gD}} = \frac{(\sin(\theta) - \mu\cos(\theta))^{1/2}}{\left(1 - \frac{\mu_N}{\mu_T}(\tan(\theta) - \mu)\right)^{3/2}} \int_{\phi_m}^{\phi_M} \left(\frac{F}{\mu_T}\right)^{1/2} \frac{\partial F}{\partial \phi} \frac{d\phi}{\phi}. \quad (8)$$

So that an important features of these flows, the total granular flux flowing down the heap $Q_{heap}$ is also determined through $F$:

$$\frac{Q_{heap}}{D\sqrt{gD}} = \frac{(\sin(\theta) - \mu\cos(\theta))^{1/2}}{\left(1 - \frac{\mu_N}{\mu_T}(\tan(\theta) - \mu)\right)^{5/2}} \int_{\phi_m}^{\phi_M} \left(\frac{F^3}{\mu_T}\right)^{1/2} \frac{\partial F}{\partial \phi} \frac{d\phi}{\phi}. \quad (9)$$

The boundary values of the integrals are such that the density at the free surface is by definition $\phi_m$ while the flow stops for $\phi = \phi_M$ in the dune for the present model.

From experimental observations, we expect the free-surface velocity and the flux to remain finite even for infinite size dunes for $\theta_{min} < \theta < \theta_{max}$. The two modelling functions $F(\phi)$ and $\mu_T(\phi)$ must be thus chosen such as to guarantee the convergence of the above integrals. $V_{heap}(0)$ and $Q_{heap}$ appear as function of $\theta$ with numerical prefactors depending on the particular properties $F$ and $\mu_T$. Different choices for these functions can be made, invoking some statistical physics argument for $F$ and convergence property for $\mu_T$. A reasonable set of functions, that we will keep further on in general, is:

$$F = F_0 \text{Log}\left(\frac{\phi_M - \phi_m}{\phi_M - \phi}\right) \quad \text{and} \quad \mu_T = \mu_{T0}\left(\frac{\phi_M - \phi_m}{\phi_M - \phi}\right)^2. \quad (10)$$

This choice of constitutive functions is discussed in more detailed in.[46] The above expression for $F$ leads to a solid fraction profile which increases

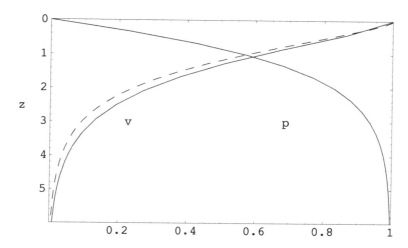

Figure 7.  Reduced velocity profile $V_{heap}/V_{heap}(0)$ versus the dimensionless distance $z/L(\theta)$ to the free surface.  The dashed curve represents the approximate expression (14). The reduced compaction profile $(\phi - \phi_m)/(\phi_M - \phi_m)$ is plotted as well.

exponentially with depth:

$$\phi_{heap}(z, \theta) = \frac{\phi_M}{1 + (\frac{\phi_M}{\phi_m} - 1)e^{-z/L(\theta)}} \tag{11}$$

where

$$L(\theta) = \frac{F_0 D}{\phi_M (1 - \frac{\mu_N}{\mu_T}(\tan(\theta) - \mu))} \tag{12}$$

represents the typical thickness of the layer flowing down the heap. Noticing that:

$$\frac{\phi_M - \phi}{\phi_M - \phi_m} = \frac{\phi_M e^{-z/L(\theta)}}{\phi_m + (\phi_M - \phi_m)e^{-z/L(\theta)}}$$

we can show that the relative velocity profile:

$$\frac{V_{heap}(z)}{V_{heap}(0)} = \frac{\int_{\phi_{heap}(z)}^{\phi_M} \left(\frac{F}{\mu_T}\right)^{1/2} \frac{\partial F}{\partial \phi} \frac{d\phi}{\phi}}{\int_{\phi_m}^{\phi_M} \left(\frac{F}{\mu_T}\right)^{1/2} \frac{\partial F}{\partial \phi} \frac{d\phi}{\phi}} \tag{13}$$

deduced from the above expressions for $F$ and $\mu_T$ has the following behavior close to the free-surface and far from it:

- for $z/L(\theta) \ll 1$, $\dfrac{V_{heap}(0)-V_{heap}(z)}{V_{heap}(0)} \propto (z/L(\theta))^{3/2}$
- for $z/L(\theta) \gg 1$, $\dfrac{V_{heap}(0)-V_{heap}(z)}{V_{heap}(0)} \propto \sqrt{z/L(\theta)}e^{-z/L(\theta)}$

Thus, this relative velocity is exponential-like for $\frac{z}{L(\theta)} \gtrsim 2$ (see Fig. 7) but displays a Bagnold-like region with a $z^{3/2}$ dependence of inverse concavity for $\frac{z}{L(\theta)} \lesssim 0.2$ (see Fig. 8). In fact, our numerical solution for the relative

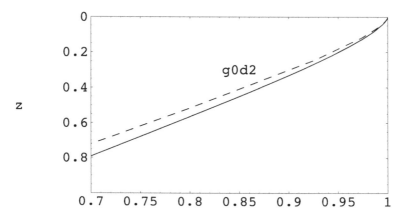

Figure 8.  Zoom of Fig. 7 showing the Bagnold like region for $z/L(\theta) < 0.2$. The dashed curve represents approximation (14).

velocity profile is quite well fitted by the analytical expression

$$1 - \frac{V_{heap}(z)}{V_{heap}(0)} = \left(\frac{\phi_{heap}(z)-\phi_m}{\phi_M-\phi_m}\right)^{\frac{3}{2}} = \left(\frac{1-e^{-z/L(\theta)}}{1+(\frac{\phi_M}{\phi_m}-1)e^{-z/L(\theta)}}\right)^{\frac{3}{2}}. \quad (14)$$

With $\phi_m = 0.5$ and $\phi_M = 0.65$ the total flux flowing down the heap reads

$$\frac{Q_{heap}}{D\sqrt{gD}} = 1.4\frac{F_0^{5/2}}{\mu_{T0}^{1/2}}\frac{(\sin(\theta)-\mu\cos(\theta))^{1/2}}{\left(1-\frac{\mu_N}{\mu_T}(\tan(\theta)-\mu)\right)^{5/2}}$$

$$= 1.4\frac{F_0^{5/2}}{(\mu_{T0}\cos(\theta_{min})^{1/2}}\left(\frac{\mu_T}{\mu_N}\right)^{5/2}\frac{\sin(\theta-\theta_{min})^{1/2}}{(\tan(\theta_{max})-\tan(\theta)))^{5/2}}.$$

The dependence on $\theta$ of $L$ and $Q_{heap}$ are represented in Fig. 9, with $\mu = 0.36$ and $\mu_N/\mu_T = 4.7$.

As suggested by the formulas, the flux diverge when approaching $\theta_{max}$ as $(\theta_{max}-\theta)^{5/2}$ indicating that above this angle the whole system is flowing.

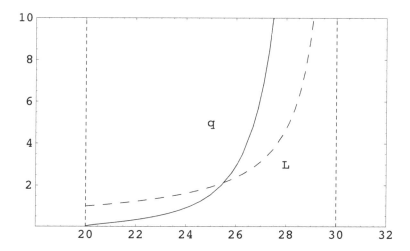

Figure 9. $\theta$-dependence of flux $Q_{heap}$ and dimensionless thickness $l(\theta) = \phi_M L(\theta)/D$.

### 3.4.1. *Flow on rough plates*

We differentiate here the flow on rough plates from the heap flow because important differences can be pointed out between the two situations: firstly, the angle for rough plate is generally imposed by the plate, except for the immature dunes as discussed below. For rough plate also, the flux of grains is imposed by the experiments while in dunes, the angle and the flux depends one on each other (it is for our model the curve in Fig. 9). The friction between the plate and the grains is still a subject of intensive research and at least, a boundary layer near the plate has to be accounted. The assumptions made above on the stress tensor are not valid anymore and we do not know at this stage how to describe the stress tensor in this layer. Indeed, the constitutive equations (4) and (5) hold in the bulk only of the dense granular medium and are likely to be modified close to the rough plate. Since the role of the plate roughness is difficult to assess quantitatively, we discard the description of the "basal layer" (boundary layer) close to the plate[29,31] and assume a slip velocity $V_s$ at some distance $\delta$ above the rough plate. Then we apply (4) and (5) to a layer of thickness $h$, so that the free surface is located at a distance $h + \delta$ above the rough incline. In the layer of thickness $h$, the solid fraction increases from $\phi_m$ at the free surface to the value $\phi_{heap}(h)$ at a distance $\delta$ from the rough plate where the velocity is $V_s$. The total flux through the core region is now given by:

$$\frac{Q_{plate}}{D\sqrt{gD}} = \frac{F(\phi_{heap}(h))}{1 - \frac{\mu_N}{\mu_T}(\tan(\theta) - \mu)} \frac{V_s}{\sqrt{gD}}$$

$$+ \frac{(\sin(\theta) - \mu\cos(\theta))^{1/2}}{\left(1 - \frac{\mu_N}{\mu_T}(\tan(\theta) - \mu)\right)^{5/2}} \int_{\phi_m}^{\phi_{heap}(h)} \left(\frac{F^3}{\mu_T}\right)^{1/2} \frac{\partial F}{\partial \phi} \frac{d\phi}{\phi}. \quad (15)$$

Although the exact structure of the basal layer is difficult to describe, we expect that a rough plate slows down the core region more efficiently than a heap would do, namely $V_s \leq V_{heap}(h)$. As a consequence, $Q_{plate}(h, \theta)$ as given in (15) should not exceed $Q_{heap}(\theta)$ given in (9). Usually, a fixed flux $Q_{plate}$ is imposed in the experiments so that the grains flow down the rough plane inclined at angle $\theta$. Therefore, two different situations can occur depending on whether the imposed flux is bigger or smaller than the one expected for the equivalent dune: when $Q_{plate}$ is larger than $Q_{heap}(\theta)$, the granular medium will in fact rearrange so as to flow down over a heap of angle $\theta + \alpha$ with $Q_{plate} = Q_{heap}(\theta + \alpha)$. We suggest here that a such scenario presents a coherent explanation for the "immature sliding flows" that were observed in some experiments.[26,27]

Due to the very large increase of $Q_{heap}$ with $\theta$ (see Fig. (9)) and because the experimental flux is limited to some maximal value, immature sliding flows were in fact observed for $\theta \sim \theta_{min}$ only. Conversely, when $Q_{plate}$ is smaller than $Q_{heap}(\theta)$ the whole layer of thickness $h$ is in motion with a velocity everywhere larger than $V_s$, the velocity and density profiles are such that the imposed flux of grains flows gently on the plate. Moreover, when $h/L(\theta) \stackrel{<}{\sim} 0.2$, the Bagnold-like velocity profile (which could hardly be observed in heap flows, see Fig. (8)) is now invading the whole core region. In fact, when expressions (10) are taken for granted and $h/L(\theta) \stackrel{<}{\sim} 0.2$, the total flux (15) has the special form

$$\frac{Q_{plate}}{D\sqrt{gD}} = \frac{\phi_m V_s h}{D\sqrt{gD}} + \frac{2}{5}\phi_m^{3/2} \left(\frac{(\sin(\theta) - \mu\cos(\theta))}{\mu_{T0}}\right)^{1/2} \left(\frac{h}{D}\right)^{5/2}. \quad (16)$$

When the role of the velocity slip can be neglected, the second contribution gives a $h^{5/2}$ scaling law for the grain flux down a rough incline. Note that this scaling stems from our particular choice (10) and varies for different choice of $\mu_T$.[22]

In the following, we will briefly show how the same model can be used in different situations where dense granular shear flows are the dominant dynamics. The model features are not changed at all but are adapted to the different geometries encountered. In fact, the changes only arise by the precise determination of the main direction of compression and the one of the shear stress. It is particularly remarkable that our approach allows a fair qualitative descriptions of all the situations we have been aware of.

### 3.4.2. *Confined shear flow*

We consider here a flows obtained by shearing a system whose coherence is maintained through an external load. The pressure load exerted on the boundaries of the granular medium is supposed to be applied along direction $z$, which is thus the direction of main compression, while shear stress is imposed along the other direction. Because gravity plays a minor role concerning the compressive forces, the constitutive relation for $\tau_{zz}$ is simply (compare with (4))

$$\tau_{zz} = \rho D^2 \mu_N(\phi) \left(\frac{dV}{dz}\right)^2 + \rho g D F(\phi), \qquad (17)$$

whatever the angle $\theta$ between the $z$ axis and gravity. The flow is along the $x$ axis and the constitutive relation for the shear stress $\tau_{xz}$ is still given by (5), without any change as compared to the free-surface case.

### 3.4.3. *Plane shear flow*

As a first type of confined shear flow, we consider the planar shear of an infinite horizontal granular layer bounded by two plates separated by a fixed distance $h$. The pressure load is then adapted for keeping $h$ constant. An alternative problem would be to consider that the external load is fixed while the separation distance adapts with the shear stress. The pressure load and the gravity are both oriented along the direction $z$ and the flow is along direction $x$ (see Fig. 5). Therefore, an important consequence of the momentum balance is that the shear stress $S$ is constant all over the volume, while the normal stress is determined by the equation:

$$\tau_{xz} = S \quad \text{and} \quad \tau_{zz}(z) = P(0) + \rho g \int_0^z \phi(\xi) d\xi$$

where $P(0)$ is the pressure load exerted on the upper plate $z = 0$ ($z = h$ stands for the lower plate). We will distinguish the situation without and with gravity, the first case being considered often in numerical simulations while the second one can be related to experiments.

### 3.4.4. *Without gravity*

In this case the normal stress is also a constant $P$ all over the granular layer and thus the only solution consists in a constant density and constant velocity gradient if any. Indeed, the constitutive equations (17) and (5) yield:

$$\rho D^2 \mu_T(\phi) \left(\frac{\partial V}{\partial z}\right)^2 = S - \mu(\phi)P$$
$$\rho D^2 \mu_N(\phi) \left(\frac{\partial V}{\partial z}\right)^2 = P - \rho g D F(\phi). \tag{18}$$

Depending on the sign of $S - \mu(\phi)P$ for all permitted $\phi$, we will have a static or a moving medium. In the static case the pressure load is noted $P_0$ and the shear is such that $S \leq \mu(\phi_0)P_0$ where $\phi_0$ is the constant compaction of the medium related to the pressure load through $P_0 = \rho g D F(\phi_0)$. In the dynamic case, the compaction is still a constant and because of mass conservation, this constant is nothing but the static value $\phi_0$. The shear $S$ is now larger than $\mu(\phi_0)P_0$. The velocity gradient is constant:

$$\rho D^2 \left(\frac{\partial V}{\partial z}\right)^2 = \frac{S - \mu(\phi_0)P_0}{\mu_T(\phi_0) + \mu(\phi_0)\mu_N(\phi_0)}.$$

Due to dilatancy effects the pressure load exerted on the plates is necessarily larger than in the static case, following:

$$P(S) = P_0 + \frac{S - \mu(\phi_0)P_0}{\mu(\phi_0) + \mu_T(\phi_0)/\mu_N(\phi_0)}.$$

Our model thus predicts a given shear threshold $S_t(\phi_0) = \mu(\phi_0)\rho g D F(\phi_0)$ for any density and an effective friction coefficient, function of $\phi_0$ and $P$:

$$\frac{S}{P} = \mu(\phi_0) + \frac{\mu_T(\phi_0)}{\mu_N(\phi_0)}\left(1 - \frac{\rho g D}{P}F(\phi_0)\right).$$

### 3.4.5. *With gravity*

In this case the normal stress increases in the downward direction so that the constitutive equation (5) results in

$$\rho D^2 \mu_T(\phi) \left(\frac{\partial V}{\partial z}\right)^2 = S - \mu(\phi)P(0) - \mu(\phi)\rho g \int_0^z \phi(\xi)d\xi.$$

If flow can occur at the upper boundary of the system, as we go down the layer, the weight of the above grains leads to a more compacted system which can eventually jam and no more motion are possible below. This shear localization of the flow, observed in experiments, is naturally a solution of our model. We will first describe the static case before considering grain motions. Because the compaction on the upper plate is necessarily different in the static and the dynamic cases, we define $P_0(0)$ as the pressure load exerted on the upper plate when the granular medium is motionless and $\phi_0(z)$ as the static compaction profile. As long as $S \leq \mu P_0(0)$, the granular slab is motionless, the compaction $\phi_0(0)$ at the upper plate satisfies $P_0(0) = \rho g D F(\phi_0(0))$ and the compaction profile is:

$$\phi_0(z) = \frac{\phi_M}{1 + \left( \frac{\phi_M}{\phi_0(0)} - 1 \right) e^{-z/L_0}} \quad \text{with} \quad L_0 = \frac{F_0}{\phi_M} D$$

When the granular medium is flowing, the compaction profile $\phi(z)$ displays larger gradients and becomes

$$\phi(z) = \frac{\phi_M}{1 + \left( \frac{\phi_M}{\phi(0)} - 1 \right) e^{-z/L}} \quad \text{with} \quad L = \frac{L_0}{1 + \mu \frac{\mu_N}{\mu_T}}$$

where $\phi(0)$ is the new compaction at the upper plate. Since mass conservation requires

$$\int_0^h [\phi(z) - \phi_0(z)] dz = 0,$$

it is clear that the inequality $L < L_0$ implies $\phi(0) < \phi_0(0)$ and $\phi(h) > \phi_0(h)$. The compaction of the moving medium is thus reduced at the upper plate as compared to its static value while it is enhanced at the lower plate. The velocity profile is then deduced from the compaction profile

$$\left( 1 + \mu \frac{\mu_N}{\mu_T} \right) \frac{D}{g} \left( \frac{\partial V}{\partial z} \right)^2 = \frac{S^* - \mu F(\phi)}{\mu_T(\phi)}$$

where $S^*$ is the dimensionless shear $\frac{S}{\rho g D}$. Let us define the volume fraction $\phi^*$ such that $S^* = \mu F(\phi^*)$. It is clear that $\phi^* > \phi_0(0)$ because $S > \mu P_0(0)$. The above equation implies that motion is present for compaction less than $\phi^*$ only. This condition leads to check the self-consistency relation $\phi(z) < \phi^*$ for $0 < z < h$. This condition is automatically satisfied in the upper part of the flow since $\phi(0) < \phi_0(0) < \phi^*$. But it may not be in the lower part, thus leading to a shear localization. This *bulk* localization is here depending on $S^*$ and $h/L$. Fig. 10 shows the compaction and velocity profiles for two

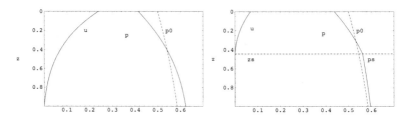

Figure 10.   Velocity profile $V(z)$ and compaction profile $\phi(z)$, left large $S$, right small $S$.

different values of $\phi^*(S^*)$ representing the two different flowing situations: shear localization or flow over the whole layer.

### 3.4.6. *Vertical chute flows*

The situation of vertical chute flow between two vertical plates shares important features with the previous case. Indeed, the flow is confined and can be stopped by imposing high enough compression. The main difference lies in the role of the gravity which here is responsible for the shear stress (see Fig. 6). The compaction is due to a pressure load $P$ exerted on the two plates along direction $z$. The flow and the gravity are oriented along direction $x$. The equations of motion result in a constant normal stress and a variable shear stress, by contrast to the previous case:

$$\tau_{zz} = P \text{ and } \tau_{xz} = \rho g \int_0^z \phi(\xi) d\xi,$$

where $z = 0$ corresponds to the symmetry plane located between the two plates at which the shear stress vanishes. The constitutive relation (5) implies

$$\rho D^2 \mu_T(\phi) \left( \frac{\partial V}{\partial z} \right)^2 = \rho g \int_0^z \phi(\xi) d\xi - \mu(\phi) P.$$

Either the right-hand side is everywhere negative (due to a very high pressure load) and the medium is motionless or there is a central region of the flow in which the shear stress does not exceed $\mu P$ and consequently where the strain rate vanishes. In this plug flow regime the solid fraction is a constant $\phi^*$ related to the pressure load as $P = \rho g D F(\phi^*)$. The thickness $z^*$ of the plug flow depends on $\phi^*$ (hence on the pressure load)

$$\frac{z^*}{D} = \frac{\mu(\phi^*)}{\phi^*} F(\phi^*).$$

Close to the vertical plates, there is a shear layer where the velocity decreases to $V_w$ dependent on the plate roughness. In this parietal shear layer, the constitutive equations (17) and (5) imply:

$$\left(\frac{D}{g}\right)^{1/2} \frac{\partial V}{\partial z} = -\left(\frac{F(\phi^*) - F(\phi)}{\mu_N(\phi)}\right)^{1/2} \tag{19}$$

and

$$D \frac{\partial \phi}{\partial z} = \frac{\phi}{\frac{\partial}{\partial \phi}\left[(\mu + \frac{\mu_T}{\mu_N})F(\phi^*) - \frac{\mu_T}{\mu_N}F(\phi)\right]}. \tag{20}$$

With the same assumptions as above for the constitutive functions, the compaction profile in the shear layer $z^* < z < z_w$ reads:

$$\phi(z) = \frac{\phi_M}{1 + \left(\frac{\phi_M}{\phi^*} - 1\right) e^{\frac{z-z^*}{L^*}}} \tag{21}$$

where $L^*$ is the typical shear layer thickness:

$$\frac{L^*}{D} = \frac{\mu_T F_0}{\mu_N \phi_M}.$$

For the flow to be dense up to the vertical plates, the wall compaction $\phi_w$ must be larger than $\phi_m$ and the shear layer thickness is

$$\frac{z_w - z^*}{L^*} = Log\left(\frac{\frac{\phi_M}{\phi_w} - 1}{\frac{\phi_M}{\phi^*} - 1}\right).$$

As a consequence, the distance $2z_w$ between the two plates is a function of $\phi^*$ (hence of $P$) and of $\phi_w$ (hence of the plate roughness). Concerning the velocity, it increases from a value $V_w$ at the wall to a value $V_{plug}$ in the central part. The computed relative velocity field is represented in Fig. 11 together with the fit

$$\frac{V(z) - V_w}{V_{plug} - V_w} = 1 - \left(\frac{\phi^* - \phi(z)}{\phi^* - \phi_w}\right)^{3/2}. \tag{22}$$

To conclude on this chapter on shear flows, we would like to insist on the consistent results obtained with this continuum model despite the important simplifications. Qualitative agreements are found with experiments[39] and for all the situations studied further on. In our model, the volume fraction and the effective friction coefficient (defined similarly as in[39] as the ratio between the shear stress and the confining pressure) are not only functions of the inertial number $I$ (dimensionless number defined by the

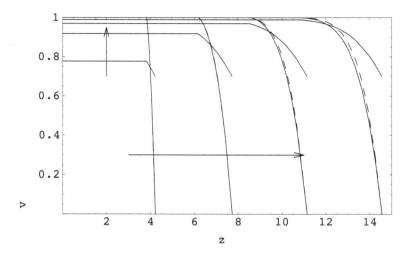

Figure 11. Reduced velocity profiles $(V(z) - V_w)/(V_{plug} - V_w)$ versus the dimension-less distance $\frac{z\mu_N\phi_M}{D\mu_T F_0}$ to the center. The arrow is in the increasing values of $\frac{P}{\rho g D F_0}$. The dashed curve represents the approximate expression (22). The reduced compaction profiles $\phi/\phi_M$ with $\phi_w/\phi_M = 0.7$ are plotted as well.

shear rate, typically $I = \dot{\gamma}D/\sqrt{P/\rho}$, where $\dot{\gamma}$ is the shear rate) but also of the ratio $P/P^*$. However, the derivation of the model as well as the deductions of the rheological parameters need to be improved in further studies. For instance, the links between a statistical descriptions of the grains contacts and the different parameters should be studied. On the other hand, this model gives well defined velocity and density profiles of stationary flows in general. Performing usual assumptions of film dynamics, it would be interesting to deduce from such laws the effective Saint-Venant like equations for granular flows.

## 4. "Built upon sand": granular solids

This title has in fact been borrowed from a review article written by L. Kadanoff.[23] The goal of this chapter is indeed to discuss the capability for an assembly of grains to support very high external load, although we all know that such materials always remain fragile under particular constraints. Actually, soil mechanics tells us that huge isotropic constraint can be imposed to granular systems, thanks to high Young's modulus and yield criteria. Grains are very hard but they can also dissipate energy very efficiently as it is used in train ballast for instance. Grain assemblies can also be very fragile: they can strongly deform or let easily slide objects when submitted to shear stresses. In fact, as already shown above, grains react very differently depending on the external driving that is imposed. Only few aspect of granular solids are rapidly discussed below and we encourage an interested readers to make use of the bibliography given here.

## 4.1. *Force distribution*

Granular system can support normal stress as usual solid do. However, while for usual solid, the force distribution is fairly homogenous in the bulk, only a few fraction of the grains participate to the effort here! The force are in fact distributed along particular network determined by the initial grains positions which concentrate all the compression forces that the system has to support. These so-called chain forces have been often compared for illustration to arches structures in building constructions. Thus, the force distribution, measured at the bottom of a cylindrical bucket submitted to high compression rate show exponential tails for high forces.[49] A random model have been derived to explain this stress focusing on preferred lines.[50,51] The model considers a triangular lattice (in 2-D first) oriented along the selected direction of the imposed compression. The normal stress and the weight are transferred from each grain to its next neighbours in the following level with random probability. The randomness of the model leads to a parabolic equation for the mean stress propagation. Such simple model, generalized in 3D, seems to reproduce qualitatively most of the experimental results on force distributions. Actually, the situation ends up to be more complex: the precise law of re-distribution of the forces between grains is in fact still questioning. Indeed, although such simple toy models where the mean stress is distributed along neighbors give stress focussing, the exact nature of the stress propagation is still difficult to determine. On one hand, random walk-like models suggest a parabolic equation for the stress evolution inside the sand pile. On the other hand stress propagation arguments argue for hyperbolic systems (see[52–55] for reviews).

## 4.2. *Sound propagation*

The propagation of sound in granular system is complex since it involves multiple diffusion and diffraction effects induced by the random network of grains. An important related question would be: can we obtain information on the granular structure from the transmitted signals? This inverse

problem is relevant in geophysical and seismic context, for the oil industry and also for medical applications but it is out the scope of our review.
Sound propagation in granular system can moreover reveal the details of
the contact forces and deformation properties of the materials. For the
interested reader, the following works provide a good introduction to the
recent results on the subject.[56–58]

## 5. Grains and glasses

Besides gas, liquid or solid state, granular have often been compared with
glasses. Glassy behaviors differ from usual thermodynamics equilibrium
because they are in out of equilibrium states which evolve very slowly. The
grains analogy with glasses is strongly supported by a sequence of recent
experimental works on granular compaction[59,60] where the very slow compaction of a grain column under tapping is studied. Similar conclusions
can also be drawn from compaction measurements of grains under cyclic
shear.[61] Moreover, there are strong similarities between the glassy transition when the viscosity of the liquid increases dramatically and the jamming
of grains.[62,63] Theoretically, one can argue that the energy phase space of
grains under gravity presents complex structures with many metastable
equilibria, similarly to spin glasses. It is then tempting to investigate the
analogy between grains and glasses, where the temperature control parameter for glasses would be replaced by some external load for grains.[62] After
a short review on some recent results on this analogy, a brief description
of the memory effects in granular compaction will be presented, following
previous work done by one of us (C.J.).[64]

### 5.1. What kind of analogy?

The analogy between grains and glasses invokes in particular the similar
complex energy landscape which contains many metastable equilibria. For
granular systems, these complexity of locally stable configuration can be
explored by a compaction experiment as performed in.[59,60] There, 1 mm–
diameter glass beads are vertically shaken in a tall, 19 mm-diameter glass
tube under vacuum. The successive vertical tapping perturbations compact
the grain assembly. By "taps", we consider here a single oscillation motion
of the whole apparatus which transmits an oscillating (one single oscillation
per tap) acceleration to the grain. The tap is thus typically defined by $\Gamma$,
the peak applied acceleration normalized by gravity, g. The packing density

of the beads are measured using capacitors mounted at four heights along the column. In these articles, it is shown that an initially loose packing of glass beads slowly compacts, asymptoting to a higher steady state packing fraction. This "equilibrium" packing fraction is lower than the random close packing limit, $\rho_{rcp} \approx 0.64$, and is a decreasing function of $\Gamma$. The relaxation dynamics are extremely slow, taking many thousands of taps for the packing fraction, $\rho$, to approach the steady state value. During this evolution, $\rho$ increases logarithmically with the number of taps, $t$. Such logarithmic behavior is typical for self–inhibiting processes where the time scale of the dynamics is the time history itself.[65] Fluctuations around the asymptoting steady state has also been investigated and their amplitudes decrease with $\Gamma$. Somehow the shaking intensity plays, at least qualitatively, the role of temperature. The spectrum of excitation itself shows however a complex behavior. It is indeed strongly non-Lorentzian,[60] revealing the existence of multiple time scales in the system. The shortest and the longest relaxation time-scales differ by as much as three order of magnitude, and the behavior of the spectrum for the intermediate frequencies is highly non-trivial involving different power laws. Similar results were also obtained for compaction monitored by periodic shear deformation of the grain assembly.[61]

These experimental observations suggested that some properties reminiscent of the glass dynamics were present also for grains.[66] In fact, glassy property of a materials is often characterized by the response of the system to sudden perturbation of the temperature, as classically studied for aging in glasses.[67,68] Thus, analogous experiments for grains have been performed on the tapping compaction,[64] on the cyclic shear experiments[61,69] and on numerical models.[70,71] Such experiments are actually helpful to test different thermodynamic theories proposed for understanding general granular properties (see for instance[45,72,73]). The difficulty for granular systems is that no temperature and in addition no entropy can be easily defined. An alternative thermodynamics function, the so-called compactivity has been proposed to allow the development of a related statistical mechanics.[45] This promising compactivity concept is based on the available volume for a grain in a well defined configuration (see for instance[73,74]). Similarly, recent experimental results have also focused on the individual motion of the grains in order to probe the microscopic properties of their dynamics.[69,75] They show in particular that the grain path alternate long and slow dynamics in cage-like structures with rapid diffusion between cages. The quest for a consistent statistical mechanics for grains is thus still a subject of intensive and questioning researches!

In the following chapter, we briefly illustrate the memory effects in granular materials. We present the experimental response obtained when a granular system under compaction is submitted to a sudden change in the amplitude of tapping $\Gamma$. The results show clearly that the grains react differently depending on the history of their compaction.

## 5.2. *Memory effects in compaction*

The dependence of granular systems on their history of preparation can be shown using a simple experiment: the system is compacted with vibration intensity $\Gamma_1$ during $t_0$ taps and the intensity is suddenly switched to $\Gamma_2$. The result depends on whether $\Gamma_2$ is bigger or smaller than $\Gamma_1$ and it shows counter-intuitive results than what would be deduced from the dependence of the asymptotic density with $\Gamma$. Namely, for $\Gamma_2 < \Gamma_1$ (Fig. 12a) we found that on short time scales the compaction rate increases. For $\Gamma_2 > \Gamma_1$ (Fig. 12b) we found that the system dilates immediately following $t_0$. These results are opposite from the long-time behavior seen in previous experiments where the asymptotic compaction rate decreases as $\Gamma$ increases increased. Note that after several taps the "anomalous" dilation ceases and there is a crossover to the "normal" behavior, with the relaxation rate becoming the same as in constant-$\Gamma$ mode.

These data show that short-term memory is important: the future evolution of $\rho$ after time $t_0$ depends not only on the density at $t_0$, $\rho(t_0)$, but also on other factors to be identified. These factors clearly depend on the way that the system has reached the density $\rho(t_0)$ which means that other variables than the mean density are needed to characterize the grains. In order to demonstrate this in a more explicit manner, we have reached the same density $\rho_0$ using three different vibrating intensities $\Gamma_1 < \Gamma_0 < \Gamma_2$. The number of taps to reach the same density is of course different for each $\Gamma$: the bigger the $\Gamma$, the smaller the number of taps. After $\rho_0$ was achieved at time $t_0$, the system was tapped with *the same intensity* $\Gamma_0$ for all three experiments. As seen in Figure 13, the evolution for $t > t_0$ strongly depends on the pre-history. The density $\rho_0$ cannot be the only order parameter describing compaction structure of a granular system. This need for extra state variables in the problem is in fact consistent with strongly non-Lorentzian behavior of the fluctuation spectrum, observed in earlier experiments.[60] It suggests that different time scales intervene during the relaxation processes.

To give a theoretical interpretation of the above results, we present

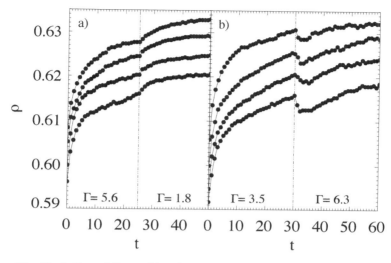

Figure 12.  Evolution of the packing fraction, $\rho$, at four heights in the column, as a function of tap number, $t$. Two different single-switch experiments: (a) $\Gamma$ was lowered from 5.6 to 1.8 at $t_0 = 25$; and (b) $\Gamma$ was increased from 3.5 to 6.3 at $t_0 = 30$. Curves are shifted vertically for clarity. Each curve is an average over 4 runs, and the measurement uncertainty in $\rho$ is $4 \times 10^{-4}$.

an illustrative model: the grain systems is composed as a set of discrete "microscopic" states corresponding to different realizations of the packing topology. These "microscopic" states can exhibit transition between different local configuration involving different local densities. The typical time scale of the transition is then a single tap.[76] On the other hand, the network of these oscillating states is slowly evolving with time as the whole system compacts. For each tap there is a possibility for a transition from one microscopic state to another one with a small different density. Since the dynamics is dissipative and the system is under external gravity, a transition to a denser configuration is typically more probable than the reverse one. Such "flip-flop" description of the system leads us to introduce a *Baseline Configuration* (BC), which plays the role of a local free energy minima for our non–thermal system. Namely, a BC may be defined as a state where any transition to a different configuration has a lower probability than the reverse one. Hence, there is a mesoscopic time scale on which the system gets trapped in the vicinity of a given BC, and its evolution is dominated by a number of flip–flop modes, i.e. local "excitations" of the baseline structure, any of which would normally relax back to the

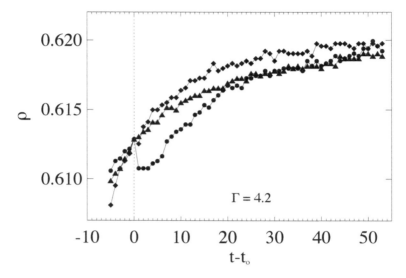

Figure 13.   The time evolution of packing fraction $\rho$ for a system which was compacted to $\rho_0 = 0.613$ at time $t_0$ using three different accelerations: $\Gamma_1 = 1.8$ (circles), $\Gamma_0 = 4.2$ (triangles), and $\Gamma_2 = 6.3$ (diamonds). After the density $\rho_0$ was achieved, the system was vibrated at acceleration $\Gamma_0$. The evolution for $t > t_0$ depended strongly on the pre-history. Each curve is an average over four experimental runs.

same BC. The evolution of the BC accounts for the slow compaction and its understanding and its description is beyond the scope of this simple model.

If we consider that the coupling between the individual flip–flop modes are accounted for by the baseline configuration evolution, we may consider the complicated configuration space as a set of independent two-state systems (similarly to[72,77–79]). It also means, that we consider only one excited states of lower mean density than the ground state. In this framework, each of these rapid modes is characterized by two transition rates, $\kappa_{e \rightarrow g} > \kappa_{g \rightarrow e}$ ("e" and "g" representing "excited" and "ground" states). $\kappa_{e \rightarrow g}/\kappa_{g \rightarrow e}$ gives the ratio of the equilibrium probabilities of populating each state and the BC corresponds to all modes at their ground state. These ground states are the one with the higher density since gravity is enhancing compaction, i.e. the volume change $v$ between the ground and the excited states is normally positive.

Because the taps introduce energy in the system and can therefore excite "open" (the one with lower density) states of the modes, the experimentally–

observed density is different from that of the current BC, $\rho_b$ and one can write the variation as:

$$\rho = \rho_b(t) \left( 1 - \frac{1}{V} \sum_n v^{(n)} \left( 1 + \frac{\kappa_{e \to g}^{(n)}}{\kappa_{g \to e}^{(n)}} \right)^{-1} \right). \tag{23}$$

The summation here is performed over all the flip–flop modes of a given BC, $V$ is the total volume, and $v^{(n)}$ is the volume difference between the excited and the ground states of the $n$-th mode. Since the excitations are due to the tapping dynamics, we expect the population of the excited states, $P(\Gamma) = (1 + \kappa_{g \to e}^{(n)} / \kappa_{e \to g}^{(n)})^{-1}$, to grow with $\Gamma$, starting from zero at $\Gamma = 0$. It is a direct consequence of the suggested qualitative analogy between $\Gamma$ and an effective temperature. Hence, for a given $\rho_b$, the total density $\rho$ will be lower at higher acceleration. With these simple ingredients only, we can explain the above experimental results. Thus, when the same density $\rho_0$ is reached with two different acceleration $\Gamma_1 > \Gamma_2$, this means that the baseline density $\rho_b$ is in fact higher for $\Gamma_1$ than for $\Gamma_2$. Thus, if one continues with $\Gamma_1$, the configuration prepared with $\Gamma_2$ will see the number of excited states increasing rapidly (on the order of few taps) leading to a rapid decrease of the density. The same reasoning can be applied to the opposite situation. Actually, after a switch from $\Gamma_1$ to $\Gamma_2$ at time $t_0 = 0$, one can determine the relaxation process using the developed model in the following way:

$$G_{\Gamma_1,\Gamma_2}(t) = \rho_b \int_0^{\kappa_{max}} F_{\Gamma_1,\Gamma_2}(v,\kappa) \left( 1 - \exp(-\kappa t) \right) dv d\kappa \tag{24}$$

Here $\kappa$ is the relaxation rate of an individual mode, and the distribution function $F_{\Gamma_1,\Gamma_2}(v,\kappa)$ is introduced as follows:

$$F_{\Gamma_1,\Gamma_2}(v,\kappa) \equiv \frac{1}{V} \sum_n \left( P^{(n)}(\Gamma_2) - P^{(n)}(\Gamma_1) \right) \delta(v - v^{(n)})$$

$$\delta \left( \kappa - \kappa_{g \to e}^{(n)}(\Gamma_2) - \kappa_{e \to g}^{(n)}(\Gamma_2) \right). \tag{25}$$

The distribution function is normalized so that $\int F_{\Gamma_1,\Gamma_2}(v,\kappa)) dv d\kappa = \rho^*(\Gamma_2) - \rho^*(\Gamma_1)$, where $\rho^*(\Gamma)$ is the equilibrium number density of the excited modes at given $\Gamma$. If we assume that $F_{\Gamma_1,\Gamma_2}$ does not vanish in the limit $\kappa \to 0$, one can show that the late stage of the relaxation of $G_{\Gamma_1,\Gamma_2}(t)$ is given by the power law:

$$G_{\Gamma_1,\Gamma_2}(t) = G_{\Gamma_1,\Gamma_2}(\infty) - \frac{\text{const}}{t}. \tag{26}$$

Therefore this model depicts two different processes: on short (in fact, mesoscopic) time scales, a fast relaxation due to the flip–flop modes is dominant, while over the long times, the dynamics are determined by the logarithmically slow evolution of the baseline density $\rho_b(t)$. The crossover between the two regimes is particularly obvious in Fig. 12b, where it results in a non-monotonic evolution. Such dynamics is unusual in spin glasses, but has been observed for conventional glasses.[80] For experiments performed at sufficiently late stages of the density relaxation we can neglect the baseline evolution although the described experiments provide us with a tool for studying the response of the system, not limited in fact to the nearly–equilibrium regime.

To investigate further on the experimental analogy with glass and also the model validity, we performed an experiment that is classical in the glassy context. A quasi-steady state for amplitude $\Gamma_0$ is reached through numerous taps and annealing procedure. "Quasi-steady state" is understood here for a very low density evolution for the constant acceleration $\Gamma_0$. We know switch to amplitude $\Gamma_1$ during a waiting time (*i.e.* a given number of taps) $\delta t$, and then switch it back to $\Gamma_0$. The issue is to study the return to the steady state as a function of the waiting time $\delta t$. Indeed, during the intermediate $\Gamma_1$–stage, the system does not have enough time to completely relax to its new equilibrium and when returning to the regular $\Gamma_0$ amplitude, the system will relax differently for different $\delta t$. Intuitively, one can estimate that only the modes whose time scale are below $\delta t$ have been perturbed by the switch while the other one remained almost in the $\Gamma_0$ frame. Following our model under simple assumptions, we can calculate the backward density relaxation similarly to Eq. (26), with $F(v, \kappa)$ effectively depleted below a minimal rate, $\kappa_0$. This cut-off frequency, $\kappa_0$, is expected to decrease monotonically with increasing perturbation duration $\delta t$. In the spirit of spin glass theories, we can characterize the density relaxation after returning to $\Gamma_0$ by the "aging" response function which now depends both on $t$ and waiting time $\delta t$. Eq. (24) gives the following form for its late–stage behavior

$$G_{\Gamma_1, \Gamma_0}(t, \delta t) = G_{\Gamma_1, \Gamma_0}(\infty) - \text{const}\left(\kappa_0 + \frac{\exp(-\kappa_0 t)}{t}\right). \qquad (27)$$

Figure 14 shows this two time function for $\delta t$ varying between 1 and 5. As predicted, the time needed to recover the steady-state density increases with the number of taps $\delta t$ spent in the "hot" regime $\Gamma_1 > \Gamma_0$. In the coordinates chosen, the relaxation curves should follow the $\delta t = \infty$ dynamics until the saturation at the cut-off time, $\kappa_0^{-1}(\delta t)$. We approximate the distribution function $F$ by a constant above this low frequency cut-off at $\kappa_0^{-1}(\delta t)$, up to a high-frequency cut-off, $\kappa_{\max} \simeq 1$. This eliminates the unphysical low-$t$ divergence in Eq. (27). Figure 14 shows fits of the data to Eq. (24), where $\kappa_0(\delta t)$ is determined from the fit. The best-fit is achieved at $\kappa_{\max} = 0.4$. We can draw few conclusions from the figure: (i) within our experimental precision, the $\delta t = \infty$ relaxation is consistent with the predicted $1/t$ law; (ii) finite–$\delta t$ relaxation curves can be parameterized by a low frequency cut-off, $\kappa_0$; and (iii) $\kappa_0$ is a decreasing function of the waiting time $\delta t$, shown in the insert of the Fig. 14.

We thus observe a range of relaxation times between one to five taps as already revealed by the fluctuation spectra of the density. We then try to relate these results to an analog of a Fluctuation-Dissipation Theorem (FDT) as it has been proposed in glassy systems (see for instance [81]). Again the difficulty here comes from the lack of a clear statistical mechanics context, so that the role of the acceleration in density fluctuations and in the dissipation is difficult to identify. However, one can propose that the density autocorrelation function can be written as follows:

$$
\begin{aligned}
\langle \delta\rho(0)\delta\rho(t)\rangle_\Gamma &\simeq \frac{\rho^2}{2V} \int \langle v^2 \rangle \exp(-\kappa t) f(\kappa) d\kappa \\
&= \frac{\rho\langle v^2 \rangle}{2V\langle v \rangle} \left( G_{0,\Gamma}(t) - G_{0,\Gamma}(\infty) \right).
\end{aligned}
\tag{28}
$$

Thus, the density correlator is simply proportional to the response function corresponding to the switch between a very low acceleration and the given one, $\Gamma$. Precision in experiments did not allow such measurements until now, although such FDT relationships have been probed in numerical models.[82]

Finally, the model gives a simple interpretation to the decreasing dependence of the steady-state density on $\Gamma$ since it can be attributed to the growth of the population of the excited states, $P(\Gamma)$. Indeed, the corresponding correction to the total density is about 1%, i.e. of the same order as the variation of the equilibrium packing fraction with $\Gamma$.[60] The slow dynamics associated with the evolution of the baseline density can also be addressed within our approach. A simple and reasonable assumption would be to consider that the relaxation of one mode to its ground state may frustrate such a transition for some of its neighbors (e.g. in 3D the

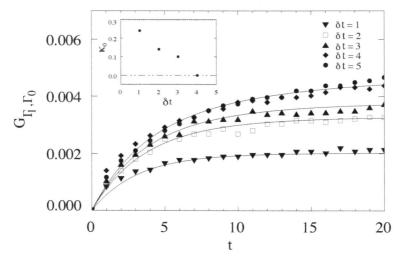

Figure 14. History-dependent density relaxation $(G_{\Gamma_1,\Gamma_0}(t,\delta t))$ of the system, prepared by tapping for a long time at $\Gamma_0 = 1.8$ and then tapping for a variable number, $\delta t$, of taps at a "hotter" intensity $\Gamma_1 = 4.2$ before being returned to $\Gamma_0$ at time $t_1$. The solid lines represent the theoretical curves, with appropriate values of the parameter $\kappa_0$. The dependence of the cut-off rate $\kappa_0$ on the waiting time $\delta t$ is shown at the insert for $\delta t \leq 4$ taps. We do not show the value for $\delta t = 8$ since we found it to be zero within the error bar, as for $\delta t = 4$. Each experimental graph is an average of 12 runs.

most compact local cluster can be created only at the expense of less dense neighboring regions). Thus, we arrive at an effective anti-ferromagnetic (AF) coupling (of an infinite strength) between the flip-flop modes. This interpretation of the slow dynamics makes a remarkable connection with the so-called reversible Parking Lot Model (PLM),[83,84] which has been successful in describing many aspects of granular compaction experiments.[59,60] A mutual frustration of individual modes is also a key ingredient of the "tetris model" (TM), another fruitful approach for modelling the dynamics of the system .[66–68,70,71] PLM, TM and our flip-flop model with AF coupling all appear to belong to the same *generic* class of frustrated spin systems.

## 6. Conclusion

In this review, the different state that granular systems can exhibit are discussed. It is always astonishing to observe that granular matter can mimic such different states of matter as gas, liquid or solid. An other striking point is that it is particularly difficult to build a consistent theory for grains that would be valid from gas to solid, saying nothing about glassy-like dynamics. This impossibility is due on one hand to the size of the grains (too big for thermal motion to be dominant, too small for not exhibiting collective motions); and on the other hand (and somehow related to the previous point), to the lack of a well defined statistical mechanics for grains.

Beside the recent results described above, the perspectives for granular research is still great. Briefly, one can cite rapidly evolving domains around the physics of powder, the sedimentation, the coupled dynamics between the grains and the surrounding fluid and the dynamics of erosion for instance.

## 7. Acknowledgements

C.J. wants to thank the Kavli Institute of Theoretical Physics at the University of Santa Barbara for its hospitality during the final achievement of this document.

## References

1. Guyon E., "Du sac de billes au tas de sable" (in french), Ed. Odile Jacob (1994).
2. Duran J., "Sables, poudres et grains" (in french), Eyrolles Sciences Ed. (1997).
3. Bagnold R.A., "The physics of blown sand and desert dunes", Dover Publications (2005).
4. Bagnold R.A., "The physics of sediment transport by wind and water: a collection of hallmark papers", ASCE (1988).
5. H. Jaeger, S. Nagel and R. Behringer, *Rev. Mod. Phys.* **68**, 1259 (1996).
6. Campbell C.S., "Rapid granular flows", *Annu. Rev. Fluid Mech.* **22**, 57 (1990).
7. Jenkins J.T. and Savage S.B., *J. Fluid Mech.* **130**, 187 (1983).
8. Goldhirsch I. and Zanetti G., "Clustering instability in dissipative gases", *Phys. Rev. Lett.* **70**, 1619 (1993).
9. Bernu B. and Mazighi R., *J. Phys. A: Math. Gen* **23**, 5745 (1990).
10. Mac Namara S. and Young W.R., "Inelastic collapse and clumping in a one-dimensional granular medium", *Phys. Fluids A* **4**, 496 (1992).
11. Mac Namara S. and Young W.R., "Kinetics of a one-dimensional granular medium in the quasielastic limit" *Phys. Fluids A* **5**, 34 (1993).
12. Mac Namara S, "Hydrodynamic modes of a uniform granular medium" *Phys. Fluids A* **5**, 3056 (1993).
13. Mac Namara S. and Young W.R., "Inelastic collapse in two dimensions" *Phys. Rev. E* **50**, R28-R31 (1994).

14. Chen S., Deng Y., Nie X. and Tu Y., "Clustering kinetics of granular media in three dimensions", *Phys. Lett.* **269**, 218 (2000).
15. Jenkins J.T. and Richman M.W., *Arch. Ration. Mech. Anal.* **87**, 355 (1985).
16. Jenkins J.T. and Richman M.W., "Kinetic theory for plane flows of a dense gas of identical, rough, inelastic circular disks", *Phys. Fluids* **28**, 3485-3494 (1985).
17. Lun C.K.K, Savage S.B. and Jeffrey D.J. and Shepurniy N., *J. Fluid Mech.* **140**, 233 (1984).
18. Schlichting H.J. and Nordmeier V., *Math. Naturwiss. Unterr* **49**, 323 (1996) (in German).
19. Eggers J., "Sand as Maxwell's demon", *Phys. Rev. Lett.* **83**, 5322-5325 (1999).
20. P. Jean, H. Bellenger, P. Burban, L. Ponson and P. Evesque, "Phase transition or Maxwell's demon in Granular gas?", *Powders and grains* **13**, 27-39 (2002).
21. D. van der Meer, K. van der Weele, and D. Lohse, "Sudden Collapse of a Granular Cluster", *Phys. Rev. Lett.* **88**, 174302 (2002).
22. Josserand C., Lagrée P.-Y. and Lhuillier D., "Stationary shear flows of dense granular materials : a tentative continuum modelling",*Eur. Phys. J. E***14**, 127-135 (2004).
23. Kadanoff L.,"Built upon sand: theoretical ideas inspired by granular flows", *Rev. Mod. Phys.* **71**, 435 (1999).
24. Azanza E., "Ecoulements granulaires bidimensionnels sur plan incliné", Doctoral Thesis from Ecole Nationale des Ponts et Chaussées, (1998).
25. Rajchenbach J., "Dense, rapid flows of inelastic grains under gravity", Phys. Rev. Letters, **90** 144302 (2003).
26. Ancey C., Coussot P. and Evesque P., "Examination of the possibility of a fluid mechanics treatment of dense granular flows", Mech. Cohesive-Frictionnal Mat. **1**, 385-403 (1996).
27. Savage S.B., "Granular flows down rough inclines - review and extension", in Proc. of US-Japan Seminar on New Models and Constitutive Relations in the Mechanics of Granular Materials, ed. by Jenkins J.T. and Satake M., Elsevier Science Publishers, Amsterdam, pp. 261-282 (1982).
28. Johnson P.C., Jackson R., "Frictional-collisional constitutive relations for granular materials, with application to plane shearing", J. Fluid Mech. **176** 67-93 (1987).
29. Ancey C.and Evesque P., "Frictionnal-collisionnal regime for granular suspension flows down an inclined channel" Phys. Rev E **62**, 8349-8360 (2000).
30. Bocquet L., Losert W., Schalk D., Lubenski T.C. and Gollub J.P., "Granular shear flow dynamics and forces: Experiment and continuum theory", Phys. Rev. E **65** 011307 (2002).
31. M.Y. Louge, "Model for dense granular flows down bumpy inclines", Phys Rev E **67**, 061303 (2003).
32. Goodman M.A., Cowin S.C., "Two problems in the gravity flow of granular materials", J. Fluid Mech. **45** 321-339 (1971).
33. Mills P., Loggia D. and Tixier M., "Model for a stationary dense granular flow along an inclined wall", Europhys. Lett. **45** 733 (1999). Mills P., Tixier M. and Loggia D., "Influence of roughness and dilatancy for dense granular

flow along an inclined wall", Eur. Phys. J. E**1** 5-8 (2000).

34. Aranson I.S., Tsimring L.S., "Continuum description of avalanches in granular media", Phys. Rev. E **64** 020301 (2001). Volfson D., Tsimring L.S. and Aranson I.S., "Order parameter description of stationary partially fluidized shear granular flows", Phys. Rev. Lett. **90** 254301 (2003).

35. Da Cruz F., Chevoir F., Roux J.-N., and Iordanoff I., "Macroscopic friction of dry granular materials", in Proceedings of the 30th Leeds-Lyon Symposium of Tribology, Dalmaz D. et al. Editors (2003).

36. Savage S.B., "Analyses of slow high concentration flows on granular materials", J. Fluid Mech. **377** 1-26 (1998).

37. Silbert L. E., Landry J. W. and Grest G. S., "Granular flow down a rough inclined plane: transition between thin and thick piles", Phys. Fluids **15**, 1-10 (2003).

38. Pouliquen O., "Scaling laws in granular flows down rough inclined planes", Phys. Fluids **11** (1999).

39. G.D.R. Midi, "On dense granular flows", Eur. Phys. J. E **14**, 341-365 (2004).

40. Jop P., Forterre Y. and Pouliquen O, "Crucial role of side walls for granular surface flows: consequence for the rheology.", J. Fluid Mech. **541**, 167-192 (2005).

41. Jop P., Forterre Y. and Pouliquen O, "A new constitutive law for dense granular flows." to be published in Nature (2006).

42. Rajchenbach J., "Granular flows", Advances in Physics **49** 229-256 (2000).

43. T.S. Komatsu, S. Inagaki, N. Nakagawa and S. Nasuno, "Creep motion in a granular pile exhibiting steady surface flow", Phys. Rev. Lett. **86**, 1757 (2001).

44. C.Ancey, "Dry granular flows down an inclined channel : Experimental investigations on the frictional-collisional regime", Phys. Rev. E**65** 011304 (2001).

45. S.F. Edwards, in *Granular Matter: An Interdisciplinary Approach*, edited by A. Mehta (Springer-Verlag, New-York, 1994).

46. C. Josserand, P.-Y. Lagrée and D. Lhuillier, "Granular pressure and the thickness of a layer jamming on a rough incline", Europhys. Lett. 73, 363-369 (2006).

47. K. Kanatani "A micropolar continuum theory for the flow of granular materials", Int. J. Engng Sci. **17**, 419-432 (1979).

48. R. Blumenfeld and S.F. Edwards "Granular entropy : Explicit calculations for planar assemblies", Phys. Rev. Lett. **90**, 114303 (2003).

49. D. Mueth, H. Jaeger and S. Nagel "Force distribution in a granular medium", Phys. Rev. E **57**, 3164-3169 (1998).

50. C. Liu, S. Nagel, D. Schecter, S. Coppersmith, S. Majumdar, O. Narayan and T. Witten "Force fluctuations in bead packs", Science **269**, 513-515 (1995).

51. S. Coppersmith, C. Liu, S. Majumdar, O. Narayan and T. Witten "Model for force fluctuations in bead packs", Phys. Rev. E **53**, 4673-4685 (1996).

52. M. Da Silva and J. Rajchenbach "Stress transmission through a model system of cohesionless elastic grains", Narure **406**, 708-710 (2000).

53. J. Rajchenbach "Stress transmission through textured granular packings", Phys. Rev. E **63**, 041301 (2001).

54. J.-P. Bouchaud, M. Cates and P. Claudin "Stress distribution in granular media and nonlinear wave equation", J. Phys. I **5** 639-656 (1995).
55. P. Claudin, J.-P. Bouchaud, M. Cates and J. Wittmer "Models of stress fluctuations in granular media", Phys. Rev. E **57**, 4441-4457 (1998).
56. E. Falcon, C. Laroche, S. Fauve and C. Coste "Collision of a 1-D column of beads with a wall", Euro. Phys. J. B **5**, 111-131 (1998).
57. C. Coste and B. Gilles "On the validity of Hertz contact law for granular material acoustics", Euro. Phys. J. B **7**, 155-168 (1999).
58. V. Tournat, V. Zaitsev, V. Gusev, V. Nazarov, P. Bequin and B. Castagnede "Probing weak forces in granular media through nonlinear dynamic dilatancy: Clapping contacts and polarization anisotropy", Phys. Rev. Lett. **92**, 085502 (2004).
59. J.B. Knight, C.G. Fandrich, C.N. Lau, H.M. Jaeger and S.R. Nagel, *Phys. Rev. E* **51**, 3957 (1995).
60. E.R. Nowak, J.B. Knight, E. Ben-Naim, H.M. Jaeger and S.R. Nagel,*Phys. Rev. E* **57**, 1971 (1998).
61. M. Nicolas, P. Duru and O. Pouliquen "Compaction of a granular material under cyclic shear", Euro. Phys. J. E **3**, 309-314 (2000).
62. A. Liu and S. Nagel "Nonlinear dynamics - Jamming is not just cool any more", Nature **396**, 21-22 (1998).
63. T. Majmudar, M. Sperl, S. Luding and R. Behringer "Jamming transition in granular systems", Phys. Rev. Lett. **98**, 058001 (2007).
64. C. Josserand, A. Tkachenko, D. Mueth and H. Jaeger: *Memory effects in granular materials*, Phys. Rev. Lett. **85**, 3632 (2000).
65. T. Boutreux and P. G. de Gennes, *Physics A* **244**, 59 (1997).
66. E. Caglioti, V. Loreto, H. J. Herrmann, and M. Nicodemi, *Phys. Rev. Lett.* **79**, 1575 (1997).
67. E. Vincent, J. Hammann and M. Ocio, in *Recent progress in random magnets*, World Scientific (1992).
68. J.-P. Bouchaud, L.F. Cugliandolo, J. Kurchan and M. Mézard, in *Spin Glasses and Random Field*, World Scientific (1997).
69. O. Pouliquen, M. Belzons and M. Nicolas "Fluctuating particle motion during shear induced granular compaction", Phys. Rev. Lett. **91**, 014301 (2003).
70. M. Nicodemi, *Phys. Rev. Lett.* **82** 3734 (1999).
71. A. Barrat and V. Loreto "Response properties in a model for granular matter", J. Phys. A **33**, 4401-4426 (2000).
72. S.F. Edwards and D. Grinev "Statistical mechanics of vibration-induced compaction of powders", *Phys. Rev. E* **58**, 4758-4762(1998).
73. A. Barrat, J. Kurchan, V. Loreto and M. Sellitto "Edwards' measures for powders and glasses", Phys. Rev. Lett. **85**, 5034-5037 (2000).
74. G. D'Anna, P. Mayor, A. Barrat, V. Loreto and F. Nori "Observing brownian motion in vibration-fluidized granular matter", Nature **424**, 909-912 (2003).
75. G. Marty and O. Dauchot "Subdiffusion and cage effect in a sheared granular material", Phys. Rev. Lett. **94**, 015701 (2005).
76. G.C. Barker and A. Mehta "Vibrated powders-structure, correlations and dynamics", *Phys. Rev. A* **45**, 3435-3446 (1992).

77. P.G. de Gennes "Tapping of granular packs: A model based on local two-level systems", J. Coll. Int. Sci. **226**, 1-4 (2000).
78. A. Mehta, and G. C. Barker "Disorder, memory and avalanches in sandpiles", *Europhys. Lett.* **27**, 501-506 (1994).
79. S. Krishnamurthy, V. Loreto and S. Roux, *Phys. Rev. Lett.* **84** 1039 (2000).
80. F. Alberici-Kious, J.P. Bouchaud, L.F. Cugliandolo, P. Doussineau and A. Levelut, *Phys. Rev. Lett.* **81** 4987 (1998).
81. L. F.Cugliandolo and J. Kurchan *Phys. Rev. Lett.* **71**, 173 (1993).
82. A. Barrat and V. Loreto "Memory in aged granular media", Europhys. Lett. **53**, 297-303 (2001).
83. P.L. Krapivsky and E. Ben-Naim, *J. Chem. Phys.* **100**, 6778 (1994)
84. A. L. Kolan, E. R. Nowak, and A. V. Tkachenko, *Phys. Rev. E* **59**, 3094 (1999).

# Chapter 5

# Relativité Générale pour Débutants (General Relativity For Beginners)

Michel Le Bellac

*INLN, Université de Nice Sophia-Antipolis, CNRS,*
*1361 route des Lucioles, 06560 Valbonne, France*
*E-mail : michel.le_bellac@inln.cnrs.fr*

Ce cours a pour objectif d'exposer à un public non initié les idées de base de la relativité générale. Il ne suppose aucun prérequis : il contient les notions nécessaires de relativité restreinte et de géométrie différentielle. Les applications traitées sont la cosmologie et les trous noirs.

This lecture offers an accessible presentation of the basic ideas of general relativity. There is no prerequisite : the background on special relativity and differential geometry are included. Applications to cosmology and black holes are considered.

## Sommaire

# 1. Introduction

## 1.1. *Bref historique*

Ce cours a pour objectif d'exposer de la façon la plus élémentaire possible les idées de la relativité générale, c'est-à-dire la théorie relativiste de la gravitation. Dans cette introduction, je commencerai par une rapide revue de l'histoire de la relativité générale. En 1905 paraît l'article d'Einstein sur la relativité restreinte, l'un des articles de "l'année miraculeuse". Même si Einstein fut le premier à les interpréter dans le cadre de l'espace-temps, les idées de la relativité restreinte étaient "dans l'air", et il est vraisemblable que Lorentz ou Poincaré (ou un autre) seraient arrivés rapidement à des conclusions identiques. Après avoir établi la relativité restreinte, Einstein commença immédiatement à réfléchir à une théorie relativiste de la gravitation. Il énonça dès 1907 le principe d'équivalence entre gravité et accélération constantes (voir section 2), mais il lui fallut encore huit ans avant d'établir à la fin de 1915 les fondements définitifs d'une théorie géométrique de la gravitation, la relativité générale. Contrairement au cas de la relativité restreinte, il est manifeste que la contribution d'Einstein à la relativité générale est unique, et que sans lui la relativité générale aurait probablement attendu quelques dizaines d'années avant d'être inventée. En effet, la relativité générale est une construction purement intellectuelle, pour laquelle il n'y avait aucune nécessité expérimentale. La théorie de Newton de la gravitation était en accord remarquable avec les mesures très précises de l'astronomie[a], contrairement à la mécanique et à l'électromagnétisme classiques qui commençaient à rencontrer des difficultés. En fait la relati-

---

[a]La seule difficulté expérimentale était l'absence d'explication de l'avance du périhélie de Mercure, de 43" (seconde d'arc) par siècle !

vité générale est restée une théorie ésotérique aux yeux de la majorité des physiciens pendant plus d'une cinquantaine d'années, et c'est seulement après la mort d'Einstein en 1955 qu'elle s'est considérablement développée, pour devenir une théorie incontournable de la physique moderne. On peut donner quelques dates clés de son histoire.

- 1916 : Einstein publie dans *Annalen der Physik* son article sur la relativité générale et explique l'avance du périhélie de Mercure. Quelques mois plus tard, Schwarzschild établit la forme (2) de la métrique qui porte son nom, et qui généralise la loi de Newton donnant le potentiel gravitationnel $\Phi(r)$ d'une masse ponctuelle $M$ sans structure et avec symétrie sphérique

$$\Phi(r) = -\frac{GM}{r} \qquad U(r) = -\frac{GMm}{r} \qquad (1)$$

où $G$ est la constante de gravitation, $r$ est la distance entre la masse $M$ et le point d'observation et $U(r)$ l'énergie potentielle d'une masse $m$ située à une distance $r$ de $M$.

L'expression de la métrique de Schwarzschild, qui sera établie à la section 6, est

$$ds^2 = \left(1 - \frac{r_S}{r}\right) dt^2 - \left(1 - \frac{r_S}{r}\right)^{-1} dr^2 - r^2 \left(d\theta^2 + \sin^2\theta \, d\varphi^2\right) \qquad (2)$$

Dans cette équation, $r_S$ est le rayon de Schwarzschild : $r_S = 2GM/c^2$, $c$ étant la vitesse de la lumière. La métrique de Schwarzschild est l'équivalent einsteinien du potentiel gravitationnel newtonien d'une masse ponctuelle $M$.

- 1919 : une éclipse de Soleil permet de mesurer la déviation d'un rayon lumineux par le Soleil, vérifiant une des prédictions majeures de la relativité générale : l'action de la gravitation sur la lumière. Malgré la précision discutable des expériences et les incertitudes expérimentales, ce résultat rend Einstein célèbre et il devient une vedette médiatique.
- 1929 : l'expansion de l'Univers devient une hypothèse plausible et Hubble établit la loi qui porte son nom (section 4). Einstein abandonne la constante cosmologique qu'il avait introduite de façon *ad hoc* pour rendre compte d'un Univers stationnaire, et qualifie cette introduction de plus grosse erreur de sa vie. En 1932 il publie avec de Sitter le premier modèle "moderne" d'Univers en expansion (voir la section 4.3.1). Puis plus rien d'important ne se passe jusqu'au début des années 1960 !

La relativité générale subit de plein fouet la concurrence de la physique quantique et le nombre d'articles consacrés à la relativité générale dans *Physical Review* est en chute libre.

- 1960 est l'année d'un progrès théorique important : la métrique de Schwarzschild (2) semble singulière à $r = r_S$, et cette singularité empêche de comprendre ce qui se passe dans la région $r \leq r_S$. Kruskal et Szekeres proposent indépendamment un système de coordonnées démontrant que la singularité est en fait un artefact du choix des variables $(t, r)$. Cette observation permettra de lancer la physique des trous noirs (section 6).

- 1960 : Pound et Rebka vérifient expérimentalement le principe d'équivalence grâce à une expérience exploitant l'effet Mössbauer.

- 1965 : Penzias et Wilson découvrent par accident le rayonnement cosmologique à 3 K. La découverte de ce rayonnement, prévu par Gamow en 1948, relance la cosmologie sur des bases expérimentales, dans un cadre qui rend indispensable l'utilisation de la relativité générale.

- 1968 : Jocelyn Bell observe le premier pulsar, qui se révèle être une étoile à neutrons. Cette découverte vaudra le prix Nobel aux deux "seniors" de l'équipe[b].

- 1974 : Hulse et Taylor observent le premier pulsar binaire, qui leur permet de vérifier l'émission d'ondes gravitationnelles par les étoiles sur orbite.

- 1976 : l'expérience mesurant le retard gravitationnel d'un écho radar, proposée par Shapiro en 1965, permet la vérification la plus précise ($\sim 10^{-3}$) à cette date de la relativité générale.

- 1998 : deux expériences indépendantes observent l'accélération de l'expansion de l'Univers : $\ddot{a} > 0$ (section 4). La constante cosmologique, qui donne une explication plausible de cette accélération, revient par la grande porte !

- 2003 : le satellite WMAP observe les fluctuations angulaires de température du rayonnement cosmologique à 3 K, ce qui permet de confirmer le jeu de paramètres du modèle standard actuel de la cosmologie (modèle $\Lambda$CDM, section 4).

---

[b]L'"oubli" de Jocelyn Bell n'est malheureusement pas un cas isolé pour les femmes scientifiques : voir Lise Meitner, Rosalind Franklin ...

## 1.2. *Plan du cours*

Le plan du cours sera le suivant

1. Principe d'équivalence
2. Espace-temps plat
3. Cosmologie
4. Boîte à outils de géométrie différentielle
5. Solutions à symétrie sphérique

Le choix des sujets a été effectué en fonction de leur intérêt physique et de la simplicité de leur traitement théorique. De plus le choix de la cosmologie (section 4) et des trous noirs (section 6) permet d'étudier la relativité générale dans des conditions où elle est essentielle, où elle n'intervient pas seulement comme une petite correction à la gravitation newtonienne. La liste ci-dessous donne un aperçu de la multitude de sujets couverts par la relativité générale, que le caractère limité de ce cours ne permet pas d'aborder ici.

- Tests classiques (i. e. dans le système solaire) de la relativité générale : précession du périhélie, déviation de la lumière, retard gravitationnel d'un écho radar, entraînement de référentiels d'inertie.
- Ondes gravitationnelles : émission et détection.
- Lentilles gravitationnelles.
- Astrophysique relativiste : pulsars.
- Solutions axisymétriques (métrique de Kerr).
- Théorèmes de singularités : Penrose, Hawking. . .
- Gravitation quantique.
- etc.

## 1.3. *Quelques références générales*

Je donne ci-dessous une bibliographie sommaire qui sera précisée pour chaque section, où les livres seront cités, à titre d'exemple, sous la forme Hartley [2003]

(1) Carrol [2004] : S. M. Mc Carrol, *Gravitation*, Addison-Wesley, New-York.
(2) Damour [2005] : T. Damour, *Relativité Générale*, contribution à l'ouvrage collectif *Actualité d'Einstein*, EDPSciences/CNRS Editions, Paris.

(3) Doubrovine [1983] : B. Doubrovine et al. *Géométrie contemporaine*, Editions Mir, Moscou

(4) Eisenstaedt [2002] : J. Eisenstaedt *Einstein et la relativité générale*, CNRS, Paris (sur l'histoire de la relativité générale).

(5) Hartle [2003] :J. Hartle *Gravity*, Addison-Wesley, New-York

(6) Ludvigsen [2000] : M. Ludvigsen *La relativité générale : une approche géométrique*, Dunod, Paris.

(7) Wald [1984] : R. M. Wald, *General Relativity*, The University of Chicago Press, Chicago.

(8) Weinberg [1972] : S. Weinberg, *Gravitation and Cosmology*, John Wiley, New-York.

## 2. Principe d'équivalence

### 2.1. *Référentiels d'inertie*

La présentation des cours élémentaires de mécanique est fondée sur la notion de référentiel d'inertie : ce sont les référentiels où les lois de Newton s'expriment simplement. Par exemple la seconde loi de Newton stipule qu'en l'absence de forces, une particule suit un mouvement rectiligne uniforme dans un référentiel d'inertie. Un référentiel en mouvement de translation uniforme par rapport à un référentiel d'inertie est aussi un référentiel d'inertie. Cependant se pose immédiatement la question : comment s'assurer qu'un référentiel est inertiel ? Le plus simple est de s'assurer que l'absence de force entraîne l'absence d'accélération : $\vec{F} = 0 \implies \vec{a} = 0$. Mais comment peut-on savoir que $\vec{F} = 0$ autrement qu'en mesurant une accélération nulle ? On tombe manifestement dans un cercle vicieux. Toutefois la situation n'est pas aussi mauvaise qu'il n'y paraît, car on peut s'appuyer sur les propriétés connues des forces. Supposons par exemple que nous voulions nous assurer de l'absence de forces électriques sur un ion $O_{16}^+$ de masse $m$ et de charge $q$. Bien sûr, dans un référentiel d'inertie, on aura $\vec{F} = q\vec{E}$, où $\vec{E}$ est le champ électrique, et on mesurera donc l'accélération correspondante $\vec{a} = q\vec{E}/m$. Cependant dans un référentiel non inertiel d'accélération $\vec{A}$ par rapport à un référentiel d'inertie (en se limitant pour simplifier à des mouvements de translation) on mesure une force $\vec{F}'$ et une accélération $\vec{a}'$ données par

$$\vec{F}' = q\vec{E} - m\vec{A} = m\vec{a}' \qquad (3)$$

où $-m\vec{A}$ est la force fictive (ou d'inertie). Il se pourrait que l'on mesure une accélération nulle, non pas parce que la force électrique est nulle, mais parce que cette force est exactement compensée par la force fictive. Il est facile de s'en sortir en utilisant un ion doublement chargé $O_{16}^{++}$, et cette fois, en négligeant une différence de masse entre les deux ions de l'ordre de $10^{-4}$, l'accélération $\vec{a}'$ est

$$m\vec{a}' = \vec{F}' = 2q\vec{E} - m\vec{A} \qquad \vec{F}' - \vec{F} = q\vec{E} \qquad (4)$$

On peut donc sans problème s'assurer de ce que la force électrique est bien nulle, car on peut modifier *indépendamment* les coefficients de $\vec{E}$ et de $\vec{A}$. Cependant cette procédure ne marche pas dans le cas de forces de gravitation. Si dans un référentiel d'inertie $\vec{F} = m\vec{g}$, dans un référentiel accéléré

$$\vec{F}' = m\vec{g} - m\vec{A} \qquad (5)$$

et on ne peut pas, comme dans l'exemple précédent, modifier *indépendamment* les coefficients de $\vec{g}$ et de $-\vec{A}$ ... sauf s'il existe deux types de masse ! Comme l'avait parfaitement compris Newton, il est *a priori* possible que la masse gravitationnelle $m_g$ intervenant dans la loi de la gravitation

$$||\vec{F}_g|| = G \, \frac{m_g m_g'}{r^2} \qquad (6)$$

où $G$ est la constante de gravitation et $r$ la distance entre les deux masses, et la masse d'inertie $m_i$ intervenant dans la force fictive $-m_i\vec{A}$ soient de caractère différent. Dans ce cas le rapport $m_i/m_g$ pourrait être différent suivant le matériau. Non seulement Newton avait envisagé cette possibilité, mais il réalisa une expérience avec un pendule pour la tester. En effet l'équation du pendule (à l'approximation des petites oscillations) s'écrit avec deux types de masse

$$m_i\ddot{\theta} + m_g g\theta = 0 \qquad (7)$$

et si l'on construit deux pendules avec des matériaux $A$ et $B$, le rapport des périodes sera

$$\frac{T^A}{T^B} = \left( \frac{m_i^A}{m_g^A} \, \frac{m_g^B}{m_i^B} \right)^{1/2} \qquad (8)$$

Newton n'observa aucun effet de ce type et conclut qu'à la précision de ses expériences le rapport $m_i/m_g$ était indépendant du matériau, ce qui impliquait qu'avec un choix convenable d'unités on pouvait écrire $m_i = m_g = m$ ; en d'autres termes, il existe un seul type de masse. Des expériences ont été réalisées ultérieurement avec une précision bien plus grande que celle qui était possible pour Newton, et on peut affirmer aujourd'hui que $m_i = m_g$ avec une précision $\sim 10^{-12}$.

## 2.2. Principe d'équivalence

Suivant le raisonnement de la section précédente, l'égalité entre masse
d'inertie et masse gravitationnelle implique que l'on ne peut pas distinguer,
*au moins localement*, entre une force de gravitation et une force fictive, et on
est donc incapable de s'assurer de l'absence de forces de gravitation, alors
que l'on peut parfaitement s'assurer de l'absence d'autres types de force,
électromagnétiques ou autres : *la gravitation joue donc un rôle particulier.*
Au lieu de s'escrimer à définir un référentiel d'inertie où l'on pourrait écrire
la loi de Newton pour une force de gravitation, il est plus économique de
décider *qu'un référentiel d'inertie est un référentiel en chute libre.* Un ascenseur en chute libre ou un satellite en orbite seront donc des référentiels
d'inertie. Avec ce choix, ce n'est pas la pomme qui tombe sur Newton dans
un référentiel d'inertie lié à la Terre, c'est Newton qui monte vers la pomme
dans le référentiel en chute libre où la pomme est au repos !

Poursuivons notre raisonnement en discutant une expérience simple : un
observateur dans une boîte fermée à la surface de la Terre lâche sans vitesse
initiale une pomme qui tombe avec une accélération $\vec{g}$. L'observateur verra
exactement le même phénomène s'il se trouve dans une fusée dans l'espace,
accélérée en direction de sa tête avec une accélération $\vec{A} = -\vec{g}$; il voit la
pomme tomber vers ses pieds avec exactement le même mouvement. Sans
regarder à l'extérieur, l'observateur est incapable de distinguer entre l'effet
d'une force de gravitation et celui d'une accélération en direction opposée[c].
Nous pouvons donc énoncer *le principe d'équivalence*

*Principe d'équivalence : aucune expérience à l'intérieur d'une boîte ne per-
met de distinguer entre une accélération constante et un champ de gravitation constant.*

On en déduit une conséquence immédiate pour les photons : dans un référentiel d'inertie, c'est-à-dire, rappelons-le, un référentiel en chute libre,
il ne peut pas exister par définition de direction privilégiée, et un photon
doit se propager en ligne droite (dans le cas contraire la direction de la
déviation indiquerait une direction privilégiée). Dans la Fig. 1, l'ascenseur
est lâché quand le photon est émis horizontalement en $A$, et la trajectoire
est une droite dans le référentiel de l'ascenseur. Comme le point d'arrivée
$B$ du photon sur la paroi de l'ascenseur ne dépend pas du référentiel, cela

---

[c]Ceci rappelle bien évidemment l'énoncé familier concernant les référentiels d'inertie
selon lequel un observateur ne peut pas détecter un mouvement uniforme par rapport
aux étoiles sans regarder à l'extérieur.

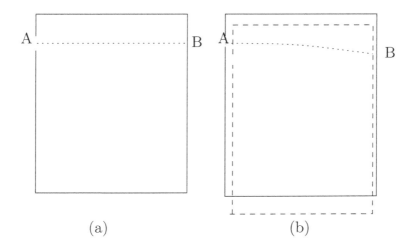

$$(a) \qquad\qquad\qquad (b)$$

FIGURE 1. Trajectoire d'un photon dans un ascenseur. (a) Référentiel en chute libre (b) Référentiel de la Terre.

veut dire que dans le référentiel de la Terre sa trajectoire est courbée : la lumière est donc sensible à un champ de gravitation. Comme application, il est tentant d'en déduire la déviation d'un rayon lumineux par une masse $M$, par exemple la déviation de la trajectoire par le Soleil d'un rayon lumineux émis par une étoile et que l'on peut mesurer au cours d'une éclipse totale de Soleil. Effectuons le calcul pour une masse $m$ en supposant la déviation petite, c'est-à-dire à l'approximation du paramètre d'impact (Fig. 2). À cette approximation, on considère que la vitesse $v$ de la masse $m$ varie peu, et l'on peut estimer la variation de l'impulsion transverse $\Delta p_y$ en intégrant la composante $F_y$ de la force

$$\Delta p_y = \int_{-\infty}^{+\infty} F_y \, \mathrm{d}t \qquad F_y = -\frac{GmM}{b^2 + v^2 t^2} \, \cos\varphi = -\frac{GmMb}{(b^2 + v^2 t^2)^{3/2}}$$

où $b$ est le paramètre d'impact. On obtient donc

$$\Delta p_y = -GmMb \int_{-\infty}^{+\infty} \frac{\mathrm{d}t}{(b^2 + v^2 t^2)^{3/2}} = -\frac{2GmM}{bv} \qquad (9)$$

soit pour l'angle de déviation[d]

$$\Delta\theta = \frac{\Delta p_y}{mv} = \frac{2GM}{bv^2} \qquad (10)$$

---

[d]Il est évidemment facile de donner une formule exacte étant donné que les orbites sont

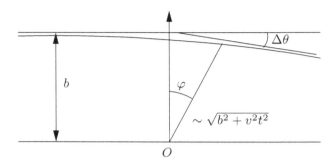

FIGURE 2.    Déviation d'un rayon lumineux par une masse $M$ placée en $O$.

Extrapolant hardiment au cas d'un photon pour lequel $v = c$, on en déduit $\Delta\theta = 2GM/(bc^2)$, ce qui correspond pour une incidence rasante sur
le Soleil ($b = R_S$ = rayon du Soleil) à une déviation de 0.87" (seconde
d'arc), résultat obtenu initialement par Einstein en 1907. Le résultat de la
relativité générale (1915) est le double : 1.75". En fait, dans ce calcul, nous
sommes allés au-delà du principe d'équivalence, valable uniquement pour
des champs de gravitation constants, et il n'est pas surprenant que notre
résultat soit quantitativement incorrect, même si le phénomène est prédit
correctement de façon qualitative.

## 2.3. Décalage vers le rouge gravitationnel

Pour simuler un champ de gravitation constant $\vec{g}$, nous allons utiliser le
principe d'équivalence et nous placer dans une fusée dont l'accélération vers
le haut est constante et égale à $-\vec{g}$. Le calcul qui va suivre est essentiellement
un calcul d'effet Doppler. Du plancher de la fusée sont émis des photons
avec une période $T$, aux temps $t = 0, t = T, \ldots, t = nT, \ldots$. Les signaux
sont reçus au plafond de la fusée aux temps $t_0, t_1, \ldots, t_n \ldots$ (Fig. 3). La
distance plancher-plafond est $h$ et le temps mis par le photon pour aller
du plancher au plafond est $\simeq h/c$; pendant ce temps la fusée acquiert une
vitesse $v \simeq gh/c$. Le calcul qui va suivre néglige les effets de la relativité
restreinte, et il ne sera valable que si $v^2/c^2 \simeq (gh/c^2)^2 \ll 1$ : le petit
paramètre du problème est en fait $gh/c^2$.

---

des hyperboles

$$\cot \frac{\Delta\theta}{2} = \frac{bv^2}{GM}$$

résultat qui coïncide avec (10) pour $\Delta\theta \ll 1$.

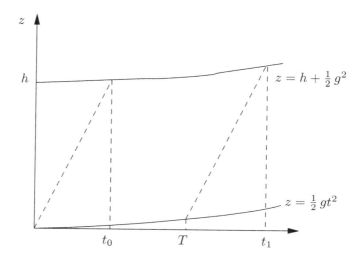

FigURE 3.   Décalage vers le rouge gravitationnel. Le plancher de la fusée suit la trajectoire $z = gt^2/2$, le plafond $z = h + gt^2/2$.

La trajectoire du plancher est $z = gt^2/2$ et celle du plafond $z = gt^2/2 + h$. Soit $t_n$ le temps de réception du signal émis à $t = nT$. Ce temps est obtenu en résolvant l'équation du second degré

$$h + \frac{1}{2}\, gt_n^2 = \frac{1}{2}\, g(nT)^2 + c(t_n - nT) \tag{11}$$

On peut bien sûr résoudre exactement, mais lorsque $gh/c \ll 1$ il est plus simple de remarquer que

$$t_n = nT + \frac{h}{c}\,(1 + \varepsilon_n) \qquad |\varepsilon_n| \ll 1$$

Dans un calcul au premier ordre en $\varepsilon_n$ on peut utiliser

$$t_n - (nT)^2 = (t_n - nT)(t_n + nT) \simeq \frac{2h}{c}\left(nT + \frac{h}{c}\right)$$

et en reportant dans (11) on obtient

$$\varepsilon_n = \frac{gnT}{c} + \frac{gh}{c^2} \qquad \Delta T = t_n - t_{n-1} - T = \frac{h}{c}\,(\varepsilon_n - \varepsilon_{n-1}) = \frac{gh}{c^2}\, T$$

Par rapport à l'émission, l'intervalle de temps entre la réception de deux photons est augmenté de $\Delta T$, avec

$$\boxed{\frac{\Delta T}{T} = \frac{gh}{c^2}} \tag{12}$$

La période mesurée par l'observateur à l'altitude $h$ est donc plus grande que celle mesurée par l'expéditeur à l'altitude zéro ; c'est le *décalage vers le rouge gravitationnel*. Il en résulte que la fréquence $\omega$ du photon reçu est décalée de $\Delta\omega$ par rapport à celle du photon émis

$$\boxed{\frac{\Delta\omega}{\omega} = -\frac{gh}{c^2}} \tag{13}$$

On peut donner une déduction plus simple de ce résultat en admettant la formule de l'effet Doppler, que nous démontrerons dans la section suivante. L'observateur qui reçoit le photon a acquis une vitesse $gh/c$ par rapport à la source pendant le temps de vol du photon, et par effet Doppler la période est allongée de

$$\frac{\Delta T}{T} = \frac{v}{c} = \frac{gh}{c^2}$$

Une dernière façon hardie de voir les choses consiste à écrire qu'entre la source et la réception le photon a perdu une énergie gravitationnelle $mgh$, avec comme précédemment $m = E/c^2$ et donc, en utilisant la relation de Planck-Einstein $E = \hbar\omega$

$$\frac{\Delta E}{E} = -\frac{1}{E}\left(\frac{E}{c^2}gh\right) = \frac{\Delta\omega}{\omega} = -\frac{gh}{c^2}$$

Cet effet fut vérifé expérimentalement par Pound et Rebka en 1960 sur une hauteur de de 20 m.

Les corrections de décalage vers le rouge gravitationnel trouvent une application dans le GPS. Dans l'espace à trois dimensions, il faut quatre satellites pour déterminer une position GPS. Nous allons nous limiter à une caricature de GPS en prenant une seule dimension d'espace, et il suffira donc de deux satellites. Dans le plan $(x, ct)$, les satellites sont supposés suivre des trajectoires (lignes d'Univers) $S_A$ et $S_B$ (Fig. 4) et l'utilisateur une trajectoire $U$. $U$ reçoit les signaux des deux satellites au temps $t$ et au point $x$, et il a accès aux positions $x_A$ et $x_B$ des satellites au moment de l'émission du signal, et aux temps d'émission $t_A$ et $t_B$ des signaux reçus car le signal GPS est codé[e]. Les deux droites de pente $\pm 1$ issues des points $(x_A, ct_A)$ et $(x_B, ct_B)$ se coupent en $(x, ct)$

---

[e]Si l'utilisateur connaissait exactement le temps, il lui suffirait d'un seul satellite (trois satellites à trois dimensions) pour repérer sa position. Le problème est qu'il n'est pas simple de transporter une horloge atomique, par exemple si l'on est en randonnée !

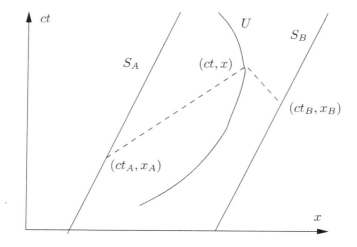

FIGURE 4. Schéma du GPS. Les lignes d'Univers $S_A$ et $S_B$ des deux satellites sont des droites, celle de l'observateur $U$ est courbe.

$$c(t - t_A) = x - x_A$$
$$c(t - t_B) = -x + x_B$$

Le système a pour solution immédiate

$$ct = \frac{1}{2}\left[c(t_A + t_B) + (x_B - x_A)\right]$$
$$x = \frac{1}{2}\left[c(t_B - t_A) + (x_B + x_A)\right]$$

(14)

Si l'on veut une précision d'un mètre[f], il faut une précision sur le temps de 3 ns, et ceci montre que les effets relativistes ne sont pas négligeables. En effet, pour un satellite de période 12 heures, et qui se trouve donc sur une orbite circulaire de rayon $R_s = 4.2\,R_T$, de vitesse $v/c \simeq 1.3 \times 10^{-5}$, l'effet de la relativité restreinte $\sim v^2/c^2$ est $\sim 10^{-10}$. En ce qui concerne le décalage vers le rouge gravitationnel, si l'on compare une horloge sur Terre et une horloge dans le satellite à une horloge hypothétique située dans un référentiel d'inertie, l'effet le plus important est en $GM_T/(R_T c^2) \simeq 1.6 \times 10^{-10}$. En une minute la dérive est $\sim 10^{-9}$ s, et il faut donc remettre les horloges à l'heure toutes les minutes pour une précision du mètre.

---

[f]Cette précision semblera modeste d'ici quelques années : on nous promet pour bientôt une précision du centimètre !

## 2.4. Interprétation géométrique

Revenons à l'expérience démontrant le décalage vers le rouge gravitationnel, en essayant de lui trouver une interprétation qui sera justifiée et affinée par la suite. Nous allons partir de l'idée que les deux observateurs ont en commun un *paramètre de temps* $t$, mais que leurs deux horloges ne mesurent pas $t$. En fait le temps mesuré par les deux horloges est un temps $\tau$, appelé *temps propre*, et relié à $t$ par

$$\tau = t \left( 1 + \frac{\Phi}{c^2} \right) \tag{15}$$

où $\Phi$ est le potentiel gravitationnel[g] au point où se trouve l'horloge : une masse test $m$ possède une énergie gravitationnelle $U = m\Phi$, et on suppose $|\Phi/c^2| \ll 1$. Dans ce cas la différence entre les temps $\tau_A$ (plancher) et $\tau_B$ (plafond) mesurés par les deux horloges est

$$\tau_B - \tau_A = \Delta T = T \left( 1 + \frac{\Phi_B}{c^2} \right) - T \left( 1 + \frac{\Phi_A}{c^2} \right) = T \frac{\Phi_B - \Phi_A}{c^2} = T \frac{gh}{c^2}$$

en accord avec (12). Une première idée pour interpréter (15) consisterait à proposer que le fonctionnement des horloges est affecté par la gravitation. Ce n'est pas l'idée retenue par Einstein, qui fait porter la responsabilité de (15) *sur la géométrie*. Pour Einstein, la distance pseudo-euclidienne (ou de Minkowski) au carré $\mathrm{d}s^2$ (voir la section 3 pour plus d'explications) entre deux événements séparés par un intervalle de temps $\mathrm{d}t$ et d'espace $\mathrm{d}\vec{r}$ est affectée par la gravitation, et lorsque $\Phi/c^2 \ll 1$ on peut écrire

$$\mathrm{d}s^2 = \left( 1 + \frac{2\Phi(\vec{r})}{c^2} \right) c^2 \mathrm{d}t^2 - \left( 1 - \frac{2\Phi(\vec{r})}{c^2} \right) \mathrm{d}\vec{r}^{\,2} \tag{16}$$

Nous verrons dans la section 6 comment cette équation peut se déduire de la métrique de Schwarzschild. Pour mieux comprendre le raisonnement d'Einstein, on peut utiliser une analogie due à Hartle : dans une projection de Mercator, les distances entre Paris et Montréal et entre Lagos et Bogota sont les mêmes. Or il est connu qu'un avion met plus de temps à voler de Lagos à Bogota que de Paris à Montréal. On peut donner deux interprétations.

1. Les règles raccourcissent quand la latidude croît (la gravité affecte les horloges).

---

[g]Il est important de noter qu'en relativité la valeur absolue de l'énergie potentielle a une signification physique : il n'y a pas de constante arbitraire dans la définition de l'énergie.

2. Les règles restent les mêmes, mais la géométrie est celle d'une sphère (la géométrie est affectée par la gravité).

Lorsque des événements se passent au même point ($d\vec{r} = 0$), on dit que l'intervalle séparant les deux événements est un *intervalle de temps propre* : $ds^2 = d\tau^2$. Suivant (16), la relation entre le paramètre de temps $t$ et le temps propre $\tau$ est donc

$$\Delta\tau = \Delta t \left(1 + \frac{2\Phi(\vec{r})}{c^2}\right)^{1/2} \simeq \Delta t \left(1 + \frac{\Phi(\vec{r})}{c^2}\right) \tag{17}$$

ce qui permet de retrouver (15). En l'absence de gravité, si ($c\Delta t, \Delta\vec{r}$) est l'intervalle d'espace-temps entre l'émission et la réception d'un photon

$$\Delta s^2 = c^2\Delta t^2 - \Delta\vec{r}^2 = 0$$

car un photon se propage à la vitesse $c$, et dans un diagramme ($ct, \vec{r}$) sa ligne d'Univers, c'est-à-dire son temps en fonction de sa position, est une droite de pente unité. En présence de gravité, nous aurons toujours par définition $ds^2 = 0$, et si nous nous limitons à une dimension d'espace, $z$, nous avons d'après (16) et compte tenu de $|\Phi/c^2| \ll 1$

$$\frac{dz}{dt} \simeq c \left(1 + \frac{2\Phi}{c^2}\right)$$

Cette équation semble indiquer que la vitesse du photon n'est plus $c$, mais c'est une illusion car $t$ est seulement un *paramètre de temps* et $z$ un paramètre d'espace. Nous pouvons par exemple définir $z'$ par

$$z' = \int^z \left(1 - \frac{2\Phi(v)}{c^2}\right) dv \qquad \frac{dz'}{dz} = 1 - \frac{2\Phi(z)}{c^2}$$

et donc

$$\frac{dz'}{dt} = \frac{dz'}{dz}\frac{dz}{dt} \simeq c$$

et avec ce nouveau paramètre d'espace les lignes d'Univers des photons sont des droites !

La forme (16) de la métrique nous a permis de rendre compte du décalage vers le rouge gravitationnel dans un cas où $d\vec{r} = 0$. Calculons maintenant l'intégrale $s_{AB}$ de la métrique sur la trajectoire d'une particule massive de vitesse $\ll c$ : comme nous le verrons dans la section suivante, cette intégrale est la "distance" minkowskienne ou le temps propre entre le point de départ $A$ et le point d'arrivée $B$. Nous avons

$$
\begin{aligned}
s_{AB} &= \int_A^B \mathrm{d}t \left[ \left( 1 + \frac{2\Phi(\vec{r})}{c^2} \right) c^2 - \left( 1 - \frac{2\Phi(\vec{r})}{c^2} \right) \left( \frac{\mathrm{d}\vec{r}}{\mathrm{d}t} \right)^2 \right]^{1/2} \\
&\simeq c \int_A^B \mathrm{d}t \left[ \left( 1 + \frac{2\Phi(\vec{r})}{c^2} \right) - \frac{1}{c^2} \left( \frac{\mathrm{d}\vec{r}}{\mathrm{d}t} \right)^2 \right]^{1/2} \\
&\simeq c \int_A^B \mathrm{d}t \left[ 1 - \frac{1}{c^2} \left( \frac{1}{2} \left( \frac{\mathrm{d}\vec{r}}{\mathrm{d}t} \right)^2 - \Phi(\vec{r}) \right) \right]
\end{aligned}
\tag{18}
$$

On reconnaît en facteur de $-1/c^2$ le lagrangien de la particule, et suivant le principe de moindre action la minimisation de $s_{AB}$ par rapport à $\vec{r}(t)$ donne tout simplement les équations du mouvement

$$
\frac{\delta s_{AB}}{\delta \vec{r}(t)} = 0 \iff \frac{\mathrm{d}^2 \vec{r}}{\mathrm{d}t^2} = -\vec{\nabla}\Phi(\vec{r})
$$

Autrement dit la particule suit une trajectoire qui extrémise le temps propre $s_{AB}$. On voit de (18) qu'en général cet extremum sera un maximum, car l'intégrale du deuxième terme dans le crochet n'est autre que l'action, qui est en général minimisée par la trajectoire. Nous verrons ultérieurement que cet énoncé est équivalent à l'affirmation selon laquelle la trajectoire d'une particule soumise uniquement à la gravitation est une géodésique de l'espace-temps. Nous voyons donc que l'interprétation géométrique nous donne à la fois le décalage vers le rouge gravitationnel et les lois de Newton.

## 2.5. *Effets de marée gravitationnels*

Le défaut principal (mais il est inévitable!) de notre définition d'un référentiel d'inertie est que cette définition est forcément locale si le champ de gravitation n'est pas uniforme. En effet, si le champ de gravitation n'est pas uniforme, on va observer des effets de marée. Prenons l'exemple d'un satellite en chute libre au-dessus de la Terre, et comparons la chute de son centre masse situé en $S$ à une distance $d$ du centre de la Terre à celle d'objets dans le satellite (Fig. 5). Si les coordonnées d'un objet dans le satellite sont $(x, z)$, en choisissant une géométrie à deux dimensions pour simplifier, l'énergie potentielle gravitationnelle d'un objet de masse $m$ est

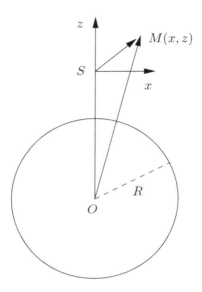

FIGURE 5. Effets de marée gravitationnels. Le centre du satellite $S$ est à une distance $d$ du centre $O$ de la Terre. La masse $m$ a pour coordonnées $(x, z)$ dans un référentiel d'origine $S$.

$$\Phi(x, z) = \frac{GmM}{[x^2 + (d + z)^2]^{1/2}}$$
$$\simeq -\frac{GmM}{d} \left[ 1 - \frac{z}{d} - \frac{x^2}{2d^2} + \frac{z^2}{d^2} \right] \tag{19}$$

où $M$ est la masse de la Terre et nous avons utilisé $|x|, |z| \ll d$ pour effectuer un développement limité. Si la masse était située en $S$, la force sur cette masse aurait pour composantes $F_x^0 = 0$, $F_z^0 = -GmM/d^2$. Pour une masse en $(x, z)$, nous avons

$$F_x - F_x^0 = F_x = -GmM \frac{x}{d^3} \tag{20}$$

$$F_z - F_z^0 = GmM \frac{2z}{d^3} \tag{21}$$

Ces équations montrent qu'au cours de la chute du satellite $x$ diminue et $z$ augmente : il y contraction suivant la direction horizontale et dilatation suivant la direction verticale. Ceci se comprend qualitativement : si une personne est en chute libre verticale, l'accélération de ses pieds est plus grande que celle de son centre de masse, et celle de sa tête plus faible. Elle a donc tendance à être dilatée verticalement dans le référentiel de sa chute. Si elle étend les bras, l'accélération de ses mains est dirigée vers le centre de forces et elle a tendance à être contractée horizontalement. C'est un *effet de marée* typique, sur lequel nous faisons les deux remarques suivantes.

- Un exemple extrême d'effet de marée est donné par la chute d'un astronaute sur une étoile en effondrement gravitationnel (section 6) : l'astronaute est dilaté de la tête aux pieds et contracté dans la direction horizontale, et finalement mis en pièces !
- Si $\vec{r}$ est la position de l'origine d'un référentiel d'inertie, et si $\vec{\xi}$ est la position d'un objet dans ce référentiel, les équations du mouvement pour $\vec{\xi}$ sont données par une généralisation facile de ce qui précède

$$\frac{\mathrm{d}^2 \xi^i}{\mathrm{d}t^2} = -\delta^{ij} \left. \frac{\partial^2 \Phi}{\partial x^i \partial x^k} \right|_{\vec{r}} \xi^k \qquad (22)$$

En résumé deux idées importantes ressortent de cette première approche.

(1) Les forces de gravitation apparaissent comme des effets de marée et on abandonne l'idée de donner une valeur absolue à la force de gravitation.
(2) Si elles sont soumises uniquement à des forces de gravitation, les particules suivent des géodésiques de l'espace-temps.

Cependant cette première approche heuristique de la relativité générale souffre pour le moment d'un défaut essentiel : nous n'avons pas considéré le cas de particules massives dont la vitesse peut être proche de celle de la lumière. Avant d'examiner ce qui se passe en relativité générale, il nous faut revenir dans la section suivante à la relativité restreinte.

## Bibliographie

Hartle [2003], chapitres 1 à 3 et 6 ; N. Ashby, *Relativity and the Global Positioning System*, Physics Today mai 2002, p. 41 ; F. Wilczek, *Total Relativity : Mach 2004*, Physics Today avril 2004 (sur la relativité et le principe de Mach).

## 3. Espace-temps plat

Cette section examine l'espace-temps plat de la relativité restreinte, non encore "déformé" par la gravité. Il s'agit donc de l'espace-temps idéal (et théorique !) d'un Univers entièrement vide de matière et d'énergie, où le temps est homogène et l'espace homogène et isotrope : c'est une arène neutre où se déroulent les processus physiques. On ne pourra donner ici qu'un survol de la relativité restreinte, et la plupart des résultats seront rendus plausibles, et non démontrés en détail. Le lecteur est renvoyé à la bibliographie pour des exposés plus complets.

**N.B.** Dans cette section et les suivantes nous utiliserons un système d'unités où la vitesse de la lumière $c = 1$ et nous utiliserons également la convention de sommation sur les indices répétés

$$\sum_i x^i y^i = x^i y^i$$

### 3.1. *Photons*

Rappelons qu'un *événement* est quelque chose qui se passe à un temps déterminé $t$ en un point déterminé $\vec{r}$ : un événement est caractérisé par une coordonnée de temps $t$, souvent notée $x^0 = t$, et trois coordonnées d'espace $\vec{r} = (x, y, z) = (x^1, x^2, x^3)$. En relativité restreinte, on a l'habitude de définir un référentiel d'inertie en quadrillant l'espace-temps par un réseau de règles rigides et d'horloges synchronisées, mais l'utilisation de règles rigides est à proscrire en relativité générale, du moins sur de grandes distances, en raison de la courbure de l'espace. En conséquence nous allons essayer de tout mesurer en utilisant uniquement des photons.

*Un observateur sera représenté par un physicien qui transporte avec lui une horloge, mesurant son temps propre, et un émetteur/détecteur de photons capable de mesurer les fréquences des photons reçus et émis. Entre d'autres termes, un observateur = une horloge et un radar.*

Nous allons maintenant énoncer l'hypothèse de base de la relativité restreinte en l'illustrant au moyen de l'explosion d'une étoile : un événement $O$ (explosion de l'étoile) se produit en émettant une bouffée de photons et de particules massives (débris). Nous allons faire l'hypothèse fondamentale que *tous les photons, indépendamment de leur couleur, arrivent au même instant à un observateur*, mais pas en général avec la même couleur que celle de l'émission. Si tel n'était pas le cas, l'explosion semblerait durer dans le temps et un événement ponctuel ne serait pas vu comme ponctuel par un observateur. Les particules massives, au contraire, n'arrivent pas au même instant et sont détectées *après* les photons. Les courbes d'espace-temps formées des positions successives de l'observateur, des photons ou

des particules et les instants correspondants, sont appelées *lignes d'Univers*
de l'observateur, des photons ou des particules. Les propriétés précédentes
doivent évidemment être valables quel que soit l'observateur. Pour en rendre
compte, on va tracer dans l'espace-temps à partir de $O$ un cône de sommet
$O$, $N(O)$, qui est une *structure géométrique absolue*, indépendante de l'observateur, et qui est appelée *cône de lumière* de $O$. Les lignes d'Univers des
photons passant par le point $O$ à $t = 0$ suivent les génératrices de ce cône.
La ligne d'Univers de l'observateur coupe $N(O)$ en un point unique $O'$, qui
définit la position de l'observateur et le temps auquel il reçoit les photons
émis par l'explosion (Fig. 6). Les lignes d'Univers des particules massives
émises au moment de l'explosion sont entièrement contenues à l'intérieur du
cône futur $N^+(O)$ de $O$. Comme $N(O)$ est une structure géométrique absolue, il ne définit aucune direction privilégiée, et tout observateur passant
par $O$ voit $N(O)$ avec une parfaite symétrie sphérique.

Considérons deux observateurs, Alice (A) et Bob (B)[h] qui échangent des
photons. Si les fréquences des photons reçus sont identiques à celles des
photons émis, on dira qu'Alice et Bob suivent des *lignes d'Univers parallèles*, et les deux observateurs peuvent alors aisément synchroniser leurs
horloges. L'échange de photons permet à Alice de définir les coordonnées
de l'événement $P$ "réflection du photon par le miroir de Bob"[i] (Fig. 7). Alice
mesure les coordonnées de l'événement $P$ comme étant $(t, x, c = 1)$

$$t = \frac{1}{2}\left(t_2 + t_1\right) \qquad x = \frac{1}{2}\left(t_2 - t_1\right) \tag{23}$$

Mais ce résultat est indépendant du fait que le photon rebondit sur le
miroir de Bob ou sur celui d'un troisième observateur, Chiara (C), qui ne
suit pas nécessairement une ligne d'Univers parallèle à celles d'Alice et de
Bob (Fig. 7). Autrement dit les coordonnées $(t, x)$ attribuées par Alice à
l'événement $P$ sont indépendantes du partenaire qui réfléchit le photon.

## 3.2. *Effet Doppler*

Nous allons maintenant établir la formule de l'effet Doppler en prenant
l'exemple de la mesure de la vitesse d'une voiture par un radar. La voiture
croise le radar au temps $t = 0$, suivant les horloges du radar et de la voiture ;
ceci est toujours possible par un choix convenable de l'origine des temps

---

[h]Ces deux héros de de l'informatique quantique sont apparus pour la première fois dans
la discussion des systèmes de cryptographie à clé secrète.
[i]Pour une version littéraire de cet échange de photons, on pourra se reporter à la nouvelle
de D. Buzzatti, "Les sept messagers", où les photons sont remplacés par des cavaliers.

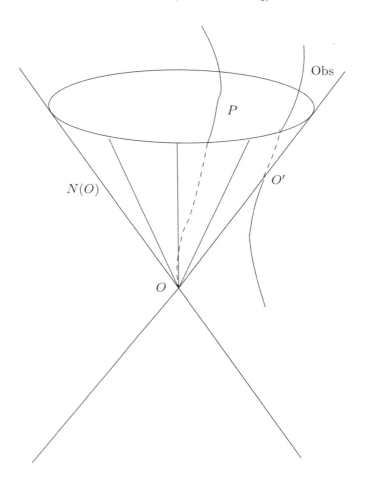

FIGURE 6. Cône de lumière $N(O)$ du point $O$. La ligne d'Univers d'une particule massive passe par $P$, celle de l'observateur coupe $N(O)$ en $O'$. Les lignes d'Univers des photons sont les génératrices du cône.

des deux horloges. Le radar émet un photon[j] au temps $t$, *suivant l'horloge du radar*. Si la voiture était immobile, elle recevrait ce photon au temps $t$, mais comme elle s'éloigne elle le recevra à un temps $Kt$ *suivant l'horloge de la voiture*, avec $K > 1$ (Fig. 8)[k]. Mais la situation radar-voiture est symétrique, car les directions $x$ et $-x$ sont équivalentes : il n'y a pas de

---

[j]En pratique un signal électromagnétique codé.
[k]En raison de l'homogénéité temporelle, $K$ ne peut pas dépendre du temps.

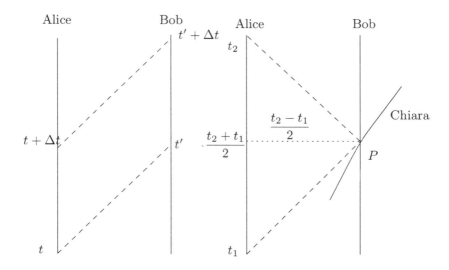

FIGURE 7. Mesure des coordonnées d'un événement $P$. Alice et Bob suivent des lignes d'Univers parallèles. Les lignes d'Univers des photons sont en tirets.

direction privilégiée dans un référentiel d'inertie. Si le photon est réfléchi au temps $Kt$ suivant l'horloge de la voiture, il atteindra le radar au temps $K^2 t$ *suivant l'horloge du radar*. Pour le radar, l'événement $P$ : le photon est réfléchi par la voiture, a pour coordonnées, avec $t_1 = t$ et $t_2 = K^2 t$ dans (23)

$$\text{temps} \qquad \frac{1}{2}\,(t_2 + t_1) = \frac{1}{2}\,t(K^2 + 1)$$

$$\text{espace} \qquad \frac{1}{2}\,(t_2 - t_1) = \frac{1}{2}\,t(K^2 - 1)$$

Effectuant le rapport espace/temps, la vitesse de la voiture mesurée par le radar est donc

$$v = \frac{K^2 - 1}{K^2 + 1} \tag{24}$$

ou encore

$$K = \sqrt{\frac{1 + v}{1 - v}} \tag{25}$$

Une autre façon d'exprimer ce résultat est de dire que la fréquence du photon mesurée par la voiture, $\omega_{\text{rec}}$ est reliée à la fréquence $\omega_{\text{em}}$ d'émission

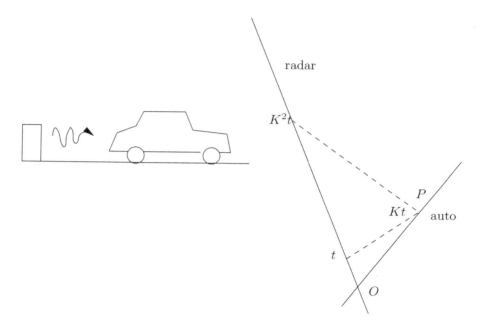

FIGURE 8. Mesure de la vitesse d'une voiture avec un radar. En traits pleins les lignes d'Univers du radar et de l'auto, en tirets celles des photons. Le signal est reçu par l'auto en $P$.

par le radar par

$$\omega_{\text{rec}} = \frac{1}{K}\,\omega_{\text{em}} = \omega_{\text{em}}\sqrt{\frac{1-v}{1+v}} \qquad (26)$$

Cette formule n'est évidemment pas autre chose que celle de l'effet Doppler relativiste.

### 3.3. Métrique de l'espace-temps plat

Pour formaliser ces considérations géométriques, on introduit une métrique sur l'espace-temps ; si

$$x^\mu = (x^0, \vec{x}) \quad \text{et} \quad y^\mu = (y^0, \vec{y})$$

sont les coordonnées spatio-temporelles de deux événements (l'exposant 0 désigne la coordonnée de temps), on pourra aussi considérer $x^\mu$ et $y^\mu$ comme des *quadrivecteurs*[l], les quadrivecteurs joignant l'origine à ces points. On définira le *produit scalaire de Minkowski* de ces deux quadrivecteurs par

$$x \cdot y = x^0 y^0 - \vec{x} \cdot \vec{y} = x^0 y^0 - x^i y^i \tag{27}$$

Par définition des quadrivecteurs sont des objets à quatre composantes tels que *leur produit scalaire (27) soit indépendant du référentiel d'inertie*. Cette condition permet de déterminer la loi de transformation des quadrivecteurs quand on passe d'un référentiel à un autre, ou transformation de Lorentz, mais nous n'aurons pas besoin de sa forme explicite. Il suffira de savoir qu'elle laisse par construction le produit scalaire invariant. De façon équivalente, nous introduisons un *tenseur métrique*[m] $\eta_{\mu\nu}$, le tenseur de Minkowski

$$\eta_{\mu\nu} = \text{diagonal } (1, -1, -1, -1) \tag{28}$$

et nous récrivons (27)

$$x \cdot y = \eta_{\mu\nu} x^\mu y^\nu = x_\mu y^\nu \quad \text{avec} \quad x_\mu = \eta_{\mu\nu} x^\nu$$

Les composantes $x^\mu$ sont appelées conventionnellement composantes *contravariantes* du quadrivecteur et les $x_\mu$ sont les composantes *covariantes*. Le produit scalaire d'un quadrivecteur par lui-même, $x^2 = x_\mu x^\mu$ est la "longueur au carré", ou norme carrée (de Minkowski) de ce vecteur[n]. On écrit en général cette longueur au carré de Minkowski pour un vecteur infinitésimal $dx^\mu$

$$ds^2 = (dx^0)^2 - d\vec{x}^2 = dt^2 - d\vec{x}^2 = \eta_{\mu\nu} dx^\mu dx^\nu \tag{29}$$

Si un photon est émis à l'origine[o] $O$ $(t = 0, \vec{x} = 0)$ et est reçu au point $x^\mu$, alors $|x^0| = ||\vec{x}||$ et

$$(x^0)^2 - \vec{x}^2 = x_\mu x^\mu = 0 \tag{30}$$

---

[l]Ce qui ne sera pas possible en Relativité Générale, où les coordonnées ne sont pas des vecteurs.

[m]Il existe deux conventions pour le tenseur $\eta_{\mu\nu}$, la convention (28) et la convention

$$\eta_{\mu\nu} = \text{diagonal } (-1, 1, 1, 1)$$

La convention utilisée dans cet exposé minimise le nombre de signe moins, car on utilise beaucoup plus d'intervalles du genre temps $ds^2 > 0$ (avec la convention (28) !) que d'intervalle du genre espace. La convention opposée à (28) est avantageuse en théorie quantique des champs, car on tombe directement sur la métrique euclidienne lorsque l'on effectue une rotation de Wick.

[n]Toutes mes excuses : $x$ peut désigner soit une coordonnée, soit le quadrivecteur $x$. J'espère que le contexte lèvera toute ambiguïté.

[o]Pour un photon émis en $x^\mu$ et reçu à l'origine, on a bien évidemment $x^0 < 0$.

Ceci n'est autre que l'équation du cône de lumière $N(O)$ défini dans la section 3.1. Pour un autre observateur qui verra l'événement réception du photon avec des coordonnées différentes $x'^{\mu}$, l'équation du cône de lumière sera $x'^0 - \vec{x}'^2 = 0$. L'intervalle d'espace-temps $x^{\mu}$ entre le point d'émission et le point de réception d'un photon vérifie donc $x_{\mu}x^{\mu} = x^2 = 0$, et on dira que le quadrivecteur $x^{\mu}$ est du *genre lumière*. Si $x_{\mu}x^{\mu} = x^2 > 0$, on dira que $x^{\mu}$ est *du genre temps*, et si $x^2 < 0$, $x^{\mu}$ est du *genre espace*.

Appliquons ces notions au cas d'un mouvement uniforme sur une droite, $x = vt$, $|v| < 1$ : par exemple une particule massive est émise en $O$ au temps $t = 0$ et est détectée en $P$ au temps $t$ et à la position $x$ (Fig. 9). La norme carrée du quadrivecteur $x^{\mu} = (t, x)$ est donnée par

$$x_{\mu}x^{\mu} = \tau^2 = t^2 - v^2t^2 = t^2(1 - v^2) = \frac{t^2}{\gamma^2}$$

avec $\gamma = (1 - v^2)^{-1/2}$, soit $t = \gamma\tau$. Mais $\tau^2$ est un produit scalaire, qui est indépendant du système de coordonnées. Si un observateur se déplace avec une vitesse $v$ en suivant la particule, les coordonnées de $P$ pour cet observateur seront $(t', x' = 0)$, et on aura

$$t'^2 = t^2 - x^2 = \tau^2$$

Le temps mesuré par une horloge liée à la particule est par définition $t'$, et c'est le temps propre de la particule : $t' = \tau$. L'expression $t = \gamma\tau$ montre que *le temps propre est toujours le plus court*.

Afin de montrer la cohérence de ce résultat avec notre raisonnement précédent sur l'effet Doppler, examinons la situation décrite dans la section 3.2 dans le référentiel où Alice est au repos (Fig. 10). D'après (25) Bob mesure sur sa montre la réception du photon au temps

$$\Delta\tau' = K\Delta t = \Delta t\sqrt{\frac{1 + v}{1 - v}}$$

Pour Alice, les coordonnées de $P$ sont données par l'intersection des droites

$$x = t - \Delta t \qquad x = vt$$

soit

$$t = \frac{\Delta t}{1 - v} \qquad x = \frac{v\Delta t}{1 - v}$$

Nous obtenons donc

$$\Delta\tau^2 = \frac{1}{(1 - v)^2}(1 - v^2)\Delta t^2 = \frac{1 + v}{1 - v}\Delta\tau'^2$$

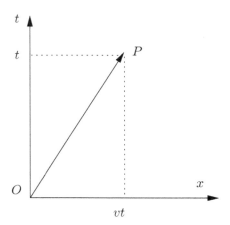

FIGURE 9. Mouvement uniforme d'une particule de vitesse $v$, qui quitte l'origine au temps $t = 0$ et arrive en $P : (t, x = vt)$ au temps $t$.

Le temps propre $\Delta\tau$ calculé à l'aide de l'invariance du produit scalaire est donc bien le même que celui $\Delta\tau'$ obtenu par le raisonnement donnant l'effet Doppler.

Examinons maintenant la ligne d'Univers $x^\mu(\tau)$ d'une particule massive paramétrée à l'aide de son temps propre. Nous appellerons *quadrivitesse* de la particule le quadrivecteur $u^\mu(\tau)$ tangent à la ligne d'Univers au point $P$ de temps propre $\tau$

$$u^\mu(\tau) = \frac{\mathrm{d}x^\mu(\tau)}{\mathrm{d}\tau} \qquad (31)$$

$u^\mu$ est bien un quadrivecteur car c'est le quotient d'un quadrivecteur $x^\mu$ par un scalaire $\tau$. Pendant un temps (propre) infinitésimal $\mathrm{d}\tau$

$$\mathrm{d}x^\mu = u^\mu \mathrm{d}\tau$$

mais comme

$$\mathrm{d}x^2 = \mathrm{d}\tau^2 = (u_\mu u^\mu)\mathrm{d}\tau^2$$

on trouve $u^2 = 1$ : *la quadrivitesse est un vecteur unitaire de genre temps*. Il est intéressant de donner l'expression de la quadrivitesse en fonction de la vitesse $\vec{v}$ dans un référentiel d'inertie (Fig. 9)

$$\frac{\mathrm{d}x^0}{\mathrm{d}\tau} = \frac{\mathrm{d}t}{\mathrm{d}\tau} = \gamma \qquad\qquad \frac{\mathrm{d}\vec{r}}{\mathrm{d}\tau} = \frac{\mathrm{d}t}{\mathrm{d}\tau}\frac{\mathrm{d}\vec{r}}{\mathrm{d}t} = \gamma\,\vec{v}$$

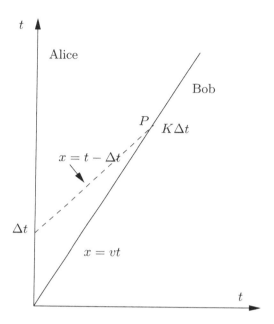

FIGURE 10. Autre déduction de l'effet Doppler. On a choisi un référentiel où Alice est au repos.

soit

$$\boxed{u^\mu = \gamma\,(1, \vec{v})} \tag{32}$$

La quadrivitesse d'une particule libre dans un référentiel d'inertie obéit à $\mathrm{d}u^\mu/\mathrm{d}\tau = 0$ : $\tau$ est ce que l'on appelle un *paramètre affine* pour la trajectoire. Ce paramètre affine n'est pas unique, $\sigma = a\tau + b$ est aussi affine. Si l'on effectue un changement de paramétrisation

$$y^\mu(\sigma) = x^\mu(\tau(\sigma)) \qquad w^\mu = \frac{\mathrm{d}y^\mu}{\mathrm{d}\sigma} = u^\mu(\tau(\sigma))\,\frac{\mathrm{d}\tau}{\mathrm{d}\sigma}$$

alors

$$\frac{\mathrm{d}^2 y^\mu}{\mathrm{d}\sigma^2} = \frac{\mathrm{d}w^\mu}{\mathrm{d}\sigma} = \frac{\mathrm{d}^2\tau}{\mathrm{d}\sigma^2}\,u^\mu \tag{33}$$

Si le changement de variable $\tau \to \sigma$ n'est pas linéaire, $\mathrm{d}w^\mu/\mathrm{d}\sigma \neq 0$, et $\sigma$ n'est pas un paramètre affine. Les choses en vont différemment pour un photon : prenons par exemple un photon se propageant suivant $Ox$ et dont la ligne d'Univers est $x^0 = x$, ce que l'on peut écrire

$$x^\mu = \lambda u^\mu \quad \text{avec} \quad u^\mu = (1, 1, 0, 0)$$

On a $dx^\mu/d\lambda = u^\mu$, $u^2 = 0$ et $du^\mu/d\lambda = 0$ : $\lambda$ est un paramètre affine. Mais on pourrait aussi bien prendre $x^\mu = \sigma^3 u^\mu$, et on vérifie que $\sigma$ n'est pas un paramètre affine.

Pour terminer cette section, nous observons que le produit scalaire de deux quadrivitesses a une interprétation intéressante. Soit Alice et Bob se croisant en $O$ avec des quadrivitesses $u_A$ et $u_B$. Dans le référentiel au repos d'Alice, $u_A = (1, \vec{0})$. Soit $v_{AB}$ la valeur absolue de la vitesse relative de $A$ et $B$, par exemple la vitesse de Bob mesurée par Alice. On a alors

$$u_A \cdot u_B = u_A^0 u_B^0 - \vec{u}_A \cdot \vec{u}_B = u_B^0 = \frac{1}{\sqrt{1 - v_{AB}^2}}$$

et donc

$$u_A \cdot u_B = \frac{1}{\sqrt{1 - v_{AB}^2}} \tag{34}$$

Cette équation permet de calculer la vitesse relative $v_{AB}$ quel que soit le référentiel où l'on connaît $u_A$ et $u_B$.

### 3.4. *Tenseur énergie-impulsion*

En multipliant la quadri-vitesse $u^\mu$ d'une particule (massive) par sa masse $m$, on obtient la *quadri-impulsion $p^\mu$*, quadri-vecteur formé de l'énergie (composante de temps) et de l'impulsion (composantes d'espace)

$$p^\mu = mu^\mu \qquad p^0 = E = \gamma m \qquad \vec{p} = \gamma m \vec{v} \tag{35}$$

où nous avons utilisé (32) ; la masse étant un scalaire, le produit $mu^\mu$ est bien un quadri-vecteur. On notera la relation importante $\vec{v} = \vec{p}/E$. Pour des faibles vitesse, $v \ll c$, on peut effectuer un développement limité en fonction de $v/c$

$$E = mc^2 + \frac{p^2}{2m} + \cdots \qquad \vec{p} = \frac{m\vec{v}}{\sqrt{1 - (v/c)^2}} = m\vec{v} + \cdots$$

où nous avons rétabli $c$. Le premier terme de $E$ est l'énergie de masse $mc^2$, le second l'énergie cinétique non relativiste habituelle.

Nous allons maintenant discuter la notion de courant en relativité. Considérons un ensemble de particules instables de vie moyenne $\tau$ dans le référentiel où elles sont au repos, et supposons pour simplifier que ces particules se désintègrent *exactement* au bout d'un temps $\tau$. Si ces particules se déplacent à une vitesse $v$ dans un référentiel d'inertie R, un observateur dans ce référentiel trouvera que les particules se désintègrent au bout d'un temps $\gamma\tau$, en raison de la relation entre le temps propre de la particule et le temps mesuré par l'observateur. Les particules parcourent donc dans R, non pas une distance $v\tau$, mais une distance[p] $\gamma v\tau$. Supposons que soit disposée dans dans R une densité linéaire de charge $\rho^*$ suivant la direction de la vitesse des particules, une particule verra, pendant un temps $\tau$, $\gamma\rho^* v\tau$ charges. En résumé, une particule voit les charges défiler à la vitesse $-v$, et elle observe une densité de charges qui n'est pas $\rho^*$, mais $\gamma\rho^*$. Un observateur lié à la particule définira donc une densité de charge $\rho = \gamma\rho^*$ et une densité de courant $\vec{j} = -\gamma\rho^*\vec{v} = -\rho\vec{v}$. Si la densité de charges est $\rho^*$ dans un référentiel où ces charges sont au repos, cet observateur de quadri-vitesse $u^\mu$ dans ce référentiel définira un quadri-courant $j^\mu$ par

$$\boxed{j^\mu = \rho^* u^\mu}\tag{36}$$

Ce quadri-courant obéit à la *loi de conservation*

$$\boxed{\partial_\mu j^\mu = \partial_0 j^0 + \vec{\nabla}\cdot\vec{j} = 0}\tag{37}$$

qui est bien connue des équations de Maxwell. Il est très important de noter qu'en relativité *une loi de conservation est obligatoirement une loi de conservation locale*. En effet, supposons que nous partions d'une situation où la charge totale est nulle, mais qu'au temps $t = 0$ une charge négative soit créée au point $O$, tandis qu'une charge positive est créée en un autre point, ce qui conserve globalement la charge. Cependant, étant donné la relativité de la simultanéité, il est facile de trouver des référentiels où la charge positive, par exemple, est créée *avant* la charge négative, et la charge n'est pas conservée. Autrement dit, la charge contenue dans un volume ne peut varier que parce qu'elle passe par la frontière de ce volume[q], ce qui n'est pas vrai par exemple pour la conservation de l'impulsion en mécanique des fluides galiléenne.

Poursuivons l'étude des lois de conservation. Une surface $t =$cste est une *3-surface de genre espace* de l'espace-temps, le vecteur orthogonal à cette surface étant le vecteur $n^\mu = (1,\vec{0})$. On peut généraliser en prenant une surface orthogonale à un vecteur unitaire de genre temps $n^\mu, n^2 = 1$. Une surface orthogonale au vecteur de genre espace $n^\mu = (0,1,0,0)$ par exemple

---

[p]Cette propriété est utilisée dans les accélérateurs de particules. Si l'on produit un faisceau de mésons-$\pi$ de vie moyenne $\sim 10^{-8}$ s, on pourrrait s'attendre qu'un méson-$\pi$ dont la vitesse est proche de celle de la lumière puisse parcourir au maximun un distance $\sim 3$ m. En fait il peut parcourir des distances bien plus grandes, ce qui permet d'éloigner la zone d'expériences de celle de production de plusieurs centaines de mètres, sans perte appréciable de mésons-$\pi$.

[q]Cette observation permet de démontrer (37) en utilisant le théorème de la divergence.

sera une *3-surface de genre temps*, et en général une telle surface sera ortho-gonale à un vecteur unitaire de genre espace $n^\mu, n^2 = -1$. On peut tracer un volume $\Delta V$ dans une 3-surface de genre temps ou de genre espace. Soit $\Delta N$ le nombre de charges dans $\Delta V$. La quantité $\Delta N$ est indépendante du référentiel, c'est un scalaire de Lorentz. Le seul scalaire que l'on puisse construire avec le courant $j^\mu$ et le vecteur $n^\mu$ caractérisant la 3-surface est $j \cdot n = j_\mu n^\mu$. Lorsque $n^\mu = (1, \vec{0})$

$$(j_\mu n^\mu)\Delta V = \rho^* \Delta V = \Delta N$$

Si le vecteur $n^\mu$ est du genre espace, par exemple $n^\mu = (0, 1, 0, 0)$, alors

$$\Delta N = (j_\mu n^\mu)\Delta t \Delta y \Delta z = j^x \Delta t \Delta y \Delta z = \frac{\Delta N}{\Delta y \Delta z \Delta t}\, \Delta t \Delta y \Delta z$$

et $j^x = \Delta N/(\Delta A \Delta t)$ est bien le flux de charges à travers la surface $\Delta A = \Delta y \Delta z$. On a donc dans tous les cas

$$\Delta N = (j_\mu n^\mu)\Delta V \tag{38}$$

Si au lieu de $\Delta N$, qui est un scalaire, on a affaire à une quantité qui est un quadrivecteur, par exemple une quadri-impulsion, on doit introduire au lieu d'un courant vectoriel $j^\mu$ un courant tensoriel $T^{\mu\nu}$. La relation correspondant à (38) est alors

$$\Delta p^\mu = (T^{\mu\nu}n_\nu)\,\Delta V \tag{39}$$

où $T^{\mu\nu}$ est le *tenseur énergie-impulsion*. Si $n^\mu = (1, \vec{0})$, cette équation donne pour la composante de temps et les trois composantes d'espace

$$\Delta p^0 = \Delta E = T^{00}\Delta V$$
$$\Delta p^i = T^{i0}\Delta V$$

$T^{00} = \epsilon$ est la densité d'énergie, $T^{i0} = \pi^i$ la densité d'impulsion suivant la direction $i$. Prenons l'exemple d'une boîte de particules de même vitesse $v$ dans R, de masse $m$ et de densité $\rho^*$ dans le référentiel où elles sont au repos. Nous aurons alors pour les densités d'énergie et d'impulsion

$$\epsilon = T^{00} = (m\gamma)(\rho^*\gamma) = m\rho^* u^0 u^0$$
$$\pi^i = T^{i0} = (m\gamma v^i)(\rho^*\gamma) = m\rho^* u^0 u^i$$

car la densité dans R est $\rho = \rho^*\gamma$.

Il nous faut maintenant interpréter $T^{0i}$ et $T^{ij}$. Suivant le même raisonnement que dans le cas scalaire, considérons une surface de genre temps orthogonale à $n^\mu = (0, 1, 0, 0)$

$$\Delta p^\mu = T^{\mu x} \Delta t \Delta y \Delta z = T^{\mu x} \Delta t \Delta A$$

$T^{\mu x} = \Delta p^\mu / (\Delta A \Delta t)$ est le flux de $p^\mu$ à travers $\Delta A$, et en particulier $T^{0x} = \Delta p^0 / (\Delta A \Delta t)$ est le flux d'énergie dans la direction $x$. Mais, en raison de la conservation locale de l'énergie-impulsion

$$\text{(flux d'energie)} \times \Delta A \Delta t = \text{(densite d'energie)} \times v^x \Delta A \Delta t$$
$$= \text{(densite d'impulsion)}^x \Delta A \Delta t$$

où nous avons utilisé $v^x = p^x / E$. En divisant par $\Delta A \Delta t$ nous obtenons $T^{0x} = T^{x0}$. D'autre part

$$T^{ix} = \frac{\Delta p^i}{\Delta A \Delta t} = \frac{\Delta p^i / \Delta t}{\Delta A}$$

qui est une force par unité de surface. Plus généralement

$$\Delta F^i = T^{ij} n_j \Delta A$$

est la composante $i$ de la force $\vec{F}$ exercée sur une surface de normale $\vec{n}$. À la limite non-relativiste, les composantes $T^{ij}$ ne sont autres que les composantes du tenseur des pressions, familier en mécanique des fluides. En résumé, le tenseur énergie-impulsion $T^{\mu\nu}$ est un tenseur symétrique : $T^{\mu\nu} = T^{\nu\mu}$. La conservation de l'énergie-impulsion se traduit par la généralisation de (37)

$$\boxed{\partial_\mu T^{\mu\nu} = 0} \tag{40}$$

Un cas particulier important est celui du fluide parfait. Dans ce cas le tenseur énergie-impulsion ne peut dépendre que de la vitesse d'ensemble $u^\mu$ du fluide, car il n'existe pas d'autre direction disponible[r] et aussi du tenseur de Minkowski $\eta_{\mu\nu}$. Dans le référentiel où le fluide est au repos, on doit avoir

$$T^{\mu\nu} = \text{diag}\,(\rho, \mathsf{P}, \mathsf{P}, \mathsf{P})$$

car la densité d'énergie est $\rho = m\rho^*$ et $T^{ij} = \mathsf{P}\delta^{ij}$, $\mathsf{P}$ étant la pression. Si l'on écrit la forme la plus générale possible de $T^{\mu\nu}$ avec des coefficients arbitraires $A$ et $B$

$$T^{\mu\nu} = Au^\mu u^\nu + B\eta^{\mu\nu}$$

---

[r]Le flux de chaleur définit une direction privilégiée dans le référentiel où le fluide est au repos, mais précisément il n'y a pas de flux de chaleur dans un fluide parfait.

on obtient dans le référentiel au repos où $u^\mu = (1, \vec{0})$

$$T^{00} = A + B = \rho \qquad T^{ij} = -B\,\delta^{ij} = \mathsf{P}\,\delta^{ij}$$

ce qui donne

$$\boxed{T^{\mu\nu} = (\rho + \mathsf{P})u^\mu u^\nu - \mathsf{P}\eta^{\mu\nu}} \tag{41}$$

Il est facile de vérifier que l'équation de conservation (40) donne l'équation de conservation de la masse et l'équation d'Euler à la limite des faibles vitesses[s]. Nous allons voir dans la section 6 que le tenseur énergie-impulsion joue un rôle fondamental dans l'écriture de l,équation d'Einstein

*courbure locale de l'espace-temps = tenseur énergie-impusion*

ou en formule

$$R_{\mu\nu} - \frac{1}{2}\,g_{\mu\nu}R = -8\pi G T_{\mu\nu} \tag{42}$$

Le tenseur de Ricci $R_{\mu\nu}$, la courbure $R$ et la métrique $g_{\mu\nu}$, qui seront définies de façon précise dans la section 5, sont des caractéristiques de la géométrie.

## Bibliographie

Hartle [2003], chapitres 4 et 5. Ludvigsen [2000], chapitres 2 à 8. Weinberg [1972], chapitre 2.

## 4. Cosmologie

Les considérations de géométrie dont nous aurons besoin dans cette section de cosmologie sont suffisamment intuitives pour qu'il ne soit pas nécessaire de faire appel aux résultats de géométrie différentielle de la section 5.

---

[s]L'équation de conservation de la masse suit de

$$u^\mu \partial_\mu T^{\mu\nu} = 0$$

et celle d'Euler de

$$(\eta_{\rho\nu} - u_\rho u_\nu)\partial_\mu T^{\mu\nu} = 0$$

Le tenseur $(\eta_{\rho\nu} - u_\rho u_\nu)$ est le projecteur, au sens de la métrique de Minkowski, sur la surface orthogonale à $u^\mu$. Les deux équations ci-dessus sont les projections de $\partial_\mu T^{\mu\nu} = 0$ sur $u^\mu$ et sur la surface perpendiculaire à $u^\mu$.

## 4.1. *Description qualitative de l'Univers*

*Matière sombre.* Il est aujourd'hui admis que la matière visible (étoiles, nuages de gaz ...) ne représente qu'une très faible partie de la masse de l'Univers, au plus quelques pour cents. Une composante essentielle est la *matière sombre.* Cette matière sombre est mise en évidence par l'étude de la rotation de nuages de gaz autour du centre de galaxies. Par exemple la Fig. 11 montre la vitesse de rotation $v(r)$ de nuages de gaz dans la galaxie d'Andromède en fonction de la distance $r$ au centre de la galaxie.

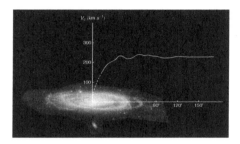

FIGURE 11. Vitesse de rotation de nuages de gaz en fonction de la distance au centre de la galaxie, ici la galaxie d'Andromède. L'échelle horizontale est en minutes d'arc et donne la taille angulaire des objets. D'après Hartle [2003].

Si le nuage est situé à une distance $r$ du centre de la galaxie, et si $M(r)$ est la masse contenue à l'intérieur de l'orbite du nuage, alors d'après la loi de Newton

$$\frac{GM(r)}{r^2} = \frac{v^2(r)}{r} \tag{43}$$

Si l'essentiel de la masse était concentrée au voisinage de $r = 0$, comme on pourrait le déduire de l'observation de la matière visible, on aurait $v(r) \propto r^{-1/2}$. En fait on trouve $v(r) \sim \text{cste}$, ce qui indique la présence d'une matière non visible importante. L'existence de cette matière sombre a reçu récemment une confirmation indépendante grâce à l'utilisation de l'effet de lentille gravitationnelle : voir Hartle [2003], chapitre 11 et l'article de Koopmans et Blanford.

*Isotropie et homogénéité de l'Univers = principe cosmologique.* Si l'on fait abstraction des fluctuations locales (galaxies...), l'observation montre que l'Univers est homogène et isotrope (en fait l'isotropie en tout point entraîne l'homogénéité). Il faut moyenner sur des distances suffisamment grandes pour que l'homogénéité soit valable et l'échelle de transition entre la répartition homogène et la structure granulaire est de l'ordre de $3 \times 10^7$ année-lumière. Un autre argument convaincant est l'isotropie du rayonnement cosmologique[t] à 3 K : les résultats des satellites COBE et WMAP montrent qu'en dehors d'une asymétrie due à l'effet Doppler provenant du mouvement de notre galaxie[u] avec une vitesse de l'ordre $10^{-3} c$, ce rayonnement est isotrope avec des fluctuations qui ne dépassent pas $10^{-5}$. En fait la qualité de la courbe de rayonnement de corps noir du rayonnement cosmologique est bien meilleure que celle de toute courbe réalisée au laboratoire ! Le point de départ de la cosmologie standard est donc l'hypothèse que *l'Univers est homogène et isotrope à grande échelle*, hypothèse qu'Einstein fut le premier à formuler en 1917, à l'époque une pure spéculation théorique sans la moindre base expérimentale.

*Expansion de l'Univers.* Comme on le sait depuis Hubble (1927), le décalage Doppler de la lumière émise par les galaxies est proportionnel à leur distance : pour une galaxie s'éloignant de la nôtre avec une vitesse $v$, le décalage Doppler est suivant (26) pour $v/c \ll 1$

$$\frac{v}{c} = \frac{\Delta \lambda}{\lambda} \equiv z \qquad (44)$$

où $\lambda$ est la longueur d'onde. Une définition plus précise de $z$ sera donnée ultérieurement. La loi établie par Hubble est $v = H_0 d$, où $H_0$ est la *constante de Hubble*, avec la valeur numérique

$$H_0 = 72 \pm 7 \, \text{km.s}^{-1}/\text{Mpc} \qquad t_H = \frac{1}{H_0} = 4.3 \times 10^{17} \, \text{s} = 13.6 \times 10^9 \text{ans}$$
$$(45)$$

Rappelons que 1 parsec (pc) vaut 3.26 année-lumière (a.l.). Le temps $t_H$ est une première approximation pour l'âge de l'Univers. En effet, en supposant que les galaxies ont eu une vitesse uniforme depuis le Big Bang, $d = v t_H$, et tenant compte d'autre part de la loi de Hubble $d = v/H_0 \Longrightarrow t_H = 1/H_0$. Pour vérifier la loi de Hubble, on a besoin d'une mesure de distance. Celle-ci est fournie en utilisant des "bougies standard", c'est-à-dire des étoiles dont la luminosité $L$ est connue (ou supposée telle !). En effet, si $f$ est le flux

---

[t]Environ 380 000 ans après le Big Bang, les photons se sont découplés de la matière et se sont retrouvés hors équilibre thermodynamique. Depuis cette époque, leur longueur d'onde a varié en proportion du facteur de dilatation $a(t)$ (voir (59)), et ils forment aujourd'hui le rayonnement cosmologique à 3 K.

[u]La vitesse de notre groupe local de galaxies est de 600 km/s par rapport au fond cosmologique. La vitesse du système solaire par rapport au centre de notre galaxie est de 270 km/s.

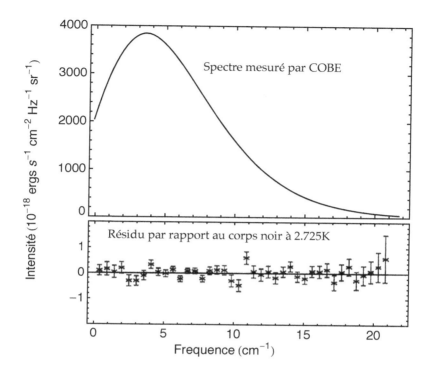

FIGURE 12.   Courbe de Planck pour le rayonnement cosmologique mesurée par le satellite COBE (1992). D'après Hartle [2003].

recueilli par un détecteur sur la Terre, ce flux est relié à $L$ et à la distance $d$ par

$$f = \frac{L}{4\pi d^2} = \frac{LH_0^2}{4\pi z^2 c^2} \tag{46}$$

d'après la loi de Hubble. Nous verrons ultérieurement une version plus précise de cette formule ; $z$ est connu par le décalage Doppler, $f$ est mesuré expérimentalement, et si l'on connaît $L$, on en déduit $d$.

Il convient de faire quelques mises en garde contre des confusions possibles.

- Il ne faut pas interpréter le Big Bang comme une explosion à partir d'un centre ! L'Univers n'a pas de centre, chaque galaxie voit les autres galaxies s'éloigner suivant la loi de Hubble. Une image classique (à deux dimensions) est celle d'un ballon que l'on gonfle : des points marqués sur le ballon s'éloignent les uns des autres, sans que l'on ait un point central sur le ballon.

- Comme on le verra dans la section 5, on ne sait pas *a priori* comparer la vitesse de deux galaxies éloignées dans un espace courbe. Pour ce faire il faudrait "transporter parallèlement" le vecteur vitesse d'une des deux galaxies au point où se trouve la seconde galaxie. Or le transport parallèle d'un vecteur dépend de la courbe choisie dans un espace de courbure non nulle. Les questions qui sont simples dans un espace plat le sont beaucoup moins dans un espace courbe !

## 4.2. *Coordonnées comobiles*

L'hypothèse d'isotropie et d'homogénéité permet de décomposer l'espace-temps $M$ en une famille de 3-surfaces de genre espace, paramétrées par un scalaire que l'on peut appeler le temps $t$. Chacune de ces 3-surfaces est une variété $\Sigma_t$ à trois dimensions homogène et isotrope : $M = \mathbb{R}^+ \times \Sigma$. On a donc un feuilletage de l'espace-temps en tranches $t =$ cste. Il existe seulement trois types de variétés à trois dimensions homogènes et isotropes

(1) L'espace plat.
(2) La sphère $S^3$ à courbure constante $> 0$.
(3) L'hyperboloïde à courbure constante $< 0$.

Ce résultat est intuitif, et il peut être montré rigoureusement. Remarquons que nous ne disons rien sur les propriétés topologiques *globales* (sauf dans le cas de la sphère) : par exemple dans le cas de l'espace plat, rien n'interdit de choisir le tore $T^3$, le tore étant une surface de courbure nulle. Nous allons nous limiter pour le moment au cas de l'espace plat, en revenant dans la section 4.4 sur les deux autres cas ; l'espace plat est le cas le plus simple, et le bonus est que c'est, semble-t-il, le cas physiquement pertinent ! Considérons les galaxies se trouvant au temps $t$ sur la variété $\Sigma_t$ : par homogénéité, ce temps, appelé *temps comobile*, peut être choisi identique pour toutes les galaxies. Au voisinage de chaque galaxie, on mesure au temps $t$ une densité de matière $\rho(t)$ qui est la même pour toutes les galaxies. L'isotropie entraîne que la métrique est de la forme[v]

$$\mathrm{d}s^2 = \mathrm{d}t^2 - a^2(t)(\mathrm{d}x^2 + \mathrm{d}y^2 + \mathrm{d}z^2) \tag{47}$$

---

[v]Anticipons sur les résultats de la section 5 : chaque galaxie suit une géodésique $G$ de l'espace-temps et on choisit comme paramètre le temps propre $t = \tau$. On a donc pour la métrique

$$g_{tt} = < \frac{\partial}{\partial t}, \frac{\partial}{\partial t} > = u^2 = 1$$

car

$$\frac{\partial}{\partial t} = \frac{\mathrm{d}}{\mathrm{d}\tau}\Big|_{\text{le long de } G}$$

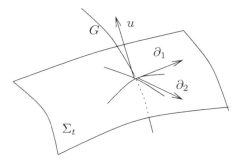

FIGURE 13.   Variété $\Sigma_t$ et géodésique dune galaxie $G$. On a représenté la quadrivitesse $u$ et deux vecteurs tangents aux axes de coordonnées $\partial_1$ et $\partial_2$ (section 5).

Les coordonnés $(x, y, z)$ repèrent la positions dans l'espace d'une galaxie et elle sont *indépendantes du temps* : ce sont des *coordonnées comobiles*. Le facteur $a(t)$ est le *facteur de dilatation* ou le *facteur d'échelle*. La métrique (47) est la généralisation la plus simple de la métrique de Minkowski (29). Une tranche d'espace-temps $t =$ cste est muni de la métrique euclidienne ordinaire pour $t =$ cste

$$-\mathrm{d}s^2 = a^2(t)(\mathrm{d}x^2 + \mathrm{d}y^2 + \mathrm{d}z^2) = a^2(t)\mathrm{d}S^2 \qquad (48)$$

Cette métrique sera évidement modifiée dans le cas de l'hyperboloïde et de la sphère. La métrique (47) correspond à un *espace* plat, mais pas à un *espace-temps* plat !

Nous allons maintenant généraliser le résultat de Hubble, qui n'est valable que pour des galaxies pas trop éloignées, en établissant la formule du *décalage vers le rouge cosmologique*. Considérons deux galaxies dont la différence de coordonnées comobiles est $(\Delta x, \Delta y, \Delta z)$. La distance qui sépare ces deux

---

Les coordonnées spatiales sont telles que les vecteurs tangents $\partial/\partial x^i$ sont orthogonaux à $\partial/\partial t$. La métrique est donc de la forme

$$\mathrm{d}s^2 = \mathrm{d}t^2 - g_{ij}\mathrm{d}x^i\mathrm{d}x^j$$

Géométriquement, les géodésiques suivies par les galaxies sont orthogonales à $\Sigma_t$, sinon elles définiraient une direction privilégiée sur $\Sigma_t$. Cela veut dire que l'on n'a pas de termes en $\mathrm{d}t\,\mathrm{d}x^i$.

galaxies au temps $t$ est d'après (48)[w]

$$d(t) = a(t) \left[\Delta x^2 + \Delta y^2 + \Delta z^2\right]^{1/2} = a(t)d_{\text{com}} \qquad (49)$$

Si $a(t)$ croît avec $t$, les deux galaxies s'éloignent : l'Univers est en expansion, ce que nous allons désormais supposer : $\dot{a}(t) > 0$. Il sera commode d'utiliser des coordonnées sphériques $(r, \theta, \varphi)$ dans l'espace et de récrire (47)

$$\mathrm{d}s^2 = \mathrm{d}t^2 - a^2(t)\left[\mathrm{d}r^2 + r^2(\mathrm{d}\theta^2 + \sin^2\theta\mathrm{d}\varphi^2)\right] = \mathrm{d}t^2 - a^2(t)\left[\mathrm{d}r^2 + r^2\,\mathrm{d}\Omega^2\right] \qquad (50)$$

Soit deux galaxies, de coordonnées comobiles $r = 0$ (nous) et $r = r_{\text{com}}$, et un photon se propageant entre les deux galaxies. Dans le cas d'un photon, on doit avoir $\mathrm{d}s^2 = 0$, mais à cause du facteur $a(t)$ dans la métrique, les lignes d'Univers d'un photon ne sont pas des droites (Fig. 14). Si les temps

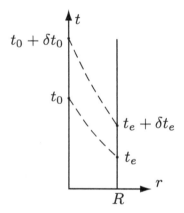

FIGURE 14. Propagation de photons entre deux galaxies et décalage vers le rouge gravitationnel, $R = r_{\text{com}}$.

d'émission par la galaxie de deux photons succcessifs à $r = r_{\text{com}}$ sont $t_e$ et $t_e + \delta t_e$ et les temps de réception sur la Terre sont $t_0$ et $t_0 + \delta t_0$, on tire de $\mathrm{d}s^2 = 0$ la relation $\mathrm{d}t^2 = a^2(t)\mathrm{d}r^2$ et

$$r_{\text{com}} = \int_{t_e}^{t_0} \frac{\mathrm{d}t}{a(t)} = \int_{t_e + \delta t_e}^{t_0 + \delta t_0} \frac{\mathrm{d}t}{a(t)} \qquad (51)$$

[w]Il convient de préciser la signification de la distance $d(t)$ : en théorie, il faudrait disposer d'un grand nombre d'observateurs disposés entre notre galaxie et la galaxie lointaine, qui mesurent chacun la distance entre eux-mêmes et leur plus proche voisin au même temps $t$. La somme de ces distances donne $d(t)$. Inutile de dire que ce n'est pas très pratique...

soit pour des $\delta t$ petits

$$\frac{\delta t_0}{a(t_0)} = \frac{\delta t_e}{a(t_e)}$$

Ceci donne pour le décalage de fréquence entre photons émis ($\omega_e$) et reçus ($\omega_0$)

$$\boxed{\frac{\omega_0}{\omega_e} = \frac{a(t_e)}{a(t_0)} \ (< 1)} \tag{52}$$

Cette équation donne l'expression générale du décalage vers le rouge cosmologique. On définit le facteur $z$ par

$$\boxed{1 + z = \frac{\lambda_0}{\lambda_e} = \frac{\omega_e}{\omega_0} = \frac{a(t_0)}{a(t_e)}} \tag{53}$$

On verra dans la section 4.4.1 que ces équations sont aussi valables pour un espace courbe. Dans ce raisonnement, toute référence à la notion hasardeuse de vitesse relative de deux galaxies a disparu. En fait il est bien préférable de définir la distance à une galaxie lointaine par la donnée de $z$, qui est une donnée *observationnelle* non ambiguë, alors que le temps d'émission d'un photon ($t_e$ dans (53)) ou bien la distance dépendent du modèle d'Univers choisi et de la définition de cette distance (voir la section 4.4.2). On dira par exemple que le découplage des photons du rayonnement cosmologique s'est produit à $z \simeq 1100$, ou que les objets les plus lointains que l'on a pu observer aujourd'hui ont $z \sim 6$.

Montrons que la loi de Hubble des équations (44) et (45) est une approximation de (53) en considérant deux galaxies proches. Le temps mis par le photon pour aller d'une galaxie à l'autre est $\Delta t \simeq a(t_0)R = d$, où $t_0$ représente l'instant actuel, mesuré à partir du Big Bang, et donc le facteur de dilatation aujourd'hui. Selon une convention habituelle en cosmologie, l'indice 0 étiquette aujourd'hui, par exemple $H_0$ est la constante de Hubble aujourd'hui et $a_0 \equiv a(t_0)$. Écrivant

$$a(t_e) \simeq a(t_0 - d) \simeq a(t_0) - d\,\dot{a}(t_0)$$

on déduit de (53)

$$1 + z \simeq \frac{a(t_0)}{a(t_0) - d\,\dot{a}(t_0)} \simeq 1 + d\,\frac{\dot{a}(t_0)}{a(t_0)} = 1 + dH_0$$

soit

$$z \simeq H_0 d \qquad H_0 \equiv \frac{\dot{a}(t_0)}{a(t_0)} = \frac{\dot{a}_0}{a_0} \tag{54}$$

L'interprétation du décalage vers le rouge cosmologique comme provenant de l'effet Doppler et de la "fuite des galaxies" n'est correcte que pour des galaxies suffisamment proches. Dans le cas général on doit utiliser (53).

Nous avons vu que l'inverse de la constante de Hubble (aujourd'hui) donne une indication sur l'âge de l'Univers $t_0$ : si la vitesse d'expansion a été uniforme depuis le Big Bang, alors $H(t) = H_0$ et l'âge de l'Univers est bien l'inverse de la constante de Hubble : $t_0 = t_H = 1/H_0$. Cependant il est intuitivement évident (et ce sera montré ci-dessous), que la gravité ne peut que ralentir l'expansion de l'Univers : une pierre lancée vers le haut est toujours freinée par la gravité, même si on la lance avec une vitesse supérieure à la vitesse de libération et $\ddot{a}(t) < 0$. La Fig. 15 montre l'évolution du facteur de dilatation $a(t)$ : on voit que $t_0 \leq t_H$. On verra que la prédiction simple et remarquable $\ddot{a}(t) < 0$ n'est pas vérifiée expérimentalement.

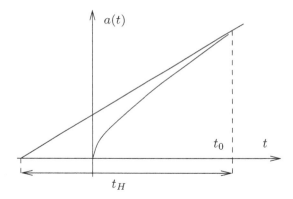

FIGURE 15. Le facteur de dilatation $a(t)$ en l'absence d'énergie du vide. La tangente au point $t = t_0$ donne la vitesse d'expansion aujourd'hui. Si cette vitesse était restée constante, l'âge de l'Univers serait $t_H$.

## 4.3. Évolution du facteur de dilatation

### Équation de Friedmann

En première approximation, l'Univers est un fluide parfait dont les molécules sont les galaxies. Un flux de chaleur ou de particules serait incompatible avec l'isotropie, car il fixerait une direction privilégiée, et on est en droit de négliger tout phénomène irréversible. Le deuxième principe ($E =$

énergie, $S$ = entropie, $P$ = pression, $V$ = volume, $T$ = température absolue)

$$dE = TdS - PdV$$

devient

$$d(\Delta E) = -Pd(\Delta V) \tag{55}$$

Des coordonnées comobiles $(\Delta x, \Delta y, \Delta z)$ définissent un covolume $\Delta V_{\text{com}} = \Delta x \Delta y \Delta z$ où le nombre de galaxies est constant, tandis que le volume $\Delta V$ est

$$\Delta V = a^3(t)\,\Delta x \Delta y \Delta z = a^3(t)\,\Delta V_{\text{com}} \tag{56}$$

Si $\rho(t)$ est la densité d'énergie (qui inclut bien évidemment l'énergie de masse), nous avons d'après (55) et (56)

$$\frac{d}{dt}\left[\rho(t)a^3(t)\Delta V_{\text{com}}\right] = -P(t)\frac{d}{dt}\left[a^3(t)\Delta V_{\text{com}}\right]$$

soit

$$\frac{d}{dt}\left[\rho(t)a^3(t)\right] = -P(t)\frac{d}{dt}\left[a^3(t)\right] \tag{57}$$

Trois cas de figure limites sont possibles.

(1) *Univers dominé par la matière.* La pression du gaz de galaxies est négligeable par rapport à l'énergie de masse et (57) devient

$$\rho(t)a^3(t) = \text{cste} \qquad \text{ou} \qquad \rho(t) = \rho(t_0)(1+z)^3 \tag{58}$$

Cette équation exprime simplement que la densité est divisée par $(a(t_0)/a(t))^3 = (1+z)^3$ quand le facteur de dilatation passe de $a(t)$ à $a(t_0)$.

(2) *Univers dominé par le rayonnement.* Dans ce cas (voir le rayonnement du corps noir), on a $P = \rho/3$ et $\rho \propto T^4$, équations qui sont aussi valides pour tout gaz de particules ultra-relativistes, quelle que soit leur statistique. On en déduit

$$\frac{\dot{\rho}}{\rho} = -3\frac{\dot{a}}{a}$$

soit

$$\rho(t) = \rho(t_0)\left[\frac{a(t_0)}{a(t)}\right]^4 = \rho(t_0)(1+z)^4$$

$$T(t) = T(t_0)\left[\frac{a(t_0)}{a(t)}\right] = T(t_0)(1+z) \tag{59}$$

Cette équation montre que la longueur d'onde des photons du rayonnement du corps noir cosmologique croît en proportion de $a(t)$ après le
découplage rayonnement-matière.

(3) *Univers dominé par le vide.* Dans ce cas $\rho$ doit être indépendant du
temps[x] : la densité d'énergie du vide ne dépend pas de l'expansion de
l'Univers. On a donc

$$\rho(t)\frac{\mathrm{d}a^3(t)}{\mathrm{d}t} = -\mathsf{P}(t)\frac{\mathrm{d}a^3(t)}{\mathrm{d}t}$$

soit

$$\mathsf{P} = -\rho \qquad\qquad (60)$$

Nous reviendrons ultérieurement sur cette "énergie du vide". Pour le moment nous prenons en compte uniquement les paramètres classiques, matière et rayonnement, et nous allons essayer d'établir l'équation d'évolution
du paramètre d'échelle $a(t)$ en utilisant un raisonnement newtonien discutable, mais qui a le mérite de la simplicité. Choisissons une origine arbitraire
$O$ dans l'Univers et considérons une galaxie $G$ de masse $m$ située à une distance $d(t)$ de $O$. En admettant la validité du théorème de Gauss pour une
distribution de matière infinie, son énergie potentielle est

$$U(d) = -G\frac{mM}{d} \qquad\qquad M = \frac{4\pi}{3}\,d^3\rho$$

soit

$$U(d) = -\frac{4\pi}{3}\,Gmd^2\rho \qquad\qquad (61)$$

Écrivons $d(t) = a(t)R$, où $R$ est la coordonnée comobile radiale de la galaxie,
l'origine $O$ ayant une coordonnée comobile $R = 0$, d'où la vitesse de la
galaxie $v(t) = R\,\dot{a}(t)$. L'énergie de la galaxie est alors

$$\begin{aligned}
E &= \frac{1}{2}\,mR^2\,\dot{a}^2(t) - \frac{4\pi}{3}mR^2G\rho(t)\,a^2(t) \\
&= \frac{1}{2}\,mR^2\left(\dot{a}^2(t) - \frac{8\pi}{3}\,G\rho(t)\,a^2(t)\right)
\end{aligned}$$

---

[x]Ceci est une simplification qui est contestée dans certains modèles. Avec l'hypothèse sim-
plificatrice $\rho_v(t) =$ cste, l'effet de l'énergie du vide est équivalent à celui d'une constante
cosmologique. Dans un référentiel d'inertie local, le seul tenseur disponible est le tenseur de Minkowski $\eta_{\mu\nu}$ et si tous les observateurs voient le même vide, on doit avoir
$T^v_{\mu\nu} \propto \eta_{\mu\nu}$. Ceci rajoute un terme $\Lambda g_{\mu\nu}$ dans l'équation d'Einstein : voir(84), où $\Lambda$ est
la constante cosmologique.

Cette énergie mécanique doit être constante, et après redéfinition convenable de $a(t)$ par un facteur multiplicatif, on obtient *l'équation de Friedmann*

$$\dot{a}^2(t) - \frac{8\pi}{3}\,G\rho(t)\,a^2(t) = -k \qquad k = 1, 0, -1 \tag{62}$$

où $k$ peut prendre les valeurs $k = -1$ : énergie mécanique positive, $k = 0$ : énergie mécanique nulle et $k = 1$ : énergie mécanique négative. Pour $k = -1\,(E > 0)$ et $k = 0\,(E = 0)$ l'expansion se poursuit indéfiniment, tandis que pour $k = 1\,(E < 0)$ l'expansion finit par s'arrêter et on revient à la singularité initiale ("Big Crunch"). La raison de la convention de signe pour $k$ apparaîtra à la section 4.4 : $k$ caractérise la courbure de l'espace, $k = +1$ pour la sphère, $k = 0$ pour l'espace plat et $k = -1$ pour l'hyperboloïde. De l'équation (62) on déduit, en la combinant avec (57)

$$\frac{\ddot{a}}{a} = -\frac{4\pi G}{3}(3\mathsf{P} + \rho) \tag{63}$$

ce qui montre que $\ddot{a} < 0$ si $(3\mathsf{P} + \rho) > 0$, en particulier si la pression et l'énergie sont positives : s'il n'y a que de la matière et/ou du rayonnement, l'expansion est ralentie par la gravité, même dans le cas où elle se poursuit indéfiniment. L'observation $\ddot{a} > 0$ implique donc une nouvelle physique !

L'Univers spatialement plat correspond à $k = 0$. De l'équation (62) prise à $t = t_0$ on déduit la densité d'énergie aujourd'hui en tenant compte de ce que $H_0 = \dot{a}_0/a_0$

$$\rho_0 = \rho_c = \frac{3H_0^2}{8\pi G} \tag{64}$$

La densité $\rho_c$ est la *densité critique* : pour $\rho \leq \rho_c$, l'expansion se poursuit indéfiniment, pour $\rho > \rho_c$ l'Univers finit dans le Big Crunch. Les cosmologistes ont l'habitude de définir les rapports des différents types de densité d'énergie à la densité d'énergie critique ($m$ = matière, $r$ = rayonnement, $v$ = vide)

$$\Omega_m = \frac{\rho_m^0}{\rho_c} \qquad \Omega_r = \frac{\rho_r^0}{\rho_c} \qquad \Omega_v = \frac{\rho_v^0}{\rho_c} \tag{65}$$

et $\Omega_m + \Omega_r + \Omega_v = 1$ pour un Univers spatialement plat ($\rho_0 = \rho_c$).

Pour fixer les idées, prenons le cas particulier d'un Univers plat dominé par la matière, $\Omega_m = 1$, $\Omega_r = \Omega_v = 0$, ce qui est en fait le premier modèle d'Univers en expansion proposé en 1932 par Einstein et de Sitter. Les deux équations (59) et (62) deviennent

$$\dot{a}^2 - \frac{8\pi G}{3a} = 0 \qquad \rho a^3 = 1$$

d'où l'on tire

$$\sqrt{a}\,\mathrm{d}a \propto \mathrm{d}t \qquad a^{3/2} \propto t$$

soit finalement une loi en $t^{2/3}$

$$a(t) = a_0 \left( \frac{t}{t_0} \right)^{2/3} \tag{66}$$

et donc $t_0 = 2t_H/3$, ce qui confirme la courbe de la Fig. 15. Pour un Univers dominé par le rayonnement on trouve $a(t) \propto t^{1/2}$ et pour un Univers dominé par le vide $a(t) \propto \exp(Ht)$ : si l'énergie du vide est non nulle, elle finit toujours par l'emporter ! Un exemple simple (non réaliste) avec $\Omega_m = \Omega_r = \Omega_v = 1/3$ est donné dans la Fig. 16.

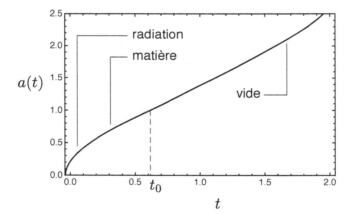

FIGURE 16.  Le paramètre d'échelle $a(t)$ en fonction du temps avec $\Omega_m = \Omega_r = \Omega_v = 1/3$. D'après Hartle [2003].

## Le problème de l'horizon

En raison de la vitesse finie de la lumière on ne peut observer qu'une fraction finie de l'univers : c'est le problème de *l'horizon*. On aurait envie de dire que l'horizon est simplement $ct_0$, mais c'est un peu plus compliqué en raison de l'expansion. Il est commode pour tracer les figures d'utiliser le "temps conforme" $\eta$ défini par

$$\mathrm{d}\eta = \frac{\mathrm{d}t}{a(t)} \tag{67}$$

En effet, avec le "temps" $\eta$, les lignes d'Univers des photons sont des droites, car la métrique devient

$$\mathrm{d}s^2 = a^2(t)[\mathrm{d}\eta^2 - (\mathrm{d}r^2 + r^2\,\mathrm{d}\Omega^2)] \tag{68}$$

Dans les coordonnées $(\eta, r)$, les lignes d'Univers des photons sont des droites à $45^o$. L'horizon $r_H(t)$ est donné par (voir aussi (51))

$$r_H(t) = \int_0^t \frac{\mathrm{d}t'}{a(t')} \tag{69}$$

La distance physique $d_H(t)$ à l'horizon au temps d'observation $t$ est

$$\boxed{d_H(t) = a(t)r_H(t) = a(t)\int_0^t \frac{\mathrm{d}t'}{a(t')}} \tag{70}$$

La région de l'Univers à laquelle on peut (en principe) accéder aujourd'hui est donc limitée par $d_H(t_0)$. Si l'on prend par exemple le cas simple d'un Univers plat et dominé par la matière, on montre immédiatement de (70) que

$$t_0 = \frac{2}{3}\,t_H \qquad d_H(t_0) = 3\,t_0$$

soit

$$d_H(t_0) = 2\,t_H = 2.7 \times 10^{10}\,\text{a.l}$$

Avec les données du modèle standard actuel ($\Lambda$CDM, section 4.5) : $\Omega_m = 30\%$, $\Omega_r \simeq 0$ et $\Omega_v = 70\%$ on trouve numériquement

$$d_H(t_0) \simeq 4.5 \times 10^{10}\,\text{a.l}$$

En fait l'Univers est opaque jusqu'au découplage des photons, qui s'effectue au bout de $4 \times 10^5$ ans environ après le Big Bang, et l'horizon effectif est plus petit (Fig. 17).

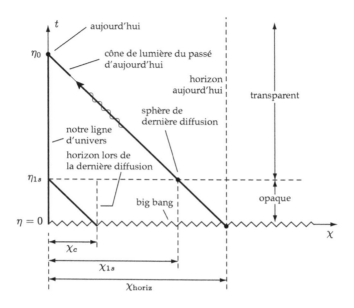

FIGURE 17. Univers visible aujourd'hui en tenant compte de l'opacité initiale. D'après Hartle [2003].

### Relation entre la luminosité et le décalage vers le rouge

Nous avons vu que pour des galaxies proches le rapport $f/L$ du flux terrestre $f$ à la luminosité $L$ est donné en fonction du décalage $z$ par (46), $c = 1$

$$\frac{f}{L} = \frac{H_0^2}{4\pi z^2}$$

Nous allons établir la généralisation de cette relation avec une première application au cas de l'espace plat, la discussion pour un courbure non nulle, légèrement plus complexe, étant renvoyée à la section 4.4.2. On définit une distance effective $d_{\text{eff}}$ entre l'étoile et le point d'observation de la façon suivante : l'aire de la sphère lieu des points ayant la même distance à la source que le point d'observation est $4\pi d_{\text{eff}}^2$, et $d_{\text{eff}} \neq d(t)$ si la courbure spatiale est $\neq 0$. On observe une réduction du flux d'énergie pour deux raisons.

(1) La fréquence (et donc l'énergie en raison de la loi de Planck-Einstein) des photons est plus faible en raison du décalage vers le rouge cosmo-

logique (53)

$$\omega_0 = \frac{\omega_e}{(1+z)}$$

(2) Comme l'intervalle $\delta t_0$ entre la réception de deux photons successifs est plus grand que l'intervalle $\delta t_e$ entre leur émission, on compte moins de photons par unité de temps

$$\delta t_0 = \delta t_e (1 + z)$$

Le rapport $f/L$ vaut dans ces conditions

$$\frac{f}{L} = \frac{1}{4\pi d_{\text{eff}}^2} \frac{1}{(1+z)^2} \qquad (71)$$

Prenons un modèle simple pour fixer les idées, celui d'Einstein et de Sitter : $\Omega_m = 1$, $\Omega_r = \Omega_v = 0$ et $k = 0$, un espace plat. Il est instructif de conduire le calcul en utilisant la variable $z$, qui est la "bonne" variable cosmologique, plutôt que $t$. Nous appellerons $z'$ la variable d'intégration et suivant (53)

$$1 + z' = \frac{a_0}{a(t)} \qquad \frac{\mathrm{d}z'}{\mathrm{d}t} = -\frac{a_0\, \dot{a}(t)}{a^2(t)}$$

On écrit d'abord en suivant (49) que $d_{\text{eff}} = a_0 d_{\text{com}}$

$$d_{\text{eff}} = a_0 \int_{t_e}^{t_0} \frac{\mathrm{d}t}{a(t)} = \int_0^z \mathrm{d}z' \frac{a(t)}{\dot{a}(t)}$$

et on exprime $a(t)/\dot{a}(t)$ en utilisant l'équation de Friedmann (62) pour $k = 0$ et $\rho = \rho_c (1 + z')^3$, ce qui donne $\dot{a}/a = H_0 (1 + z')^{3/2}$, soit

$$d_{\text{eff}} = \frac{1}{H_0} \int_0^z \frac{\mathrm{d}z'}{(1+z')^{3/2}} = \frac{2}{H_0} \left( 1 - \frac{1}{\sqrt{1+z}} \right)$$

Cette expression se réduit à $z/H_0$ pour $z \ll 1$ et on retrouve (46). Écrivons $f/L$ sous sa forme finale

$$\frac{f}{L} = \frac{H_0^2}{16\pi} \frac{1}{(1+z)[(1+z)^{1/2} - 1]^2} \qquad (72)$$

Le cas général est examiné au § 4.4.2. La Fig. 18 illustre le calcul précédent dans le cas à deux dimensions d'espace.

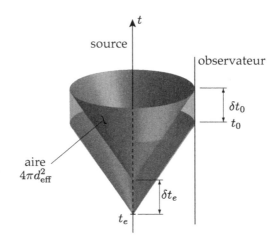

FIGURE 18.    Relation flux/luminosité. D'après Hartle [2003].

## 4.4. *Cas de la courbure spatiale non nulle*

*Métrique de Friedmann–Robertson–Walker*

Traitons brièvement le cas de courbure spatiale non nulle, puisque la nature semble avoir choisi la courbure nulle! Nous avons déjà mentionné qu'il n'existe que trois types de variétés à trois dimensions homogènes et isotropes

(1) L'espace plat.
(2) La sphère $S^3$ à courbure constante $> 0$.
(3) L'hyperboloïde à courbure constante $< 0$.

L'espace plat vient d'être étudié; passons au cas de la sphère $S^3$, qui est sans doute le plus intuitif. La sphère unité $S^3$ peut être paramétrée par trois angles $(\chi, \theta, \varphi)$

$$0 \leq \chi \leq \pi \qquad 0 \leq \theta \leq \pi \qquad 0 \leq \varphi \leq 2\pi$$

Les coordonnés cartésiennes $(X, Y, Z, W)$ sont données par

$$
\begin{aligned}
X &= \sin\chi \sin\theta \cos\varphi & Z &= \sin\chi \cos\theta \\
Y &= \sin\chi \sin\theta \sin\varphi & W &= \cos\chi
\end{aligned}
\tag{73}
$$

La métrique spatiale se calcule immédiatement à partir de (73)

$$dS^2 = d\chi^2 + \sin^2\chi(d\theta^2 + \sin^2\theta d\varphi^2) \tag{74}$$

Le volume de l'espace est fini au temps $t$ et est proportionnel à $2\pi^2 a^3(t)$. Le cas de l'hyperboloïde s'obtient en remplaçant $\sin\chi$ par $\sinh\chi$

$$\mathrm{d}S^2 = \mathrm{d}\chi^2 + \sinh^2\chi(\mathrm{d}\theta^2 + \sin^2\theta\,\mathrm{d}\varphi^2) \tag{75}$$

Pour regrouper les trois cas de figure possibles, on écrit

$$\mathrm{d}S^2 = \mathrm{d}\chi^2 + S_k^2(\chi)(\mathrm{d}\theta^2 + \sin^2\theta\,\mathrm{d}\varphi^2) \tag{76}$$

le facteur $S_k(\chi)$ étant donné par :

(1) sphère : $S_1(\chi) = \sin\chi$ ;
(2) espace plat : $S_0(\chi) = \chi$ ;
(3) hyperboloïde : $S_{-1}(\chi) = \sinh\chi$.

Contrairement aux apparences, on voit que $r$ est en fait une coordonnée angulaire ! L'équation de propagation d'un photon est (voir (48))

$$\mathrm{d}s^2 = \mathrm{d}t^2 - a^2(t)\mathrm{d}S^2 = 0$$

Si un photon reçu au temps $t_0$ a été émis par une galaxie dont la coordonnée est $\chi_{\mathrm{com}}$, le temps d'émission $t_e$ est donné par

$$\chi_{\mathrm{com}} = \int_{t_e}^{t_0} \frac{\mathrm{d}t}{a(t)}$$

et on en déduit que (53) reste valable pour un espace courbe. L'aire d'une sphère à deux dimensions dont les points sont à une coordonnée comobile $\chi_{\mathrm{com}}$ de l'origine est

$$4\pi a_0^2\, S_k^2(\chi_{\mathrm{com}})$$

On trouve souvent la métrique de l'espace-temps écrite sous la forme de Friedmann–Robertson–Walker (FRW), obtenue grâce à un changement de variables élémentaire

$$\mathrm{d}s^2 = \mathrm{d}t^2 - a^2(t)\left[\frac{\mathrm{d}r^2}{1-kr^2} + r^2(\mathrm{d}\theta^2 + \sin^2\theta\,\mathrm{d}\varphi^2)\right]$$

mais nous ne nous servirons pas de cette forme.

Dans un traitement rigoureux, l'évolution du facteur de dilatation $a(t)$ est fixée par l'équation d'Einstein (42)

$$R_{\mu\nu} - \frac{1}{2}Rg_{\mu\nu} = -8\pi G T_{\mu\nu}$$

et par l'expression (41) de $T_{\mu\nu}$ du fluide parfait de galaxies

$$T_{\mu\nu} = (\mathsf{P} + \rho)u_\mu u_\nu - \mathsf{P}g_{\mu\nu}$$

La combinaison de ces deux équations permet de démontrer rigoureusement l'équation de Friedmann (62). Ensuite, étant donné une équation d'état, on peut en déduire la loi d'évolution de $a(t)$. Pour $k = +1$, l'expansion s'arrête au bout d'un certain temps et l'Univers se met à se contracter : c'est le Big Crunch. Pour $k = 0$ et $k = -1$, l'expansion se poursuit indéfiniment. Toutefois, si une énergie du vide est présente, on observe toujours une expansion indéfinie. Une prédiction remarquable de la relativité générale est que le type d'Univers homogène et isotrope, caractérisé par sa courbure, est lié à la densité d'énergie qu'il contient.

*Distance de luminosité dans le cas général*

Revenons sur la distance de luminosité lorsque la courbure spatiale est non nulle, en écrivant l'équation de Friedmann sous la forme

$$\left(\frac{\dot{a}}{a}\right)^2 = H_0^2 \left[\frac{\rho}{\rho_c} + (1 - \Omega_t)\frac{a_0^2}{a^2}\right] \tag{77}$$

où nous avons utilisé l'expression (64) de la densité critique $\rho_c$ et défini $\Omega_t = \rho_0/\rho_c$ ; il est instructif de vérifier que cette équation est bien correcte à $t = 0$ en retrouvant (64). On décompose $\rho$ en une partie matière (non relativiste) $\rho_m$, une partie rayonnement $\rho_r$ et une partie vide $\rho_v$ suivant (65). Dans le cas de la matière par exemple on écrit

$$\frac{\rho_m}{\rho_c} = \frac{\rho_m^0 \, a_0^3}{\rho_c \, a^3} = \Omega_m(1 + z')^3$$

et en procédant de la même manière avec $\rho_r$ et $\rho_v$ on aboutit à

$$\frac{\dot{a}}{a} = H_0 \left[\Omega_m(1 + z')^3 + \Omega_r(1 + z')^4 + \Omega_v + (1 - \Omega_t)(1 + z')^2\right]^{1/2} \tag{78}$$

On en déduit $\chi_{\text{com}}$

$$\chi_{\text{com}} = \frac{1}{a_0 H_0} \int_0^z \frac{dz'}{[\Omega_m(1 + z')^3 + \Omega_r(1 + z')^4 + \Omega_v + (1 - \Omega_t)(1 + z')^2]^{1/2}} \tag{79}$$

Ainsi que nous l'avons déjà mentionné, la surface de la sphère à utiliser dans le calcul est

$$4\pi d_{\text{eff}}^2 = 4\pi a_0^2 S_k^2(\chi_{\text{com}})$$

et la distance effective est $d_{\text{eff}} = a_0 S_k(\chi_{\text{com}})$ où l'on rappelle que

$$S_0(\chi) = \chi \qquad S_1(\chi) = \sin\chi \qquad S_{-1}(\chi) = \sinh\chi$$

Ceci donne le rapport $f/L$

$$\frac{f}{L} = \frac{1}{4\pi a_0^2 S_k^2(\chi_{\text{com}})(1+z)^2} = \frac{1}{4\pi d_{\text{eff}}^2(1+z)^2} = \frac{1}{4\pi d_L^2} \qquad (80)$$

Cette expression permet d'identifier la *distance de luminosité*

$$d_L = d_{\text{eff}}(1+z) = a_0 S_k(\chi_{\text{com}})(1+z)$$

Une autre distance comunément utilisée est la *distance angulaire* $d_A$ : considérons un objet que nous voyons sous un angle $\Delta\theta$. Les photons se sont propagés depuis cet objet jusqu'à nous à $\theta$ et $\varphi$ constants. La taille de l'objet au moment de l'émission était $\Delta L = a(t)S_k(\chi_{\text{com}})\Delta\theta$, et comme l'angle n'a pas varié, on a aujourd'hui

$$\Delta\theta = \frac{\Delta L}{a(t)S_k(\chi_{\text{com}})} = \frac{\Delta L(1+z)}{a_0 S_k(\chi_{\text{com}})} = \frac{\Delta L}{d_A}$$

par définition de la distance angulaire $d_A$. On a donc la relation suivante entre $d_A$ et $d_L$

$$d_A = a_0 S_k(\chi)(1+z)^{-1} = d_L(1+z)^{-2}$$

## 4.5. *Le modèle* $\Lambda$*CDM*

Le rapport $f/L$ dépend des paramètres $\Omega_m, \ldots \Omega_t$ du modèle d'Univers, et sa mesure en fonction de $z$, qui est une donné observationnelle indépendante, nous donne donc accès à ces paramètres. Il est commode de se placer à $t = t_0$ est de définir le *paramètre de décélération* $q_0$ par

$$q_0 = -\left.\frac{a(t)\,\ddot{a}(t)}{\dot{a}^2(t)}\right|_{t_0} \qquad (81)$$

Ce paramètre a été initialement défini avec un signe moins, car tout le monde s'attendait d'après l'équation (63) à ce que $q_0$ soit positif, c'est-à-dire que l'expansion de l'Univers ralentisse.

En 1998, deux groupes indépendants réussirent à mesurer la décélération de l'Univers en utilisant comme bougies standard des supernovae de type Ia et la grande surprise fut l'observation d'une valeur négative de ce paramètre $q_0$, soit une expansion de l'Univers qui s'accélère ! En fait les supernovae à grand $z$ sont moins lumineuses que dans un Univers où l'expansion ralentirait. Comme nous l'avons vu, un tel comportement ne peut s'expliquer que par la présence d'une énergie du vide. Les résultats sont résumés dans la Fig. 19.

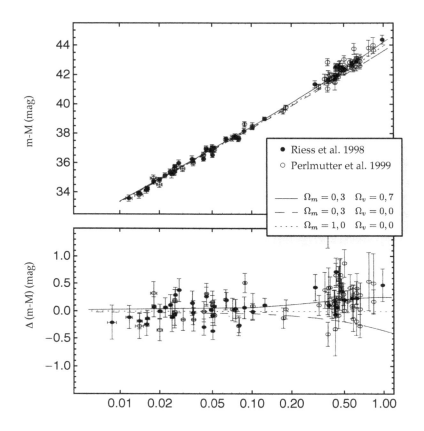

FIGURE 19.   Accélération de l'expansion de l'Univers. $z$ est défini en (53).

Il est intéressant de calculer le rapport $\ddot{a}/a$. En différentiant (78), ou mieux en utilisant directement (63), on trouve

$$\frac{\ddot{a}}{a} = -H_0^2 \left[ \frac{1}{2} \Omega_m (1+z)^3 + \Omega_r (1+z)^4 - \Omega_v \right] \qquad (82)$$

soit aujourd'hui, en négligeant le rayonnnement

$$\frac{\ddot{a}}{a} = -H_0^2 \left[ \frac{1}{2} \Omega_m - \Omega_v \right] \qquad q_0 = \frac{1}{2} \Omega_m - \Omega_v \qquad (83)$$

En l'absence d'énergie du vide on aurait bien $\ddot{a} < 0$.

Nous avons déjà mentionné que la matière visible n'est qu'une faible faction de la matière contenue dans l'Univers. Mais il y a pire : on pourrait imaginer que la matière non visible soit ordinaire, c'est-à-dire constituée de protons et de neutrons, une matière sombre baryonique. Malheureusement, les observations sont incompatibles avec cette hypothèse. La matière sombre doit être pour l'essentiel non baryonique, car l'étude des grandes structures de l'Univers montre qu'elle doit interagir très faiblement avec la matière ordinaire. De plus cette matière sombre doit être froide, ou non relativiste[y] : l'énergie cinétique des particules qui composent la matière sombre doit être petite par rapport à leur énergie de masse. Les deux données précédentes conduisent au modèle standard actuel de la cosmologie, modèle appelé ΛCDM : CDM = Cold Dark Matter, et Λ fait référence à la constante cosmologique introduite par Einstein. En effet une modification possible de (42) est

$$R_{\mu\nu} - \frac{1}{2}\,g_{\mu\nu} = -8\pi G T_{\mu\nu} - \Lambda g_{\mu\nu} \qquad (84)$$

Cette modification est équivalente à l'introduction d'une énergie du vide constante (voir cependant la note 5) avec la correspondance

$$\rho_v = \frac{c^4 \Lambda}{8\pi G} \qquad (85)$$

| Paramètres | 2002 | WMAP |
|---|---|---|
| $H_0$ (Hubble) [km/s/Mpc] | $72 \pm 7$ | $71 \pm 4$ |
| Décélération $q_0 = -a\ddot{a}/\dot{a}^2\vert_{t_0}$ | $-0.67 \pm 0.25$ | $-0.66 \pm 0.10$ |
| Âge de l'Univers [Gan] | $13 \pm 1.5$ | $13.7 \pm 0.2$ |
| $\Omega_t$ | $1.03 \pm 0.03$ | $1.02 \pm 0.02$ |
| $\Omega_b$ | $0.039 \pm 0.008$ | $0.044 \pm 0.004$ |
| $\Omega_{cdm}$ | $0.29 \pm 0.04$ | $0.23 \pm 0.04$ |
| $\Omega_v$ | $0.67 \pm 0.06$ | $0.73 \pm 0.06$ |

La table 4.1 compare les résultats disponibles en 2002 à ceux du satellite WMAP, qui a mesuré avec une grande précision les anisotropies du rayonnement cosmologique. Ces anisotropies sont liées aux fluctuations qui ont donné naissance aux grandes structures par instabilité gravitationnelle

[y]Le scenario CDM prédit que les petites structures se forment avant les grandes (scenario bottom up), alors que le scenario HDM (Hot Dark Matter) prédit au contraire que les petites structures s'obtiennent par fractionnement des grandes (scenario top-down). C'est le scenario bottom up qui est favorisé par l'observation.

et leur mesure permet de remonter à un grand nombre de paramètres du modèle de Big Bang.

Comment comprendre l'énergie du vide ? Il n'y a pas pour le moment d'autre solution que de faire appel aux fluctuations quantiques. Prenons comme exemple le cas familier du champ électromagnétique quantifié dans une cavité de volume $V$. Chaque mode normal de la cavité est un oscillateur harmonique de fréquence $\omega_k = c|\vec{k}|$, où $\vec{k}$ est le vecteur d'onde. Il lui correspond en physique quantique un oscillateur harmonique quantifié dont l'énergie de point zéro, ou énergie de l'état fondamental, est $\hbar\omega_k/2$. La somme des énergies de point zéro de ces modes, ou énergie du vide, est infinie et vaut

$$V\rho_v = \sum_{\vec{k}} \hbar\omega_k = \frac{\hbar c V}{2\pi^2} \int_0^\infty k^3 \mathrm{d}k \tag{86}$$

Par une curieuse revanche de l'histoire, la quantification du rayonnement, postulée par Planck et Einstein pour se débarrasser des infinis dans le rayonnement du corps noir, introduit une autre source d'infinis, cette fois à température nulle ! En théorie quantique des champs on "renormalise" en décidant de prendre comme zéro d'énergie l'énergie du vide, qui est inobservable en valeur absolue, sauf en relativité générale, où l'on doit prendre en compte *toutes* les formes d'énergie. En revanche on observe des *différences* d'énergie du vide en disposant par exemple dans le vide deux plaques conductrices qui se font face. La modification des modes normaux due à la présence des plaques change l'énergie du vide par une quantité *finie*, et se traduit physiquement par une attraction entre les plaques. C'est *l'effet Casimir*, qui a été vérifié récemment avec une précision $\sim 10^{-3}$. Si l'on essaie de deviner un cut-off pour limiter l'intégrale dans (80), le seul cut-off naturel est fixé par l'échelle de Planck $l_P = \sqrt{G\hbar/c^3} \sim 10^{-35}$ m, et on trouve que la densité d'énergie du vide ainsi estimée est d'un facteur $10^{120}$ supérieure à celle mesurée en cosmologie ! Cependant ce calcul pose le problème du référentiel : il n'est pas invariant de Lorentz.

En résumé le modèle $\Lambda$CDM est en excellent accord avec les observations. En particulier la concordance des résultats pré-WMAP avec ceux de WMAP est impressionnante. Cependant il reste deux questions fondamentales, auxquelles les physiciens des particules et de la théorie quantique des champs n'ont pour le moment aucune réponse. Quelle est la particule (ou les particules) qui entrent dans la composition de la matière sombre froide ? Quelle est l'origine de l'énergie du vide ? En l'absence d'une réponse satisfaisante à ces deux questions, un doute continuera à planer sur la pertinence du modèle $\Lambda$CDM, et on ne pourra pas empêcher les astrophysiciens de spéculer sur des solutions encore plus radicales, comme la variabilité dans le temps des constantes fondamentales ou bien la faillite de la relativité générale aux très grandes distances.

## Bibliographie

Hartle [2003], chapitres 17 à 19 ; F. Bouchet, *Relativité, cosmologie et évolution de l'Univers*, contribution à l'ouvrage collectif *Actualité d'Einstein*, EDPSciences/Éditions du CNRS, à paraître en 2005 ; J. Rich *Principes de la cosmologie*, Éditions de l'École Polytechnique, (2002) ; P. Coles et F. Lucchini *Cosmology : the origin and evolution of cosmic structure*, John Wiley, New-York (2002) ; W. Freeman et M. Turner, *Rev. Mod.Phys*, **75**, 1433 (2003) ; P. Peebles et B. Ratra, *Rev. Mod.Phys*, **75**, 559 (2003) ; S. Perlemutter, *Supernovae, Dark Energy and the Accelerating Universe*, Physics Today, avril 2003, p. 53 ; L. Koopmans et R. Blanford, *Gravitational Lenses*, Physics Today, juin 2004, p. 45.

## 5. Boîte à outils de géométrie différentielle

**N.B.** Cette section n'est évidemment pas une introduction, même succincte, à la géométrie différentielle. Il se borne à rassembler les notions qui sont strictement indispensables à la compréhension des équations fondamentales de la relativité générale.

### 5.1. *Espace tangent à une variété*

Une *variété* $M$ de dimension $N$ est un espace topologique qui est *localement* isomorphe à $\mathbb{R}^N$. Cela ne dit rien sur les propriétés topologiques globales de la variété. Ainsi le plan réel à deux dimensions $\mathbb{R}^2$, le tore à deux dimensions $T^2$ et la sphère $S^2$ sont tous trois des variétés de dimension 2, bien que leurs propriétés topologiques globales soient très différentes. Nous n'aurons besoin dans ce cours que des propriété locales. On peut introduire localement sur la variété des coordonnées $x^i$ qui repèrent un point de la variété. Toutefois, on ne peut pas en général repérer tout point $P$ de la variété par un système de coordonnées unique. Il faut en général un système de cartes dont chacune recouvre un partie de la variété et qui se raccordent entre elles, l'ensemble des ces cartes formant un *atlas*.

Soit une courbe $C$ tracée sur $M$ paramétrée par $t$ : les coordonnées d'un point de $C$ sont $x^i(t)$. On appelle *dérivée directionnelle* $\xi$ par rapport à la courbe $C$ au point $P \in C$

$$\xi = \left.\frac{\mathrm{d}}{\mathrm{d}t}\right|_{\text{le long de } C \text{ en } P} \equiv \left.\frac{\mathrm{d}}{\mathrm{d}t}\right|_{C,P}$$

l'application qui fait correspondre à toute fonction $f(t)$ sa dérivée $\mathrm{d}f/\mathrm{d}t$. L'ensemble de ces dérivées directionnelles forme *l'espace tangent $T_P$ en $P$ à la variété*; $T_P$ est donc l'espace des dérivées directionnelles, et les dérivées directionnelles sont les vecteurs de $T_P$. Nous verrons ci-dessous que $T_P$ est un espace vectoriel de dimension $N$. Montrons la linéarité : si $\Gamma$ est une autre courbe paramétrée par $t$ et coupant $C$ en $P$ (Fig. 20) avec $\eta = \mathrm{d}/\mathrm{d}t|_{\Gamma,P}$, on peut former la combinaison linéaire avec des coefficients $a$ et $b^z$

$$a\left.\frac{\mathrm{d}}{\mathrm{d}t}\right|_{C,P} + b\left.\frac{\mathrm{d}}{\mathrm{d}t}\right|_{\Gamma,P}$$

et cette combinaison est bien une *dérivation*, c'est-à-dire qu'elle obéit bien à la règle de Leibniz

$$(a\xi + b\eta)(fg) = g(a\xi + b\eta)f + f(a\xi + b\eta)g$$

On peut trouver aisément une base de $T_P$ en utilisant un système de coordonnées $x^i(t)$ valable dans le voisinage du point $P$. Un point sur $C$ étant alors repéré par $x^i(t)$, on peut écrire

$$
\begin{aligned}
\left.\frac{\mathrm{d}f}{\mathrm{d}t}\right|_{C,P} &= \frac{\mathrm{d}x^i}{\mathrm{d}t}\frac{\partial f}{\partial x^i} = \frac{\mathrm{d}x^i}{\mathrm{d}t}\,\partial_i f \\
\xi = \left.\frac{\mathrm{d}}{\mathrm{d}t}\right|_{C,P} &= \frac{\mathrm{d}x^i}{\mathrm{d}t}\,\partial_i = \xi^i \partial_i
\end{aligned}
\tag{87}
$$

Si l'on interprète le paramètre $t$ comme un temps, $x^i(t)$ décrit le parcours d'un point sur $C$ en fonction du temps, et $\mathrm{d}x^i/\mathrm{d}t$ n'est autre que le vecteur vitesse de ce point en $P$ : Fig. 20. Ceci donne une interprétation géométrique utile de la dérivée directionnelle, mais cette interprétation a l'inconvénient de masquer le caractère intrinsèque de cette dérivée : il n'est pas nécessaire de plonger $M$ dans un espace de dimension $> N$, même si la Fig. 20 peut être utile pour l'intuition.

---

$^z$Pour le lecteur qui souhaite des notations plus explicites : soit $x^i(t) = c^i(t)$ la paramétrisation de la courbe $C$ par $N$ fonctions $c^i(t)$ et $\gamma^i(t)$ celle de $\Gamma$ : $x^i(t) = \gamma^i(t)$. Alors $\xi = \dot{c}^i\partial_i$ et $\eta = \dot{\gamma}^i\partial_i$. La combinaison linéaire des deux vecteurs est

$$a\xi + b\eta = (a\dot{c}^i + b\dot{\gamma}^i)\partial_i$$

.

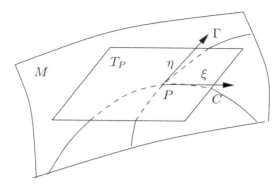

FIGURE 20.    (a) Courbes $C(t)$ et $\Gamma(\tau)$, vecteur vitesse en $P$ et interprétation géométrique du plan tangent en $P$.

L'équation (87) montre que le vecteur $\xi \in T_P$, qui, soulignons le une fois de plus, *existe indépendamment de tout système de coordonnées*, a pour composantes $\xi^i = \mathrm{d}x^i/\mathrm{d}t$ dans la base $e_i = \partial_i$ de $T_P$. L'ensemble des $e_i$ forme une base de $T_P$, appelée *base de coordonnées*, et comme il y a $N$ dérivées partielles $\partial_i$ indépendantes, $T_P$ est manifestement de dimension $N$. Il est crucial de comprendre que l'espace tangent à $M$ est associé au point $P$, et on ne saura pas *a priori* comparer les vecteurs de deux plans tangents $T_P$ et $T_{P'}$ associés à deux points $P$ et $P'$ différents. En résumé, dans une base de coordonnées, un vecteur $\xi$ s'écrit comme

$$\xi = \xi^i e_i = \xi^i \partial_i \qquad e_i^k = \delta_i^k \tag{88}$$

Le vecteur $\xi$ est appelé *vecteur contravariant*, ses composantes sont en exposant. En ce qui concerne les vecteurs de base contravariants $e_k$, $k$ étiquette les vecteurs de base, de composantes $e_k^i = \delta_k^i$. Il est très important de comprendre que, contrairement à ce que pourrait laisser supposer la notation $x^i$, la coordonnée $x^i$ *n'est pas* la composante $i$ d'un vecteur contravariant. Dans le cas de l'espace-temps plat de la section 3, $x^\mu$ était *à la fois* une coordonnée et la composante $\mu$ d'un vecteur qui relie l'origine au point de coordonnées $x^\mu$. Ce n'est plus le cas dans l'espace-temps de la relativité générale. En fait on a noté $x^i$ par commodité, mais on aurait aussi bien pu utiliser $x_i$ ; la notation $x^i$ a l'avantage de la cohérence, car $u^i = \mathrm{d}x^i/\mathrm{d}t|_{C,P}$ est une composante de vecteur, contrairement à $x^i$, alors que dans le cas de la section 3 on distingue $x^\mu$ et $x_\mu$ (*cf.* (29)) : voir l'exemple de la section 5.3.3 où $x^1 = \theta$ et $x^2 = \varphi$. On n'insistera jamais assez sur le fait que les vecteurs sur une variété sont *attachés à l'espace tangent en un point de ladite variété.*

La *dérivée directionnelle* $\partial_\xi$ d'une fonction $f(x^i)$ est par définition

$$\partial_\xi f = \xi^i \partial_i f \tag{89}$$

Si l'on choisit pour $\xi$ un vecteur de base $e_k$, $\xi = e_k$

$$e_k^i \partial_i f = \partial_{e_k} f = \delta_k^i \partial_i f = \partial_k f$$

$\partial_k$ est la dérivée dans la direction $k$, les autres coordonnées restant constantes : Fig. 21. Dans un changement de système de coordonnées

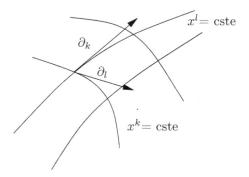

FIGURE 21.    Dérivées directionnelles $\partial_k$ et $\partial_l$.

$x^i(x'^j)$, la loi de transformation des dérivées directionnelles est

$$\frac{\mathrm{d}x^i}{\mathrm{d}t}\bigg|_P = \frac{\partial x^i}{\partial x'^j}\bigg|_P \frac{\mathrm{d}x'^j}{\mathrm{d}t}\bigg|_{C,P} \ \text{ou} \ \ \xi^i = \frac{\partial x^i}{\partial x'^j}\bigg|_P \xi'^j \tag{90}$$

On peut donc écrire une forme matricielle de cette loi de transformation

$$\xi^i = A^i{}_j(P)\xi'^j \qquad A^i{}_j(P) = \frac{\partial x^i}{\partial x'^j}\bigg|_P \tag{91}$$

C'est bien sûr une transformation linéaire sur des vecteurs, mais cette transformation est spécifique du point $P$ : la matrice $A^i{}_j(P)$ dépend du point $P$.

Passons ensuite à la notion de *covecteur = vecteur covariant* ou encore *1-forme*. Un covecteur est simplement une forme linéaire sur les vecteurs, c'est-à-dire une application $T_P \to \mathbb{R}$ qui à tout vecteur fait correspondre un nombre réel en respectant la propriété de linéarité. Nous noterons $\Lambda$ un covecteur et $< \Lambda, \xi >$ l'application linéaire. La linéarité implique l'action suivante sur la combinaison $a\xi + b\eta$ des vecteurs $\xi$ et $\eta$

$$< \Lambda, a\xi + b\eta >= a < \Lambda, \xi > +b < \Lambda, \eta >$$

L'application $df$ qui au vecteur $\xi = d/dt|_{C,P}$ fait correspondre le nombre $df/dt|_{C,P}$

$$< df, \frac{d}{dt}\Big|_{C,P} >\equiv \frac{df}{dt}\Big|_{C,P} \tag{92}$$

est bien un covecteur. En effet

$$< df, a\xi + b\eta >= a\frac{df}{dt}\Big|_{C,P} + b\frac{df}{dt}\Big|_{\Gamma,P}$$

Soit $e^i$ la base duale de $e_i$, c'est-à-dire la base telle que $< e^i, e_k >= \delta^i_k$. Les composantes de $df$ ne sont autres que $\partial_i f$. En effet

$$< \frac{\partial f}{\partial x^i}e^i, \frac{d}{dt}\Big|_{C,P} >=< \partial_i f e^i, \xi^k e_k >= \frac{dx^i}{dt}\partial_i f = \frac{df}{dt}\Big|_{C,P}$$

La base $e^i$ est souvent notée $e^i \equiv dx^i$. La loi de transformation des composantes $\partial_i f$ d'un covecteur sont *différentes* de celles d'un vecteur. Écrivons les explicitement

$$\frac{\partial f}{\partial x^i} = \frac{\partial f}{\partial x'^j}\frac{\partial x'^j}{\partial x^i}\Big|_P \quad \text{soit} \quad \Lambda_i = \frac{\partial x'^j}{\partial x^i}\Lambda'_j\Big|_P \tag{93}$$

## 5.2. *Champs de tenseurs*

Un champ de vecteurs est un ensemble de vecteurs dépendant du point $P$, de composantes $T^i(P) \equiv \xi^i(P) = \xi^i(x^j)$ dans la base de coordonnées $e_i$, et se transformant en chaque point selon (90). On définit de même des champs de covecteurs $T_i(P) \equiv \Lambda_i(P) = \Lambda_i(x^j)$ se transformant suivant (93). Un tenseur de type $(m, n)$ aura $m$ composantes contravariantes et $n$ composantes covariantes

$$T^{i_1 \ldots i_m}{}_{j_1 \ldots j_n} \tag{94}$$

se transformant par (90) pour les composantes contravariantes et suivant (93) pour les composantes covariantes. Par exemple le tenseur de type (0,2) $T_{ij}$ se transformera comme

$$T_{ij} = \frac{\partial x'^k}{\partial x^i}\frac{\partial x'^l}{\partial x^j}T'_{kl} \tag{95}$$

Tout comme le vecteur $\xi$, le tenseur $T$ existe indépendamment de ses composantes

$$T = T^{i_1 \cdots i_m}{}_{j_1 \ldots j_n}\, e_{i_1} \otimes \cdots \otimes e_{i_m}\, e^{j_1} \otimes \cdots \otimes e^{j_n} \tag{96}$$

où $\otimes$ indique le produit tensoriel. Le type de tenseur est défini par la donnée du couple $(m, n)$. Il existe deux opérations sur les champs de tenseurs qui sont intrinsèques à la variété, c'est-à-dire qui ne dépendent d'aucune structure additionnelle

(1) *La différentiation extérieure* : nous nous limiterons à en donner la définition dans le cas d'un covecteur $T_i$, dont la dérivée extérieure est[a]

$$(\mathrm{d}T)_{ij} = \frac{\partial T_i}{\partial x^j} - \frac{\partial T_j}{\partial x^i} \tag{97}$$

C'est un bon exercice que de vérifier que $(\mathrm{d}T)_{ij}$ se transforme comme un tenseur (0,2).

(2) *La dérivée de Lie $L_\xi$* d'un champ de tenseurs, qui est définie de la façon suivante. Pour une fonction scalaire $f(x^i)$, $L_\xi f$ est simplement la dérivée directionnelle $\partial_\xi f$

$$L_\xi f = \partial_\xi f = \xi^i \partial_i f \tag{98}$$

Pour un champ de vecteurs $\eta^i$

$$L_\xi \eta^i = \xi^j \partial_j \eta^i - \eta^j \partial_j \xi^i = \partial_\xi \eta^i - \partial_\eta \xi^i = -L_\eta \xi^i \tag{99}$$

Un résultat important est que $[\partial_\xi, \partial_\eta]$ définit un champ de vecteurs. En effet

$$[\partial_\xi, \partial_\eta]f = \partial_\xi \partial_\eta f - \partial_\eta \partial_\xi f = \partial_{L_\xi \eta} f \tag{100}$$

La démonstration de ce résultat est immédiate

$$[\partial_\xi, \partial_\eta]f = \xi^j \partial_j(\eta^i \partial_i f) - \eta^j \partial_j(\xi^i \partial_i f)$$
$$= (\xi^j \partial_j \eta^i)(\partial_i f) - (\eta^j \partial_j \xi^i)(\partial_i f) = \partial_{L_\xi \eta} f$$

Autrement dit, $[\partial_\xi, \partial_\eta]$ est un opérateur différentiel du premier ordre. Le champ de vecteurs $L_\xi \eta = [\xi, \eta]$ est le *crochet de Lie* des champs de vecteurs $\xi$ et $\eta$.

---

[a]Nous n'aurons pas l'occasion de nous servir de la différentielle extérieure, ni de la dérivée de Lie. Cependant ces deux notions donnent une bonne occasion de se familiariser avec les champs de vecteurs.

### 5.3. *Connexions*

*Dérivée covariante*

On pourrait penser qu'un objet comme $\partial_i T_j$ se comporte comme un tenseur (0,2) dans un changement de coordonnées, mais tel n'est pas le cas. En revanche, les termes indésirables, ceux qui ne correspondent pas à une loi de transformation de type (0,2) disparaissent dans le cas de la différentielle extérieure $(dT)_{ij}$, qui est bien un tenseur de type (0,2). De même $\partial_i T^j$ n'est pas un tenseur de type (1,1). On va donc chercher au lieu de la dérivée partielle $\partial_i$ une opération $\nabla_i$, qui, appliquée sur un vecteur $T^j$, donne bien un tenseur (1,1). Cette opération sera appelée la *dérivée covariante*. Nous allons exiger de cette opération les propriétés suivantes

(1) C'est un opération linéaire.
(2) Elle se réduit à la dérivée ordinaire pour une fonction scalaire $f$

$$\nabla_i f = \partial_i f \tag{101}$$

(3) C'est une dérivation : elle obéit à la règle de Leibniz.

Soit $T^j$ un vecteur (contravariant). Des propriétés (2) et (3) on tire

$$\partial_i(fT^j) - \nabla_i(fT^j) = f[\partial_i T^j - \nabla_i T^j]$$

Cette équation montre que la quantité $[\partial_i T^j - \nabla_i T^j]$ ne dépend que de la valeur de $T^j$ au point $P$, $T^j(P)$. En effet, si les champs de vecteurs $T'^j$ et $T^j$ coïncident au point $P$ : $T'^j(P) = T^j(P)$, on peut écrire un développement du type

$$T'^j - T^j = \sum_\alpha f^\alpha \omega_\alpha^j \ \text{ avec } \ f^\alpha(P) = 0$$

On a donc

$$\partial_i(T'^j - T^j) - \nabla_i(T'^j - T^j) = \partial_i\left(\sum_\alpha f^\alpha \omega_\alpha^j\right) - \nabla_i\left(\sum_\alpha f^\alpha \omega_\alpha^j\right) \propto f^\alpha(P) = 0$$

Enfin, d'après la propriété (1), $[\partial_i T^j - \nabla_i T^j]$ est une combinaison linéaire des $T^l$, soit

$$\partial_i T^j - \nabla_i T^j = \Gamma_{li}^j T^l$$

La quantité $\Gamma_{li}^j$ est un *coefficient de connexion*, ou simplement *connexion*, ou encore un symbole de Christoffel. Une même variété peut être munie de connexions différentes. La seule condition est que la connexion doit obéir aux propriétés $(1-3)$. En résumé, l'action de la dérivée covariante sur un vecteur $T^j$ est

$$\boxed{\nabla_i T^j = \partial_i T^j + \Gamma^j_{li} T^l} \tag{102}$$

L'action de la dérivée covariante sur un covecteur s'obtient aisément en remarquant que la quantité $T_i S^i$, construite à partir du covecteur $T_i$ et du vecteur $S^i$ est une fonction scalaire, comme on le vérifie à partir de (91) et (93)

$$T_i S^i = T'_i S'^i$$

et par conséquent

$$\nabla_i (T_j S^j) = \partial_j (T_j S^j)$$

On en déduit

$$\boxed{\nabla_i T_j = \partial_i T_j - \Gamma^l_{ij} T_l} \tag{103}$$

Attention cependant : contrairement à ce que la notation pourrait laisser croire, $\Gamma^l_{ij}$ *n'est pas* un tenseur de type (1,2) ! La *torsion* $\Theta^l_{ij}$ est définie par

$$\Theta^l_{ij} = \Gamma^l_{ij} - \Gamma^l_{ji} = \Gamma^l_{[ij]} \tag{104}$$

où nous avons introduit une notation standard

$$A_{[ij]} = A_{ij} - A_{ji} \qquad A_{(ij)} = A_{ij} + A_{ji} \tag{105}$$

Contrairement au cas de la connexion, on peut montrer que la torsion est bien un tenseur de type (1,2) : comme dans le cas de différentielle extérieure, les termes indésirables disparaissent à cause de l'antisymétrisation dans (104).

L'action sur les vecteurs d'une base de coordonnées donne directement la connexion. En effet, d'après (102)

$$(\nabla_i e_m)^j = \partial_i e^j_m + \Gamma^j_{li} e^l_m = \Gamma^j_{mi}$$

car $e^j_m = \delta^j_m$. On peut donc écrire

$$\boxed{\nabla_i e_m = \Gamma^l_{mi} e_l} \tag{106}$$

*Transport parallèle et géodésiques*

La notion qui suit logiquement celle de connexion est celle de *transport parallèle*. Examinons d'abord cette notion dans le cas d'une fonction. Soit une courbe $C$ paramétrée par $x^i(t)$. Si une fonction $f(x^i)$ est constante le long de la courbe $C$, on dira que la fonction $f$ est transportée parallèlement le long de la courbe $C$

$$f(x^i(t + \mathrm{d}t)) \simeq f\left(x^i + \frac{\mathrm{d}x^i}{\mathrm{d}t}\,\mathrm{d}t\right) \simeq f(x^i) + \partial_i f \frac{\mathrm{d}x^i}{\mathrm{d}t}\,\mathrm{d}t$$

et $f(x^i(t)) = f(x^i(t + \mathrm{d}t))$ implique

$$\frac{\mathrm{d}x^i}{\mathrm{d}t}\,\partial_i f = 0 \tag{107}$$

En termes imagés : "le gradient $\partial_i f$ est perpendiculaire au vecteur vitesse". Un vecteur $T^i$ sera transporté parallèlement le long de $C$ si sa dérivée covariante est "perpendiculaire" au vecteur vitesse

$$0 = \frac{\mathrm{d}x^k}{\mathrm{d}t}\,\nabla_k T^i = \frac{\mathrm{d}x^k}{\mathrm{d}t}\left(\partial_k T^i + \Gamma^i_{jk} T^j\right)$$

soit

$$\boxed{\frac{\mathrm{d}T^i}{\mathrm{d}t} + u^k \Gamma^i_{jk} T^j = 0 \qquad u^k = \frac{\mathrm{d}x^k}{\mathrm{d}t}} \tag{108}$$

Il est essentiel de remarquer que l'équation (108) est une équation tensorielle admissible, valable dans tout système de coordonnées si elle est valable dans un système particulier, car

$$\frac{\mathrm{d}x^k}{\mathrm{d}t}\,\nabla_k T^i$$

est un vecteur, alors que

$$\frac{\mathrm{d}x^k}{\mathrm{d}t}\,\partial_k T^i = \frac{\mathrm{d}T^i}{\mathrm{d}t}$$

*n'est pas un vecteur*. Si $\mathrm{d}T^i/\mathrm{d}t = 0$ dans un système de coordonnées particulier, rien ne garantit qu'il en sera de même dans un autre système ; cette équation n'a pas une structure tensorielle admissible. Le transport parallèle dépend de la courbe (Fig. 22), mais non du système de coordonnées. Si l'on se donne une courbe $C$ entre deux points $A$ et $B$ paramétrés par exemple par $t = 0$ et $t = 1$ et une condition initiale $T^i(t = 0)$, alors $T^i(t = 1)$ est déterminé de façon unique car (108) est un système d'équations différentielles du premier ordre.

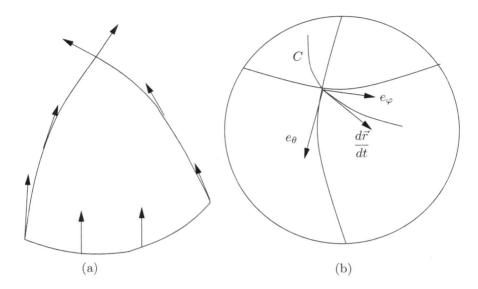

FIGURE 22. (a) Transport parallèle sur une sphère suivant deux courbes (des arcs de cercle) différentes. (b) Vecteurs $e_\theta$ et $e_\varphi$ du plan tangent.

La notion de *géodésique* est étroitement associée à celle de transport parallèle : une courbe $G$ est une géodésique si son vecteur vitesse est transporté parallèlement à lui-même le long de $G$, soit (rappelons que $u^i = \mathrm{d}x^i/\mathrm{d}t$)

$$u^i \nabla_i u^j = u^i(\partial_i u^j + \Gamma^j_{ki} u^k) = 0$$

ce qui donne l'équation des géodésiques

$$\boxed{\frac{\mathrm{d}^2 x^j}{\mathrm{d}t^2} + \Gamma^j_{ki} \frac{\mathrm{d}x^i}{\mathrm{d}t} \frac{\mathrm{d}x^k}{\mathrm{d}t} = 0} \tag{109}$$

où nous avons utilisé

$$\frac{\mathrm{d}}{\mathrm{d}t} = \frac{\mathrm{d}x^i}{\mathrm{d}t}\,\partial_i \ \text{ soit } \ \frac{\mathrm{d}x^i}{\mathrm{d}t}\,\partial_i\left(\frac{\mathrm{d}x^j}{\mathrm{d}t}\right) = \frac{\mathrm{d}^2 x^j}{\mathrm{d}t^2}$$

Deux remarques sur les géodésiques.

(1) Seule la partie symétrique dans les deux indices inférieurs de la connexion $\Gamma$ intervient dans l'équation des géodésiques ; en d'autres termes, cette équation ne dépend pas de la torsion.

(2) Une géodésique $x^i(t)$ est déterminée de façon unique par la donnée d'une position initiale $x^i(t=0)$ et d'une vitesse initiale $u^i(t=0)$. En effet (109) est un système d'équations différentielles du second ordre.

*Exemple de la sphère*

Nous allons illustrer les notions de connexion et de transport parallèle sur le cas familier de la sphère $S^2$, choisie de rayon unité. Dans un premier temps il sera commode de considérer $S^2$ comme une variété plongée dans l'espace $\mathbb{R}^3$. Le mouvement $\vec{r}(t) = (x(t), y(t), z(t))$ d'un point sur $S^2$ est repéré par exemple par ses coordonnées polaires $\theta(t)$ et $\varphi(t)$

$$
\begin{aligned}
x(t) &= \sin\theta(t)\cos\varphi(t) \\
y(t) &= \sin\theta(t)\sin\varphi(t) \\
z(t) &= \cos\theta(t)
\end{aligned}
\tag{110}
$$

En différentiant ces équations par rapport à $t$ on obtient le vecteur vitesse $u$

$$
u = \dot\theta\, e_\theta + \dot\varphi\, e_\varphi
\tag{111}
$$

où les vecteurs $e_\theta$ et $e_\varphi$ sont des vecteurs du plan tangent $T_P$ à la sphère

$$
e_\theta = (\cos\theta\cos\varphi, \cos\theta\sin\varphi, -\sin\theta) = \frac{\partial}{\partial\theta}
\tag{112}
$$

$$
e_\varphi = (-\sin\theta\sin\varphi, \sin\theta\cos\varphi, 0) = \frac{\partial}{\partial\varphi}
\tag{113}
$$

Par convention, nous noterons les vecteurs sans flèche lorsqu'ils sont considérés comme des vecteurs du plan tangent : $(e_\theta, e_\varphi)$, et avec une flèche lorsqu'ils sont considérés comme vecteurs de $\mathbb{R}^3$ : $(\vec{e}_\theta, \vec{e}_\varphi)$. Tout champ de vecteurs sur la sphère $X(\theta, \varphi)$ peut se décomposer sur les vecteurs de base $e_\theta$ et $e_\varphi$

$$
X(\theta, \varphi) = X^\theta(\theta, \varphi)\, e_\theta + X^\varphi(\theta, \varphi)\, e_\varphi
\tag{114}
$$

Si l'on considère $\vec{e}_\theta$ et $\vec{e}_\varphi$ comme des vecteurs de $\mathbb{R}^3$, on peut les différentier par rapport à $\theta$ et $\varphi$ $(\hat{r} = \vec{r}/r)$

$$
\begin{aligned}
\partial_\theta \vec{e}_\theta &= (-\sin\theta\cos\varphi, -\sin\theta\sin\varphi, -\cos\theta) = -\hat{r} \\
\partial_\varphi \vec{e}_\theta &= (-\cos\theta\sin\varphi, \cos\theta\cos\varphi, 0) = \cot\theta\, \vec{e}_\varphi \\
\partial_\theta \vec{e}_\varphi &= (-\cos\theta\sin\varphi, \cos\theta\cos\varphi, 0) = \cot\theta\, \vec{e}_\varphi \\
\partial_\varphi \vec{e}_\varphi &= (-\sin\theta\cos\varphi, -\sin\theta\sin\varphi, 0) = -\sin^2\theta\,\hat{r} - \sin\theta\cos\theta\, \vec{e}_\theta
\end{aligned}
\tag{115}
$$

On constate que $\partial_\theta\vec{e}_\theta$ est orthogonal à $T_P$, que $\partial_\varphi\vec{e}_\theta$ et $\partial_\theta\vec{e}_\varphi$ sont contenus dans $T_P$ et que $\partial_\varphi\vec{e}_\varphi$ possède une composante dans $T_P$ et une composante orthogonale à $T_P$. Soit

$$
\partial_X = X^\theta\partial_\theta + X^\varphi\partial_\varphi
$$

la dérivé directionnelle suivant $X$. Si $Y(\theta, \varphi)$ est un autre champ de vecteurs, l'expression $(\vec{X} \cdot \vec{\partial})\vec{Y}$ ne définit pas un vecteur de $T_P$ : c'est bien un vecteur de $\mathbb{R}^3$, mais ce n'est pas un vecteur de $T_P$. Pour obtenir un vecteur de $T_P$, il faut projeter sur ce plan à l'aide du projecteur $\mathsf{P}$

$$\mathsf{P}\left[(\vec{X} \cdot \vec{\partial})\vec{Y}\right] = (\vec{X} \cdot \vec{\nabla})\vec{Y}$$

et cette projection définit une dérivée covariante. En effet

- c'est une opération linéaire
- elle se réduit à la dérivée ordinaire pour une fonction
- c'est une dérivation

Pour faire le lien avec les notations utilisées jusqu'ici, on se donne la correspondance $\theta \to x^1$ et $\varphi \to x^2$ : $e_\theta \to e_1$, $e_\varphi \to e_2$ De l'équation (106) $\nabla_i e_m = \Gamma^l_{mi} e_l$ on tire par identification avec (115)

$$
\begin{aligned}
\nabla_\theta e_\theta &= 0 & \nabla_\varphi e_\theta &= \cot\theta \, e_\varphi \\
\nabla_\theta e_\varphi &= \cot\theta \, e_\varphi & \nabla_\varphi e_\varphi &= -\sin\theta\cos\theta \, e_\theta
\end{aligned}
\tag{116}
$$

d'où la connexion

$$\Gamma^\varphi_{\theta\varphi} = \Gamma^\varphi_{\varphi\theta} = \cot\theta \qquad \Gamma^\theta_{\varphi\varphi} = -\sin\theta\cos\theta \tag{117}$$

les autres composantes étant nulles. À partir de ce point on peut oublier que $S^2$ est plongée dans $\mathbb{R}^3$ : les équations (117) définissent de façon *intrinsèque* une connexion sur $S^2$. Ceci donne l'équation des géodésiques

$$
\begin{aligned}
\ddot{\theta} - \sin\theta\cos\theta \, \dot{\varphi}^2 &= 0 \\
\ddot{\varphi} + 2\cot\theta \, \dot{\theta}\dot{\varphi} &= 0
\end{aligned}
\tag{118}
$$

C'est un exercice instructif, mais pas entièrement trivial, de montrer à partir de ces équations, que les géodésiques de la sphère sont des grands cercles parcourus à vitesse constante.

### 5.4. *Métrique et courbure*

*Connexion associée à une métrique*

Dans l'exemple de la sphère $S^2$ donné ci-dessus, on sait que la métrique euclidienne sur $\mathbb{R}^3$

$$\mathrm{d}s^2 = \mathrm{d}x^2 + \mathrm{d}y^2 + \mathrm{d}z^2 \tag{119}$$

induit sur la sphère unité une métrique

$$ds^2 = d\theta^2 + \sin^2\theta \, d\varphi^2 = g_{ij}(\theta,\varphi) \, dx^i dx^j \qquad (120)$$

avec comme précédemment $x^1 = \theta$ et $x^2 = \varphi$ ; $g_{ij}$ est appelé le *tenseur métrique*, et sous forme matricielle

$$g(\theta,\varphi) = \begin{pmatrix} 1 & 0 \\ 0 & \sin^2\theta \end{pmatrix} \qquad (121)$$

La longueur d'un arc de courbe $C : (\theta(t), \varphi(t))$ sur la sphère est

$$\ell = \int_{t_A}^{t_B} \left[ \left(\frac{d\theta}{dt}\right)^2 + \sin^2\theta \left(\frac{d\varphi}{dt}\right)^2 \right]^{1/2} dt \qquad (122)$$

Une métrique $g_{ij}(x)$ sur une variété $M$ est un tenseur symétrique $(g_{ij} = g_{ji})$ (0,2) qui définit une forme bilinéaire $(\bullet, \bullet)$ sur les vecteurs : étant donné deux vecteurs $\xi$ et $\eta$, cette forme bilinéaire est

$$(\xi, \eta) = g_{ij}(x)\, \xi^i \eta^j \qquad (123)$$

De plus la matrice $g_{ij}$ doit être une matrice positive. Si l'on sait munir la variété $M$ d'une métrique, le tenseur métrique $g_{ij}(x)$ permet de calculer la longueur d'un arc de courbe sur la variété[b] par une généralisation immédiate de (122)

$$\ell = \int_{t_A}^{t_B} \left[ g_{ij}(x) \frac{dx^i}{dt} \frac{dx^j}{dt} \right]^{1/2} dt \qquad (124)$$

Définissons le covecteur $\overline{\xi}$ associé au vecteur $\xi$ par

$$\overline{\xi}_i = g_{ij}(x)\, \xi^j \qquad (125)$$

Alors la forme bilinéaire $(\xi, \eta)$ peut s'écrire

$$(\xi, \eta) = <\overline{\xi}, \eta> = \overline{\xi}_i \eta^i$$

et peut être interprétée comme le produit scalaire des vecteurs $\xi$ et $\eta$. On voit que le tenseur métrique permet de "descendre les indices" suivant (125), en associant par exemple un covecteur à un vecteur. On note $g^{ij}(x)$ la matrice inverse de $g_{ij}(x)$

$$g_{ij}(x)\, g^{jk}(x) = \delta_i^k$$

---

[b]Nous nous plaçons pour l'instant dans le cas d'une métrique euclidienne, cas où la matrice $g_{ij}$ est définie positive.

qui existe car $g_{ij}$ est supposée définie positive. Le tenseur $g^{ij}$ permet de "monter les indices", par exemple

$$\xi^i = g^{ij}(x)\,\overline{\xi}_j$$

Cette équation, tout comme (125), se généralisent trivialement à un tenseur quelconque, par exemple on passe d'un tenseur (2,0) à un tenseur (1,1) par

$$T^i_j = g_{jk}\,T^{ik}$$

La *trace* d'un tenseur (1,1) est définie par

$$\mathrm{Tr}\,T = T^i{}_i = g_{ik}\,T^{ik} = g^{ik}\,T_{ik} \qquad (126)$$

Les notions de connexion et de métrique sont *a priori* des notions indépendantes. Cependant, ainsi que nous allons le voir, on peut déduire d'une métrique une connexion *unique, la connexion associée à la métrique*, pourvu que l'on exige les deux conditions suivantes

(1) La torsion est nulle est dans une base de coordonnées : $\Gamma^i_{jk} = \Gamma^i_{kj}$
(2) La dérivée covariante "tue la métrique" : $\nabla_k g_{ij} = 0$

En manipulant les indices dans l'équation $\nabla_k g_{ij} = 0$ et en utilisant la symétrie $\Gamma^i_{jk} = \Gamma^i_{kj}$, on trouve l'expression explicite de la connexion associée à la métrique

$$\Gamma^k_{ij} = \Gamma^k_{ji} = \frac{1}{2}\,g^{kl}\,(\partial_j\,g_{li} + \partial_i\,g_{lj} - \partial_l g_{ij}) \qquad (127)$$

Le fait que l'on obtienne cette formule explicite montre que la connexion associée à la métrique est bien unique.

Pour illustrer ces concepts, revenons à l'exemple de la sphère où

$$\mathrm{d}s^2 = \mathrm{d}\theta^2 + \sin^2\theta\,\mathrm{d}\varphi^2,$$
$$g_{\theta\theta} = 1, \qquad g_{\theta\varphi} = g_{\varphi\theta} = 0, \qquad g_{\varphi\varphi} = \sin^2\theta$$

On vérifie imédiatement que la connexion trouvée en (117) est symétrique et tue la métrique (120)

$$\nabla_i g_{jk} = 0$$

La connexion $\Gamma$ (117) est bien la connexion associée à la métrique, comme on peut le vérifier explicitement en évaluant (127) dans le cas de la métrique de la sphère.

Terminons cette brève revue de la connexion associée à la métrique par l'énoncé de quelques propriétés utiles. Tout d'abord on remarque que si l'on utilise dans la dérivée covariante (102) la connexion associée à la métrique, alors cette dérivée commute avec l'opération de montée ou de descente des indices

$$\nabla_k T_i = \nabla_k(g_{ij} T^j) = g_{ij}(\nabla_k T^j) \tag{128}$$

où nous avons utilisé $\nabla_k g_{ij} = 0$. Ensuite, si $T^i(t)$ et $S^i(t)$ sont des vecteurs transportés parallèlement le long d'une courbe $C$, leur produit scalaire est invariant le long de cette courbe

$$u^k \nabla_k(g_{ij} T^i S^j) = u^k \left[ g_{ij}(\nabla_k T^i)S^j + g_{ij}T^i(\nabla_k S^j) \right] = 0 \tag{129}$$

car par définition du transport parallèle

$$u^k \nabla_k T^i = u^k \nabla_k S^j = 0$$

La divergence (ordinaire) d'un champ de vecteurs $\partial_i T^i$ n'est pas un scalaire, mais $\nabla_i T^i$ *est* un scalaire

$$\begin{aligned} \nabla_i T^i &= \partial_i T^i + \Gamma_{ki}^i T^k \\ &= \frac{1}{\sqrt{g}} \, \partial_i(\sqrt{g} \, T^i) \end{aligned} \tag{130}$$

avec $g = \det g_{ij}$. Enfin une remarque importante concerne les géodésiques. Nous avons donné en (124) une expression de la longueur d'une courbe entre deux points de paramètres $t_A$ et $t_B$. On peut montrer par des techniques standard de calcul variationnel que l'on retrouve l'équation (109) des géodésiques avec la connexion associée à la métrique (127) en cherchant la courbe qui minimise (plus généralement extrémise) la longueur $\ell$ entre deux points $A$ et $B$ fixés : $\delta\ell = 0$. Mais en fait l'équation (109) $u^k \nabla_k u^i = 0$ donne en plus la façon dont est parcourue la géodésique. Ceci s'explique par le fait que $t$ est un paramètre affine (*cf.* (33). Si l'on choisit un paramètre non affine $\lambda$, alors $u^k \nabla_k u^i = \alpha u^i$, où $\alpha$ est une constante. On le voit dans le cas d'un mouvement rectiligne uniforme dans l'espace euclidien ordinaire, où les géodésiques sont des droites.

(1) Géodésique avec paramètre affine : $\dfrac{\mathrm{d}^2 \vec{r}}{\mathrm{d}t^2} = 0$

(2) Géodésiques avec paramètre non affine : $\dfrac{\mathrm{d}^2 \vec{r}}{\mathrm{d}\lambda^2} = \alpha \dfrac{\mathrm{d}\vec{r}}{\mathrm{d}\lambda}$

*Tenseur de courbure*

Il nous reste à définir la notion de *courbure*. À partir d'une connexion quelconque (non nécesairement associée à une métrique) et d'une base de coordonnées $e_i$ on définit le tenseur de torsion $\Theta^i_{kl}$

$$\nabla_k e_l - \nabla_l e_k = \Theta^i_{kl} e_i$$

et le *tenseur de courbure* $R^i_{\ qkl}{}^c$

$$\boxed{[\nabla_k, \nabla_l] e_q = -R^i_{\ qkl}\, e_i} \tag{131}$$

La définition de la torsion suit de (104), compte tenu de (106). En utilisant une approche analogue à celle qui suit (101), on peut montrer que $[\nabla_k, \nabla_l] e_q$ est une fonction linéaire des $e_i$, ce qui justifie (131). Si la connexion est symétrique, une manipulation d'indices permet de calculer explicitement $R^i_{\ qkl}$

$$R^i_{\ qkl} = -\partial_k \Gamma^i_{ql} + \partial_l \Gamma^i_{qk} - \Gamma^i_{pk} \Gamma^p_{ql} + \Gamma^i_{pl} \Gamma^p_{qk} \tag{132}$$

L'interprétation géométrique la plus parlante du tenseur de courbure est la suivante : considérons un covecteur $T_i$ transporté parallèlement le long d'une courbe fermée $C$. Après avoir parcouru une fois la courbe fermée, le covecteur $T_i$ est différent du covecteur initial par $\Delta T_i$, qui vaut

$$\Delta T_i = \frac{1}{2} R^j_{\ ikl} T_j \oint x^k \mathrm{d}x^l \tag{133}$$

---

[c] Il existe une correspondance intéressante avec les théories de jauge, abéliennes et non abéliennes. Pour simplifier la discussion, je me limiterai aux théories abéliennes dans un situation indépendante du temps. Dans le formalisme de l'intégrale de chemin, le poids statistique d'une trajectoire d'une particule chargée de charge $q$ est donnée par

$$\exp\left(-\mathrm{i}\frac{q}{\hbar} \int_{1(C)}^{2} \mathrm{d}\vec{r} \cdot \vec{A}(\vec{r})\right)$$

où $\vec{A}$ est le potentiel vecteur. Cette expression peut être interprétée comme le transport parallèle de la fonction d'onde du point 1 au point 2 en suivant la courbe $C$. L'analogue de la courbure est le champ magnétique

$$\vec{B} = \vec{\nabla} \times \vec{A}$$

et si la courbure est non nulle, le transport parallèle le long d'une courbe fermée est non trivial

$$\oint \vec{A} \cdot \mathrm{d}\vec{r} = \iint \vec{B} \cdot \mathrm{d}\vec{S}$$

d'après le théorème de Stokes. L'analogue de la dérivée covariante est

$$\vec{D} = \vec{\nabla} - \mathrm{i}\frac{q}{\hbar} \vec{A}$$

Si un vecteur est transporté parallèlement à lui-même le long d'une courbe fermée et qu'il ne coïncide pas avec le vecteur initial, la courbure est nécessairement $\neq 0$

En relativité générale, la situation est très souvent la suivante. On dispose d'une métrique $g$, à partir de laquelle on calcule la connexion associée à la métrique via (127), puis le tenseur de courbure via (132). Le schéma est donc le suivant[d]

$$\text{métrique} \rightarrow \text{connexion} \rightarrow \text{courbure}$$

Les expressions (127) et (132) montrent que le tenseur de courbure dépend *non linéairement* de la métrique. Le tenseur $R^i{}_{qkl}$ vérifie plusieurs relations de symétrie, par exemple

$$R^i{}_{qkl} = -R^i{}_{qlk}$$

évidente d'après (131). En raison de ces propriétés de symétrie, on montre que dans une variété de dimension $N$ le tenseur de courbure possède

$$d_N = \frac{1}{12} N^2(N^2 - 1)$$

composantes indépendantes, soit

$$\begin{aligned}
N &= 2 & d_2 &= 1 \\
N &= 3 & d_3 &= 6 \\
N &= 4 & d_4 &= 20
\end{aligned}$$

Le tenseur de courbure obéit aussi à *l'identité de Bianchi*, qui est en fait reliée à l'identité de Jacobi

$$\nabla_{[i} R_{jk]lq} = 0 \tag{134}$$

où [ ] indique l'antisymétrisation comme dans (105). À partir du tenseur de courbure on construit le *tenseur de Ricci* $R_{ij}$, qui est un tenseur symétrique

$$R_{ij} = R^q{}_{iqj} = R_{ji} = \partial_j \Gamma^k_{ik} - \partial_k \Gamma^k_{ij} + \Gamma^k_{lj}\Gamma^l_{ik} - \Gamma^k_{lk}\Gamma^l_{ij} \tag{135}$$

et la courbure $R = R^i{}_i$, qui est la trace du tenseur de Ricci, et est donc un scalaire. Le tenseur symétrique $G_{ij}$

$$G_{ij} = R_{ij} - \frac{1}{2} R\, g_{ij} \tag{136}$$

---

[d]Les calculs explicites de (127) et (131) étant donné une métrique sont en général longs et pénibles et ils ne sont pas instructifs. Ils servent juste à se convaincre que l'on serait encore capable de passer le concours de l'Ecole Polytechnique. Il existe des programme *Mathematica* pour effectuer ces calculs de façon automatique (voir le site WEB du livre de Hartle).

vérifie, en raison de l'identité de Bianchi, l'identité importante

$$\boxed{\nabla^i\, G_{ij} = 0} \tag{137}$$

Pour $N = 2$ et $N = 3$, le tenseur de courbure peut être construit à partir du tenseur de Ricci et de la métrique, et si $R_{ij} = 0$, le tenseur de courbure s'annule. Nous verrons que les équations d'Einstein impliquent que le tenseur de Ricci s'annule en l'absence de matière, et donc la courbure doit s'annuler dans le vide en dimension 2 et en dimension 3, ce qui entraîne l'absence de forces de gravitation : selon Einstein, la gravitation ne peut exister que pour des dimensions d'espace-temps $\geq 4$ !

## 5.5.  *Adaptation à la relativité générale*

La dimension de l'espace-temps étant $N = 4$, la relativité générale utilisera une variété à quatre dimensions. Un point de cette variété sera repéré par quatre coordonnées $x^\mu$, $\mu = (0, 1, 2, 3)$. Comme dans la section 3, $x^0 = 0$ est une coordonnée de type temps, et $x^i$ une coordonnée de type espace. Cependant on doit tenir compte de ce que la métrique $g_{\mu\nu}$ n'est pas définie positive, mais que la matrice $g_{\mu\nu}$, tout comme $\eta_{\mu\nu}$, a une valeur propre positive et trois valeurs propres négatives. Il est facile de se convaincre que le nombre de valeurs propres négatives est inchangé dans un changement de coordonnées (91), et la seule conséquence pratique est que, tout comme dans l'espace de Minkowski (*cf.* (29)) , la quantité d$s^2$ ("l'élément de longueur")

$$\mathrm{d}s^2 = g_{\mu\nu}\, x^\mu\, x^\nu \tag{138}$$

peut être positive, négative ou nulle[e]. Comme la matrice $g_{\mu\nu}$ est symétrique, on peut la diagonaliser par une transformation orthogonale en un point donné $x_0^\mu$ ; évidemment la diagonalisation ne sera en général pas valable en un autre point. On peut ensuite effectuer une dilatation sur chacune des quatre coordonnés de façon à se ramener à la situation de l'espace- temps plat : $g_{\mu\nu}(x_0) \rightarrow \eta_{\mu\nu}$. Un nouveau changement de coordonnées permet d'éliminer les termes linéaires $\mathrm{O}(x^\mu - x_0^\mu)$ avec pour résultat

$$g_{\mu\nu}(x) = \eta_{\mu\nu} + \mathrm{O}(x^\mu - x_0^\mu)^2 \tag{139}$$

D'après (127) la connexion $\Gamma$ s'annule en $x_0$

$$\Gamma^\mu_{\rho\sigma}(x_0) = 0 \tag{140}$$

L'équation (139) (ou (140)) définit un *référentiel d'inertie local* (RIL) : c'est le référentiel en chute libre de la section 2. On peut associer à ce RIL une

---

[e]On doit remplacer dans (130) $\sqrt{g}$ par $\sqrt{|g|}$.

base pseudo-orthogonale comme celle de l'espace-temps plat. Cette base ne doit pas être confondue avec une base de coordonnées.

Dans le RIL, l'équation d'une géodésique est *localement* celle d'un mouvement rectiligne uniforme (*cf.* (33))

$$\frac{\mathrm{d}^2 x^\mu}{\mathrm{d}\tau^2} = 0 \quad \text{ou} \quad \frac{\mathrm{d}u^\mu}{\mathrm{d}\tau} = 0 \tag{141}$$

où $u^\mu$ est la quadrivitesse et $\tau$ le temps propre, mais (141) ne peut pas être valable partout : ce n'est pas une équation tensoriellement admissible. L'équation tensoriellement admissible qui se réduit à (141) est l'équation d'une géodésique

$$u^\nu \nabla_\nu u^\mu = \frac{\mathrm{d}^2 x^\mu}{\mathrm{d}\tau^2} + \Gamma^\mu_{\rho\sigma} \frac{\mathrm{d}u^\rho}{\mathrm{d}\tau} \frac{\mathrm{d}u^\sigma}{\mathrm{d}\tau} = 0 \tag{142}$$

Les particules, massives ou non, suivent des géodésiques de l'espace-temps : (142) se réduit à (141) dans un RIL.

Une dernière remarque concerne la loi de conservation du tenseur énergie-impulsion. L'équation de conservation (40) $\partial_\mu T^{\mu\nu} = 0$ n'est pas une équation tensoriellement admissible. L'équation tensoriellement admissible qui se réduit à celle-ci dans un RIL est

$$\boxed{\nabla_\mu T^{\mu\nu} = \partial_\mu T^{\mu\nu} + \Gamma^\mu_{\sigma\mu} T^{\sigma\nu} + \Gamma^\nu_{\sigma\mu} T^{\mu\sigma} = 0} \tag{143}$$

## Bibliographie

Tous les livres de relativité générale contiennent une introduction à la géométrie différentielle, par exemple Hartle [2003], chapitres 7, 8 et 20. À mon avis aucune de ces introductions ne vaut l'exposé de Doubrovine *et al.* [1983], chapitres 3 et 4.

## 6. Solutions à symétrie sphérique

### 6.1. *Équation d'Einstein*

Le principe qui sous-tend l'équation d'Einstein consiste à écrire une relation entre la géométrie et l'énergie-impulsion. On ne peut pas démontrer l'équation d'Einstein, tout comme on ne peut pas démontrer les équations de Maxwell ou de Newton. Toutefois on peut argumenter de la façon suivante.

(1) C'est l'équation la plus simple possible satisfaisant au principe précédent.

(2) Elle est mathématiquement cohérente et définit un problème de valeurs initiales.

(3) Elle redonne l'équation de Newton dans une limite appropriée.

L'équation la plus simple possible est la suivante

$$G_{\mu\nu} \equiv R_{\mu\nu} - \frac{1}{2} R g_{\mu\nu} = -\kappa T_{\mu\nu} \tag{144}$$

où $\kappa$ est une constante de proportionnalité à déterminer. En effet, si l'on veut avoir dans le membre de droite le tenseur énergie-impulsion, le membre de gauche doit aussi être un tenseur symétrique d'ordre deux construit avec la courbure. Les deux tenseurs symétriques les plus simples à notre disposition sont le tenseur de Ricci $R_{\mu\nu}$ et le tenseur métrique $g_{\mu\nu}$. Comme le tenseur énergie-impulsion obéit à l'équation de conservation

$$\nabla^{\mu} T_{\mu\nu} = 0$$

il doit en être de même du membre de gauche de (144), et compte tenu de l'identité de Bianchi (137), la seule combinaison possible de $R_{\mu\nu}$ et de $g_{\mu\nu}$ convenable est précisément $G_{\mu\nu}$. Prenant la trace de (144) on déduit la courbure scalaire sous la forme

$$R = -\kappa T^{\mu}{}_{\mu} = -\kappa T$$

ce qui permet de récrire (144) sous la forme souvent utile

$$R_{\mu\nu} = -\kappa \left( T_{\mu\nu} - \frac{1}{2} T g_{\mu\nu} \right) \tag{145}$$

Sous cette forme, on voit immédiatement qu'en l'absence de matière ($T_{\mu\nu} = 0$), le tenseur de Ricci et la courbure sont nuls. Dans un espace-temps de dimension $\leq 3$, ceci implique que le tenseur de courbure est nul, et il ne peut y avoir de gravitation dans un sens usuel que pour une dimension $\geq 4$ !

Il reste à fixer la constante $\kappa$. Pour ce faire, nous allons nous placer en champ gravitationnel faible indépendant du temps et considérer une situation où les vitesses des particules contribuant à $T_{\mu\nu}$ sont petites par rapport à $c$. Dans cette limite $T_{00} \simeq T \simeq \rho$, où $\rho$ est la densité de matière, et toutes les autres composantes du tenseur sont négligeables par rapport à $T_{00}$. L'hypothèse du champ faible permet d'écrire $g_{00} = 1 + h_{00}$, avec $|h_{00}| \ll 1$, et la composante (00) de (145) devient

$$R_{00} = -\frac{1}{2} \kappa T_{00}$$

Nous devons évaluer la composante $R_{00}$ du tenseur de Ricci à partir de la métrique. D'après (135)

$$R_{00} = \partial_0 \Gamma^i_{0i} - \partial_i \Gamma^i_{00} + O(\Gamma^2)$$

et les termes $O(\Gamma^2)$ peuvent être négligés car ils sont d'ordre $(h_{00})^2$. La contribution de $\partial_0$ s'annule dans une situation stationnaire et il ne reste que $-\partial_i \Gamma^i_{00}$. En utilisant (127) on évalue $\Gamma^i_{00}$

$$\Gamma^i_{00} = \frac{1}{2} g^{il} (\partial_0 g_{l0} + \partial_0 g_{0l} - \partial_l g_{00}) \simeq -\frac{1}{2} \eta^{il} \partial_l h_{00}$$

où $\eta^{il}$ représente les composantes spatiales du tenseur de Minkowski (rappelons que les lettres latines vont de 1 à 3 et étiquettent les composantes spatiales). On en déduit

$$R_{00} = -\frac{1}{2} \eta^{ij} \partial_i \partial_j h_{00} = \frac{1}{2} \nabla^2 h_{00}$$

soit

$$\nabla^2 h_{00} = -\kappa T_{00} = -\kappa \rho$$

Mais nous avons vu Ĺ la section 1 que $h_{00} = 2\Phi(\vec{r})$, où $\Phi$ est le potentiel gravitationnel, et donc

$$\nabla^2 \Phi(\vec{r}) = \frac{\kappa}{2} \rho = 4\pi G \rho$$

d'après l'équation de Poisson pour $\Phi$. Nous pouvons donc faire l'identification $\kappa = 8\pi G$, ce qui donne pour (144)

$$\boxed{G_{\mu\nu} \equiv R_{\mu\nu} - \frac{1}{2} R g_{\mu\nu} = -8\pi G \, T_{\mu\nu}} \tag{146}$$

L'équation d'Einstein peut être vue comme un ensemble d'équations aux dérivées partielles non linéaires et du second ordre pour la métrique $g_{\mu\nu}$. Les dix équations sont en fait réduites à 6 équations indépendantes en raison de l'identité de Bianchi $\nabla^\mu G_{\mu\nu} = 0$. En principe, si l'on se donne $g_{\mu\nu}$ et ses dérivées premières par rapport au temps sur une surface du genre espace (par exemple sur une surface $t = t_0 = $ cste), on doit pouvoir calculer $g_{\mu\nu}(x)$ pour tout temps $t > t_0$. En fait la situation est plus complexe en raison de la possibilité de transformations de jauge, et en pratique on essaiera plutôt de construire une métrique compatible avec les symétries du problème.

Afin de discuter commodément des généralisations possibles de l'équation d'Einstein (146), il est utile d'introduire *l'action d'Einstein-Hilbert $S_H$* : en effet il est possible de déduire (144) d'un principe de moindre action, en définissant

$$S_H = \int \mathrm{d}^4 x \sqrt{|g|}\, R \tag{147}$$

Un calcul un peu long montre alors que

$$\frac{1}{\sqrt{|g|}} \frac{\delta S_H}{\delta g_{\mu\nu}} = -G_{\mu\nu} \tag{148}$$

Pour obtenir le second membre de (144), il faut se donner une action $S_M$ pour la matière, et *par définition* le tenseur énergie-impulsion est

$$T^{\mu\nu} = \frac{1}{\sqrt{|g|}} \frac{\delta S_M}{\delta g_{\mu\nu}} \tag{149}$$

Par exemple $S_M$ pourrait être l'action d'un champ scalaire. Dans ces conditions l'équation d'Einstein se déduit du principe de moindre action

$$\frac{\delta}{\delta g_{\mu\nu}} \left( \frac{1}{8\pi G} S_H + S_M \right) = 0 \tag{150}$$

On peut géneraliser (144) en ajoutant un terme de constante cosmologique

$$G_{\mu\nu} = -8\pi G T_{\mu\nu} - \Lambda g_{\mu\nu} \tag{151}$$

Ceci est bien compatible avec la structure tensorielle, mais ne redonne pas la gravité newtonienne dans les limites convenables. On peut réinterpréter cette constante cosmologique comme une contribution à $T_{\mu\nu}$ provenant de l'énergie du vide[f]. Une autre généralisation possible consiste à ajouter à $S_H$ des dérivées de $g_{\mu\nu}$ d'ordre plus élevé que deux, par exemple en introduisant des termes non linéaires en $R_{\mu\nu}$

$$S_H = \int \mathrm{d}^4 x \sqrt{|g|}\, \left( R + \alpha_1 R^2 + \alpha_2 R_{\mu\nu} R^{\mu\nu} + \ldots \right)$$

ou à ajouter un mélange avec un champ scalaire $\lambda$ couplé à la courbure, comme dans la théorie de Brans–Dicke

$$S_{\mathrm{BD}} = \int \mathrm{d}^4 x \sqrt{|g|}\, \left[ f(\lambda R) - \frac{1}{2}\, g^{\mu\nu} \partial_\mu \lambda \partial_\nu \lambda - V(\lambda) \right].$$

À ce point, il vaut la peine de bien caractériser la relativité générale par rapport à la relativité restreinte. Cette caractérisation est parfaitement résumée par T. Damour (Damour [2005]). "Le principe de relativité générale

---

[f]Voir cependant la note 4 de la section 4.

a un statut physique différent du principe de relativité restreinte. Le principe de relativité restreinte est un principe de symétrie de la structure de l'espace-temps qui affirme que la physique est *la même* dans une classe particulière de référentiels, et que donc certains phénomènes "correspondants" se déroulent de la même façon dans des référentiels différents ("transformations actives"). En revanche, le principe de relativité générale est un *principe d'indifférence* : les phénomènes ne se déroulent (en général) pas de la même façon dans des systèmes de coordonnées différents, mais aucun des systèmes de coordonnées (étendu) n'a de statut privilégié par rapport aux autres."

Nous n'insisterons pas sur les vérifications de la relativité générale et renvoyons à la revue récente de T. Damour (Damour [2005]) : en résumé, la relativité générale a été testée aujourd'hui dans des situations très diverses, depuis notre environnement immédiat (système GPS) jusqu'aux confins de l'Univers (lentilles gravitationnelles). Les tests les plus précis atteignent une précision relative de $10^{-5}$.

## 6.2. *Métrique de Schwarzschild*

Lorsqu'une métrique est invariante dans un changement de coordonnées particulier, on dit que l'on a une symétrie de la métrique. Dans le cas simple où la métrique est invariante par une translation, par exemple $x^1 \to x^1 + a$, on associe à cette invariance un *vecteur de Killing*

$$\xi = (0, 1, 0, 0) \tag{152}$$

Les conséquences d'une symétrie de la métrique se déduisent commodément de l'étude des vecteurs de Killing. En particulier on montre que $\xi \cdot u$ est invariant le long d'une géodésique, $u^\mu = \mathrm{d}x^\mu/\mathrm{d}\tau$ étant la quadrivitesse le long de cette géodésique.

Nous allons nous intéresser au cas de la *symétrie sphérique*. Il est intuitif, mais long à montrer rigoureusement, que la partie spatiale de la métrique $g_{ij} \propto \delta_{ij}$. En effet ceci assure que la partie spatiale du $\mathrm{d}s^2$ a la forme

$$\mathrm{d}x^2 + \mathrm{d}y^2 + \mathrm{d}z^2$$

les termes tels que $\mathrm{d}x\mathrm{d}y$ étant incompatibles avec la symétrie sphérique. Écrivant

$$\mathrm{d}x^2 + \mathrm{d}y^2 + \mathrm{d}z^2 = r^2(\mathrm{d}\theta^2 + \sin^2\theta\,\mathrm{d}\varphi^2) = r^2\mathrm{d}\Omega^2$$

la forme la plus générale de la métrique compatible avec la symétrie sphérique est, en fonction de deux coordonnées $a$ et $r$

$$\mathrm{d}s^2 = g_{aa}(a,r)\mathrm{d}a^2 + 2g_{ar}\mathrm{d}a\mathrm{d}r - g_{rr}(a,r)\mathrm{d}r^2 - r^2\mathrm{d}\Omega^2 \qquad (153)$$

Des termes croisés $(a,\theta)$, $(a,\varphi)$, $(r,\theta)$ et $(r,\varphi)$ sont exclus par la symétrie sphérique. On peut donc se limiter à l'étude des coordonnées $(a,r)$. On effectue un changement de coordonnées $a \to t(a,r)$

$$\mathrm{d}t = \frac{\partial t}{\partial a}\,\mathrm{d}a + \frac{\partial t}{\partial r}\,\mathrm{d}r$$

et on souhaiterait avoir

$$\mathrm{d}s^2 = A(t,r)\mathrm{d}t^2 - B(t,r)\mathrm{d}r^2 - r^2\mathrm{d}\Omega^2$$

ce qui donne 3 équations pour 3 fonctions inconnues, $t$, $A$ et $B$. On peut donc toujours mettre la métrique sous la forme

$$\mathrm{d}s^2 = \mathrm{e}^{2\alpha(r,t)}\,\mathrm{d}t^2 - \mathrm{e}^{2\beta(r,t)}\,\mathrm{d}r^2 - r^2\,\mathrm{d}\Omega^2 \qquad (154)$$

Il faut évidemment prendre garde au fait que la relation entre $t$ et le temps effectivement mesuré par un observateur est indirecte ($cf.$ (158)), et de même $r$ n'est pas la distance au centre de symétrie. En revanche la surface d'une sphère de rayon $r$ est bien $4\pi r^2$. On va s'intéresser pour l'instant uniquement à la région extérieure à la source du champ de gravitation, où le tenseur de Ricci $R_{\mu\nu} = 0$ d'après (146). Notre programme consiste maintenant à calculer le tenseur de Ricci à partir de la métrique (154) et à l'annuler. On trouve à partir de (132)

$$R_{tr} = \frac{2}{r}\,\partial_t\beta = 0$$

$$R_{\theta\theta} = \mathrm{e}^{-2\beta}\Big[r(\partial_r\beta - \partial_r\alpha) - 1\Big] + 1 = 0$$

On en déduit que $\beta$ est seulement une fonction de $r$, $\beta(r)$ ; de plus $\partial_t R_{\theta\theta} = 0 \implies \partial_t\partial_r\alpha = 0$ soit

$$\alpha(r,t) = f(r) + g(t)$$

ce qui donne pour la métrique

$$\mathrm{d}s^2 = \mathrm{e}^{2f(r)}\,\mathrm{e}^{2g(t)}\,\mathrm{d}t^2 - \mathrm{e}^{2\beta(r)}\mathrm{d}r^2 - r^2\mathrm{d}\Omega^2 \qquad (155)$$

On effectue un changement de variables $t \to t'$ tel que

$$\mathrm{d}t' = \mathrm{e}^{g(t)}\mathrm{d}t$$

et en réétiquetant $t' \to t$ on obtient le *théorème de Birkhoff*

$$ds^2 = e^{2\alpha(r)} dt^2 - e^{2\beta(r)} dr^2 - r^2 d\Omega^2 \tag{156}$$

En mots, le théorème de Birkhoff nous dit que dans le cas d'une symétrie sphérique, *la métrique est stationnaire*, ce qui implique l'existence d'un vecteur de Killing $\xi = (1,0,0,0)$. Une distribution de matière à symétrie sphérique ne rayonne pas d'ondes gravitationnelles, de même qu'en électromagnétisme une distribution de charges à symétrie sphérique ne rayonne pas : il n'existe pas de rayonnement monopolaire !

Pour finir de déterminer la métrique, examinons la combinaison suivante de $R_{tt}$ et de $R_{rr}$

$$e^{2(\beta-\alpha)} R_{tt} - R_{rr} = \frac{2}{r}(\partial_r \alpha + \partial_r \beta) = 0$$

d'où $\alpha = -\beta + \text{cste}$. Enfin $R_{\theta\theta} = 0$ devient, compte tenu des résultats précédents

$$e^{2\alpha}(2r\partial_r\alpha + 1) = 1 \implies \partial_r\left(re^{2\alpha}\right) = 1 \implies e^{2\alpha} = 1 + \frac{C}{r}$$

Ceci conduit à la forme suivante de la métrique

$$ds^2 = \left(1 + \frac{C}{r}\right) dt^2 - \left(1 + \frac{C}{r}\right)^{-1} dr^2 - r^2 d\Omega^2$$

Pour déterminer la constante $C$, il reste à examiner la limite de champ faible où

$$g_{tt} \simeq 1 + 2\Phi(r) = 1 - \frac{2GM}{r}$$

ce qui donne la forme finale de la métrique se Schwarzschild

$$\boxed{ds^2 = \left(1 - \frac{2GM}{r}\right) dt^2 - \left(1 - \frac{2GM}{r}\right)^{-1} dr^2 - r^2 d\Omega^2} \tag{157}$$

La métrique de Schwarzschild donne accès à tous les résultats classiques de la relativité générale

- déviation de la lumière par une masse ;
- précession du périhélie de Mercure ;
- décalage vers le rouge gravitationnel ;
- retard de l'écho radar
- *etc.*

Il faut calculer les géodésiques et utiliser le fait que $\xi \cdot u$ est une constante le long de ces géodésiques, si $\xi$ est un vecteur de Killing. Les calculs exacts n'étant pas possibles en général, on doit effectuer un développement en puissances de $v/c$ appelé approximation post-newtonienne.

## 6.3. *Trous noirs*

La métrique de Schwarzschild semble singulière lorsque la coordonnée $r = r_S = 2GM$ ($r_S = 2GM/c^2$ si l'on rétablit $c$) ; $r_S$ est apelé le *rayon de Schwarzschild*. Dans le cas du Soleil $r_S \simeq 3\,\text{km}$, ce qui est évidemment négligeable par rapport au rayon du Soleil, que l'on peut donc assimiler en pratique à une masse ponctuelle. En revanche le rayon d'une étoile à neutrons est du même ordre de grandeur que son rayon de Schwarzschild[g], et la question de savoir ce qui se passe lorsque $r \to r_S$ est intéressante. Le calcul de la courbure scalaire $R$ pour $r = r_S$ donne un résultat fini, ce qui suggère que la singularité de la métrique de Schwarzschild à $r = r_S$ est un artefact du système de coordonnées, ce que nous allons montrer en exhibant un système de coordonnées manifestement non singulier à $r = r_S$. Nous allons toujours nous placer à $(\theta, \varphi)$ fixés, et prendre en compte uniquement les coordonnées $t$ et $r$. Examinons d'abord le décalage vers le rouge gravitationnel (Fig. 23). L'observateur $O_1$ en $r_1$ émet des signaux à des intervalles $\Delta t$ réguliers dans une direction fixée ($\mathrm{d}\theta = \mathrm{d}\varphi = 0$), mais pour lui les intervalles de temps propre entre l'émission de deux signaux consécutifs sont

$$\Delta\tau_1 = \left(1 - \frac{2GM}{r_1}\right)\Delta t \tag{158}$$

Compte tenu d'une relation similaire entre $\Delta t$ et les intervalles de temps propre de l'observateur $O_2$ en $r_2$ entre la réception de deux signaux consécutifs, on obtient pour le rapport entre fréquences reçues et émises

$$\frac{\omega_2}{\omega_1} = \frac{\Delta\tau_1}{\Delta\tau_2} = \left(\frac{1 - 2GM/r_1}{1 - 2GM/r_2}\right)^{1/2} \xrightarrow[r_2 \gg r_S]{} \left(1 - \frac{2GM}{r_1}\right)^{1/2}$$

et pour un observateur $O_2$ à l'infini, avec $r_1 = r_{\text{em}}$

$$\boxed{\omega_\infty = \omega_{\text{em}}\left(1 - \frac{2GM}{r_{\text{em}}}\right)^{1/2}} \tag{159}$$

---

[g]$r_S \simeq 0.4r$ si $r \simeq 10\,\text{km}$ est le rayon d'une étoile à neutrons dont la masse est égale à $\simeq 1.5$ fois celle du Soleil.

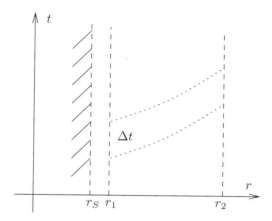

FIGURE 23.  Décalage vers le rouge gravitationnel. La trajectoire des photons est en pointillés.

Lorsque $r_{\text{em}} \rightarrow r_S$, le rythme d'émission des signaux et la fréquence des photons tendent vers zéro pour un observateur à l'infini, et bientôt il ne verra plus rien : au fur et à mesure que $r_{\text{em}} \rightarrow r_S$, le rythme des signaux est de plus en plus ralenti et l'énergie des photons devient de plus en plus petite. Il est aussi intéressant d'examiner le cône de lumière

$$\frac{\mathrm{d}t}{\mathrm{d}r} = \pm \left(1 - \frac{2GM}{r}\right)^{-1} \tag{160}$$

ce qui montre que l'angle au sommet du cône devient de plus en plus aigu quand $r \rightarrow r_S$ (Fig. 24).

Pour comprendre ce qui se passe lorsque $r < r_S$, il faut faire appel à un autre système de coordonnées. Le plus simple est sans doute celui d'Eddington-Finkelstein (EF) où $(t, r) \rightarrow (v, r)$, avec

$$t = v - r - 2GM \ln\left|\frac{r}{2GM} - 1\right| = v - r - r_S \ln\left|\frac{r}{r_S} - 1\right| \tag{161}$$

À partir de (161) on obtient immédiatement

$$\mathrm{d}t = \mathrm{d}v - \frac{1}{1 - r_S/r}$$

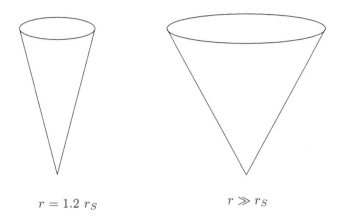

$$r = 1.2\, r_S \qquad\qquad\qquad r \gg r_S$$

FIGURE 24.   Évolution des cônes de lumière dans les coordonnées $(t, r)$.

d'où le d$s^2$

$$\mathrm{d}s^2 = \left(1 - \frac{r_S}{r}\right)\left(\mathrm{d}v - \frac{1}{1 - r_S/r}\right)^2 - \left(1 - \frac{r_S}{r}\right)^{-1}\mathrm{d}r^2$$

$$= \left(1 - \frac{r_S}{r}\right)\mathrm{d}v^2 + 2\mathrm{d}v\mathrm{d}r - r^2\mathrm{d}\Omega^2 \tag{162}$$

On constate que la métrique n'est plus singulière pour $r = r_S$ ! De plus le déterminant $g$ de la métrique vaut

$$g|_{r=r_S} = -r_S^4\,\sin^2\theta$$

et $g^{-1}$ existe pour $r = r_S$. Étudions le cône de lumière ; il faut résoudre l'équation d$s^2 = 0$

$$\mathrm{d}s^2 = \left(1 - \frac{r_S}{r}\right)\mathrm{d}v^2 + 2\mathrm{d}v\mathrm{d}r = 0 \tag{163}$$

Il y a deux possibilités

(1)  d$v = 0$, $v = $ cste. Pour $r \gg r_S$, cela correspond à $t + r = $ cste, c'est-à-dire à des rayons lumineux entrants : $r$ décroît si $t$ croît.

(2)

$$\left(1 - \frac{r_S}{r}\right)\mathrm{d}v + 2\mathrm{d}r = 0$$

soit

$$v - 2\left(r + r_S\ln\left|\frac{r}{r_S} - 1\right|\right) = \text{cste}$$

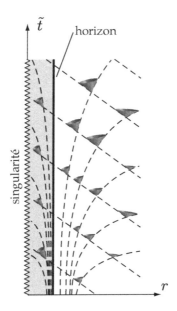

FIGURE 25. Les rayons lumineux entrants et sortants dans les variables $(\tilde{t}, r)$. D'après Hartle [2003].

Si $r \gg r_S$

$$v - 2r \simeq t + r - 2r = t - r = \text{cste}$$

ce qui correspond à des rayons lumineux sortants : $r$ croît si $t$ croît.

Pour la discussion générale, il est commode d'introduire la variable $\tilde{t} = v - r$. Lorsque l'on se trouve dans le cas (1), on aura $\tilde{t} + r = \text{cste}$, ce qui correspond à des droites entrantes pour $r > r_S$ aussi bien que pour $r < r_S$. Dans le cas (2)

$$\tilde{t} = \text{cste} - r + 2 \left( r + r_S \ln \left| \frac{r}{r_S} - 1 \right| \right)$$

soit

$$\frac{\mathrm{d}\tilde{t}}{\mathrm{d}r} = 1 + \frac{2}{r/r_S - 1} \tag{164}$$

ce qui veut dire que

$$\frac{\mathrm{d}\tilde{t}}{\mathrm{d}r} \xrightarrow[r \to r_S^+]{} +\infty \qquad\qquad \frac{\mathrm{d}\tilde{t}}{\mathrm{d}r} \xrightarrow[r \to r_S^-]{} -\infty$$

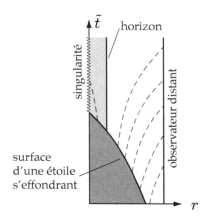

FIGURE 26.    Schéma de l'effondrement gravitationnel d'une étoile. D'après Hartle [2003].

d'où le schéma de la Fig. 25. La surface $r = r_S$ est une surface à trois dimensions du genre lumière ; les rayons lumineux se propagent le long de cette surface, qui est appelée *l'horizon du trou noir*. Aucun rayon lumineux émis depuis $r \leq r_S$ ne peut se propager à l'infini. La Fig. 26 donne le schéma de l'effondrement gravitationnel d'une étoile qui termine sa vie comme un trou noir. Les rayons émis depuis la surface de l'étoile sont d'abord reçus par un observateur à l'infini, puis leur fréquence diminue progressivement au fur et à mesure de l'effondrement, jusqu'au moment où l'observateur ne reçoit plus rien.

Comme par définition on ne peut pas voir un trou noir, on ne peut avoir que des présomptions sur leur existence. La première possibilité de trou noir vient de l'effondrement gravitationnel d'une étoile dont la masse est supérieure à 3 fois la masse solaire environ. L'existence d'un tel trou noir peut être "mise en évidence" s'il a un compagnon dont il perturbe l'orbite. On a ainsi "détecté" dans la Galaxie une dizaine de trous noirs plausibles, dont celui du Cygne. Une autre possibilité est que l'on ait des trous noirs au centre de certaines galaxies, dont la masse pourrait atteindre un million voire un milliard de masses solaires. Il est à peu près certain qu'il existe un trou noir d'environ un million de fois la masse solaire au centre de notre Galaxie. Enfin les quasars sont probablement des trous noirs géants, dont la masse est de l'ordre de $10^9$ masses solaires, et qui "avalent" une quantité de matière énorme autour d'eux. Cette matière rayonne des rayons X en abondance, ce qui permet de conclure à l'existence plausible d'un trou noir.

Toutes les galaxies ont probablement en leur centre un tel trou noir. Mais comme il n'y a plus de gaz à aspirer, le trou noir devient "invisible" faute d'être alimenté.

**Bibliographie**

Hartle [2003], chapitres 9, 12 et 13; Carroll [2004], chapitres 4 et 7, Wald [1984], chapitre 6; Damour [2005]; R. Blanford et N. Gehrels, *Revisiting the black hole*, Physics Today, juin 1999, p. 40.

## 6.4. *Remerciements*

Je suis très reconnaissant à Pierre Coullet et Yves Pomeau qui m'ont incité à donner ce cours aux rencontres non linéaires de Peyresq. Je remercie également Thierry Grandou et Mathieu Le Bellac pour leur lecture attentive et critique du manuscrit.

# Chapter 6

# Bioadhesion

Eric Perez and Frédéric Pincet

*Laboratoire de Physique Statistique*
*UMR 8550*
*24 rue Lhomond*
*75005 Paris, France*

## Contents

In living matter, interactions and bonds are unceasingly formed and broken between various molecules or organized aggregates and cells.[1] These interactions occur either spontaneously or after a signal which triggers production of the molecular species corresponding to the programmed biological phenomenon. Cell adhesion is involved in a large number of biological processes such as the defenses of organism, transport, metastasis invasion.[2] A majority of cells can divide and multiply only when they adhere to other cells or tissues. The adhesion between cells if often a way to exchange signals, for instance mediated by a cross-talk between different families of cell adhesion molecules. The first stages of inflammation correspond to several different cell adhesion processes after an external microorganism is detected.[3] As a consequence, white blood cells start having a weak adhesion to the vessel walls which restricts their movement to rolling along the wall. This first adhesive contact due to selectins and their ligand triggers another process which increases the adhesion and blocks the cell in a fixed position. Then, other adhesion processes drive the white blood cells across the vessel walls by letting them pass between the cells that make up the wall. Eventually, the white blood cells reach the microorganisms and adhere to them in order to destroy them. There is an increasing interest in understanding biological adhesion in views of designing new anti-inflammatory drugs, new biomaterials, understanding some processes linked to metastasis proliferation, and also control the properties of numerous materials nowadays designed to have increasing sophisticated functions. There is a large difference between the adhesion of inert materials and biological ones. In general, the latter is triggered by a signal while the former just depends on more static physico-chemical features. In contrast to inert materials, the adhesion of cells always rely on receptor/ligand (often named key/lock) interactions which consist in molecules from one cell that recognize those of another cell. This molecular recognition always results from known physico-chemical interactions (van der Waals forces, electrostatic forces, hydrophobic interactions, hydrogen bonds...) combined with a particular geometry of the molecules. In spite of these complications, several physical methods allow to investigate this biological adhesion either by studying simplified models with a limited number of parameters, or by studying a single bond between cell adhesion molecules, or by investigating realistic biological systems such as cells for which it is still possible to obtain some physical description.

A first section deals with the adhesion of living cells and some theoretical description of adhering spheres and shells. An experimental procedure for the measurement of the adhesion strength is given as well as its application to a biologically relevant problem involving receptors and ligands located at the cell surface. The second section describes theoretical models for cell adhesion involving surfaces bearing adhesion sites either immobile or mobile and also reports experimental tests of these models: adhesion energies are measured and related to the surface density of adhesion molecules and their binding energy. In a third section, the dynamics of rupture of a single receptor/ligand bond under force is described from a statistical mechanics point of view and the experimental methods to measure such forces are reported. The case of a single and multiple potential barriers is examined.

## 1. Adhesion of living cells

### 1.1. *Living cells: an active material*

A living cell can usually be schematized by a viscous medium wrapped in a membrane. Its usual size is about 10 $\mu$m. This medium has the specific feature of containing a cytoskeleton that provides its peculiar mechanical properties. It can reorganize with the various constraints applied to the cell. The characteristic time of reorganization is about a few minutes. Three main types of cytoskeleton are present in the cell: the cortical cytoskeleton (made of actin filaments), the microtubules (tubulin) and the intermediate filaments. Pictures of these types of cytoskeleton are shown in Fig. 1. The cortical cytoskeleton has a structure which is mainly bidimensional and is bound to the cell membrane. Red blood cells, for instance, have no cytoskeleton except this cortical one. Microtubules and intermediate filaments have a three-dimensional structure. Most of the time, there is no cytoskeleton in the nucleus of the cell.

When the forces are applied fast enough for the cytoskeleton not to reorganize, and when the resulting deformations are small compared to the cell size, it can be imagined that the cell will behave like a classical elastic material. The type of cytoskeleton (cortical or three-dimensional) will then determine whether the cell can be seen as a spherical shell or as a full elastic sphere. In this part, we will see how to determine which is the best adapted model for two adhering cells.

### 1.2. *Adhering spherical shells*

To simplify, we shall work on the model presented in Fig. 2 where a spherical shell adheres on a flat substrate. The procedure is the same for two adhering cells (for which the substrate is not flat, but a spherical shell) and the result

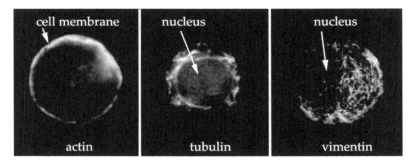

Figure 1.   Three of the main cytoskeleton types: actin filaments, microtubules and intermediate filaments (from left to right). They are visualized with fluorescent antibodies specific to a cytoskeleton component (here, actin, tubulin and vimentin).

will be given at the end of this paragraph.  Let us consider an adhering spherical shell on which a pulling force normal to the contact zone is applied with a very small pipe (micropipette).[4]  The problem has a cylindrical symmetry.  When no force is applied, the shell looks like a truncated sphere. The work to displace one unit area of membrane under tension (see Fig. 2) is: $\tau\left(1-\cos\theta_e\right)$. The change in interfacial energy is $W_a$.  At equilibrium the free energy F is minimal: $\delta$F=0.  Therefore, the contact angle, $\theta_e$, between the surface and the shell is related to the adhesion energy, $W_a$, through the Young-Dupré equation:

$$W_a = \tau\left(1-\cos\theta_e\right) \tag{1}$$

where $\tau$ is the membrane tension which is uniform and given by:

$$\Delta P = 2\tau(1/r_p - 1/R) \tag{2}$$

$\Delta$P is the aspiration in the micropipette, $r_p$, the micropipette radius and R, the shell radius.  Equation 2 can be obtained by applying twice the Laplace Eq. on the shell inside the micropipette and in the "free" region (cf. Fig. 3). When a finite non-zero force is applied, the mechanical equilibrium can be written on every section normal to the axis of symmetry in the "free" region of the cell.

The pressure inside the cell being higher than the ones outside, one gets: f+pressure force = tension force.  Which gives, explicitly:

$$2\pi r\tau\sin\theta - \pi r^2\tau C = f \tag{3}$$

$r$ is the radius of the cell at the level of the cross section and $C$ the curvature of the surface (which is constant because of Laplace law).  Applying Eq. 3 at the equator and at the contact area:

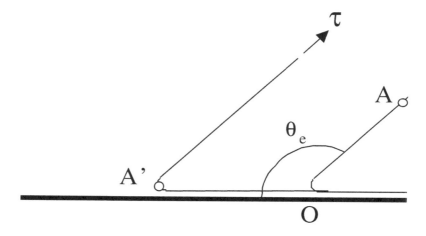

Figure 2. A membrane under tension $\tau$ adhering to a substrate is displaced by one unit area.

$2\pi R_c\tau\sin\theta_e - \pi R_c^2\tau C = f$
$2\pi R\tau - \pi R^2\tau C = f$

By eliminating $C$, it yields:

$$f\left(1 - \frac{R_c}{R}\right) = 2\pi R\tau\left(\frac{R_c}{R}\sin(\theta_e) - \left(\frac{R_c}{R}\right)^2\right). \tag{4}$$

In general, it can be assumed that $R_c/R \ll 1$, which simplifies 4 to:

$$f = 2\pi R\tau\left(\frac{R_c}{R}\sin(\theta_e) - \left(\frac{R_c}{R}\right)^2\right). \tag{5}$$

The shape of f as a function of $R_c$ is given in Fig. 4. An instability occurs when $f > f_{sep}$, with:

$$f_{sep} = \pi R W_a. \tag{6}$$

This is the expression of the separation force. In the case of two adhering shells, Eq. 6 is still valid by taking R as the harmonic mean of the shells radii. If they are identical, it simplifies to:

$$f_{sep} = \frac{\pi R W_a}{2} \tag{7}$$

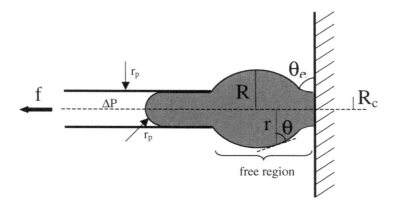

Figure 3.    Schematic of the system. A cell is held by a micropipette and adheres on the substrate. A pulling force (generated by the aspiration pressure) is applied to separate the cell and the substrate.

### 1.3.  *Adhering elastic spheres*

In the case of a full sphere with a given elastic modulus, K, the mechanical equilibrium can be written by minimizing the sum of the different contributions to the energy: elastic energy, mechanical energy due to the displacement of the spheres and surface energy.[5,6] When the sphere adheres on a substrate, the result is similar to that obtained with the spherical shells with a factor $3/2$:

$$f_{sep} = \frac{3\pi R W_a}{2}. \tag{8}$$

As for the shells, Eq. 8 remains valid for two cells by taking R as the harmonic mean of the shells radii. If they are identical, Eq. 8 becomes:

$$f_{sep} = \frac{3\pi R W_a}{4}. \tag{9}$$

It is worth noting that to obtain the same separation force, the required adhesion energy for spherical shells is higher than that of elastic spheres.

The radius $R_c$, of the contact area for a given force f verifies:

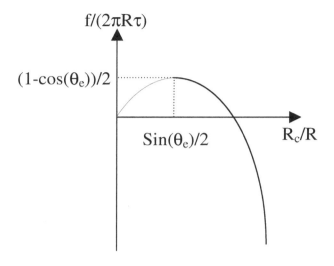

Figure 4.  f as a function of Rc, as given by Eq. 5.

$$(R_c)^3 = \frac{R}{2K}\left[f + \frac{3\pi R W_a}{2} + \sqrt{3\pi R W_a f + \left(\frac{3\pi R W_a}{2}\right)^2}\right]. \qquad (10)$$

From Eq. 10, it is easy to deduce a simple relation between the contact radius under no force, $R_{c0}$ and the contact radius at separation, $R_{csep}$ :

$$R_{csep} = \frac{1}{\sqrt[3]{4}} R_{c0}. \qquad (11)$$

### 1.4. *Experimental characterization of the mechanical properties of living cells*

In order to know whether a cell has indeed an elastic behavior, a possible approach is to apply a controlled adhesion and to compare $f_{sep}$ and $W_a$. Equations 6 and 8 show that the following relation is expected:

$$\frac{f_{sep}}{R} \propto W_a \qquad (12)$$

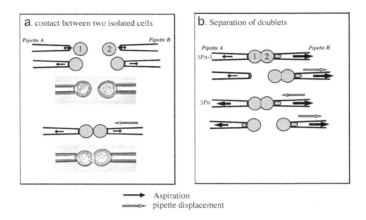

Figure 5. Experimental protocol: Two cells are grabbed by micropipettes, forced in contact (a) and mechanically separated (b) for increasing values of the aspiration in the left micropipette until the rupture of the adhesion.

If this relation is experimentally verified, the prefactor will allow determining whether the studied cells behave more like spherical shells or like spheres.

Such a study can be conducted by forcing initially non-adhering cells to adhere through non-specific forces such as a depletion effect induced by long polymers in solution. The principle of this effect is well known. When the cells approach each other, the excluded volume for the polymer decreases generating an attractive entropic force between the cells. More quantitatively, the expression of the adhesion energy as a function of the volume fraction of the polymer $\phi$ is well-known[7]:

$$W_a = \frac{k_B T}{\xi^2} = \frac{k_B T}{a^2} \phi^{1.5} \tag{13}$$

where a is the size of one monomer.

In the case of dextrane, Eq. 13 has been experimentally checked on lipid vesicles.[8] It is therefore possible to test the mechanical behavior of cells by forcing them to adhere through depletion forces at different volume fractions $\phi$. A suitable experimental technique consists in holding two cells in micropipettes with a controlled aspiration. The following experimental procedure can be used (cf. Fig. 5): the aspiration is high in one of the pipettes (the right one on the figure), and lower in the other one. A mechanical traction (movement of one of the pipette along its axis) is applied to the cell doublet in order to separate them. Two cases can occur: If the adhesion is stronger than the lowest aspiration, the doublet remains intact and detaches from the pipette. If, on the contrary, the aspiration is predominant, the cells will separate. In the first case, the doublet is reacquired in the pipette and the aspiration is increased to try again to separate the cells. The cycle is repeated until the breakage of the adhesion during the $n^{th}$ cycle. The separation force is then well approximated by:

$$F_s = (\Delta P_{n-1} + \Delta P_n)\pi r_p^2/2 \tag{14}$$

where $\Delta P_{n-1}$ and $\Delta P_n$ are respectively the aspirations for cycle n-1 and cycle n and $r_p$ the inner radius of the pipette (left pipette in Fig. 5).

The relation 14 was experimentally checked on murine sarcoma S180 cells with a microdynamometer (here, a microneedle).

The results obtained with these S180 cells show that they behave more like elastic sphere than like spherical shells.[9] This conclusion was cross-checked by measuring the variation of the contact radius with the pulling force and deducing an elastic modulus through equation 10. The obtained value is in good agreement with the ones from the literature.[10] Also, these cells verified Eq. 11.

The chemical disruption of the cytoskeleton lowered dramatically the measured separation force, showing that the mechanical behavior of the cell was indeed due to the cytoskeleton. It turns out to be difficult to systematically predict what behavior a given cell will have. For instance, red blood cells that only have a cortical cytoskeleton behave more like spherical shells.

Finally, it is important to mention that if the characteristic time of the experiment is longer than the time required for the cytoskeleton to reorganize, the cells would not have an elastic behavior (viscoelastic behavior) and the results would not be valid anymore.

## 1.5. *Application of adhesion measurements to biologically relevant problems*

Measurements of the adhesion of cells as presented in the previous paragraph can be used to try to solve biological or medical problems.[11] The value of the adhesion energy may not be the most relevant parameter. However, the comparison between separation forces that represent the strength of the adhesive junction can bring interesting information. For example, such an approach allowed to better understand the origin of an epidemiologic observation with a chemokine called fractalkine and its receptor, called CX3CR1. This receptor has two natural humane forms: wild and mutated. The latter one is associated with a significant decrease of cardiovascular risks. The fractalkine/CX3CR1 pair has two functions, a first one connected to the chemotraction, which is cell migration in a gradient of solute (here, the fractalkine molecules), and a second one connected to cell adhesion. Force measurements between a cell expressing CX3CR1 and a cell expressing fractalkine have shown that the difference between the two CX3CR1 forms is due to the adhesive function (cf. Fig. 6).

Complementary force measurements allowed to understand part of the underlying intracellular signals resulting from the attachment of the fractalkine on CX3CR1, and to show some difference in the signals between the two CX3CR1 types. This reveals the relation between these signals and the epidemiologic observations.

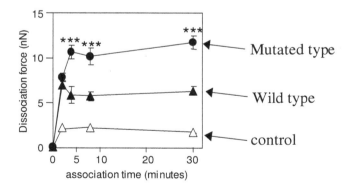

Figure 6. Dissociation force as a function of the association time for the mutated and wild type CX3CR1.

## 2. Model systems for bio-adhesion

### 2.1. *Surface forces and adhesion energy measurement techniques*

The surface force apparatus (SFA)

The surface force apparatus was developed in the early seventies[12] to probe colloidal forces. These forces are closely related to the nature of the interface between a colloidal particle and the dispersing liquid. These forces, called surface forces, can be van der Waals forces, electrostatic double-layer forces, steric forces, undulation forces, solvation forces etc... They govern the stability of colloids: repulsive forces will make the colloid stable while attractive forces will make it coagulate. This technique is based on a mechanical apparatus shown in Fig. 7 which can move two surfaces relative to each other and change their distances, a measurement of their actual distance at the angstrom level, and molecularly smooth surfaces (mica) which give a sense to the precision of the distance measurement.[13] The force is measured with the spring supporting one of the surfaces (the lower one). The mechanical part is a translation device with several stages including micrometric translations (accuracy $1\mu m$), differential spring systems (accuracy $\sim 1nm$) and piezoelectric crystals (accuracy $\sim 0.1nm$). Distance is measured by multiple beam interferometry (accuracy $\sim 0.1nm$).

The mica sheets are glued onto two cylindrical discs and arranged in a cross cylinder geometry in order to avoid alignment problems. This is equivalent to a sphere/plane geometry when the distance between the surfaces is much smaller than their radius of curvature.

The multiple beam interferometric technique involves a Fabry-Perot formed by two thin mica sheets (of equal thickness $2 - 5\mu m$) free from crystal steps and silvered on one side and facing each other with the two silver layers outside (see Fig. 8). When a white light beam is sent to the interferometer, the various wavelengths are reflected and transmitted across the different interfaces and produce a transmitted spectrum made of fringes of equal chromatic order (FECO). These FECOs allow to measure simultaneously and independently the distance and refractive index of the medium between the mica surfaces.

According to the Derjaguin approximation[14]: $F(D)/R = 2\pi E(D)$, the force between the curved surfaces is proportional to the energy between plane surfaces if the distance is much smaller than the radius of curvature

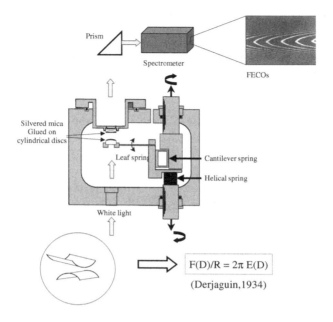

Figure 7. The surface force apparatus. Here, the upper and lower cylinders are driven by a micrometric translation. The lower one is connected to a differential spring system made of one helical spring pushing a cantilever spring with whose stiffness is 1000-fold higher; $1\mu m$ move of the lower cylinder will move the lower surface by $1nm$. The measurement of the leaf spring bending when multiplied by its stiffness will be equal to the forces between the surfaces.

of the surfaces and if the interaction is decreasing with distance sufficiently rapidly (at least in $1/d^2$).

Vesicle micromanipulation

The adhesion free energy of two giant lipid vesicles (size: several micrometers) can be measured with the micropipette aspiration technique and the use of Eq. 1. The adhesion energy can be obtained by determining the contact angle $\theta$ of the two vesicles (see Fig. 9) and the tension $\tau$ of the flaccid vesicle membrane. Two osmotically controlled vesicles are aspirated in micropipettes and brought into contact. They are observed in interference contrast microscopy. One of them is pressurized into a tight-rigid sphere with large bilayer tension, whereas the adherent vesicle is held with low pressure and remains deformable.

The (negative) pressure $\Delta P$ in each pipette controls the (positive) hydrostatic pressure in the vesicle and thus the mechanical tension $\tau$ in its

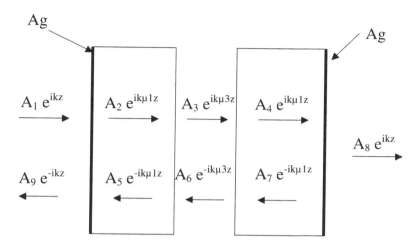

Figure 8. Multiple beam interferometry: two mica sheets silvered on one side and separated by a third medium are placed in order to form a Fabry-Perot interferometer. The transmitted spectrum is made of fringes of equal chromatic order (FECO) which can be observed in a spectrometer and whose wavelength is related to the distance D between the surfaces and the refractive index $\mu$ of the intervening medium. By following two successive fringes, one can obtain simultaneously and independently D and $\mu$.

membrane according to Eq. 2, which, combined with Eq. 1 allows to relate $\Delta P$ and $W_a$:

$\Delta P = C.W_a$ where $C$ depends only on geometrical parameters:

$$C = \frac{2\left(\frac{1}{r_p} - \frac{1}{r_v}\right)}{1 - \cos\theta_e}. \tag{15}$$

The measurement of $\theta_e$ is numerically deduced from geometrical parameters as indicated in Ref. 15. It is measured for several tension values of the flaccid vesicle membrane by decreasing the aspiration and then increasing it in order to check the reversibility of the adhesion. The adhesion results are displayed as plots of $\Delta P$ versus $C$, $W_a$ being the slope of the line.

## 2.2. *Adhesion of surfaces coated with a compact layer of adhesion sites*

### Theory

When two surfaces $S_1$ and $S_2$ bear a compact layer of adhesion sites with densities $\delta_1$ and $\delta_2$ with $\delta_1 < \delta_2$, and a binding energy $e_b$ per site, each site of $S_1$ faces a site of $S_2$. The adhesion energy of the two surfaces is the energy

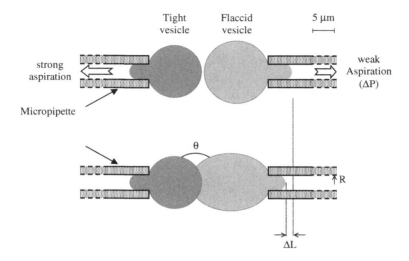

Figure 9.    Micromanipulation of two vesicles and measurement of their contact angle.

necessary to separate both planar surfaces per unit area. The contribution of the bonds between sites is equal to $e_b$ times the number of bonds per unit area. The probability for two sites to be bound is: $exp(e_b)/(1 + exp(e_b))$. Therefore the adhesion free energy is equal to:

$$W_a = \delta_1 e_b \exp(e_b\, k_B\, T)/(1 + exp(e_b k_B T)) \tag{16}$$

It is possible to coat one surface with sites of one kind and another surface with complementary sites which will form specific bonds with the former ones. These sites can be the complementary DNA bases (shown in Fig. 10) which are well suited to this kind of experiment. There are four such bases: adenine (A), thymine (T), cytosine (C) and guanine (G). They form hydrogen bonds in such a way that A binds specifically to T and C to G.

If these bases make up the polar headgroup of a lipid (see Fig. 11), it becomes possible to form monolayers of this lipid and compress them in a film balance, shown in Fig. 12, to make them compact. The mica surfaces can then be covered with these bases with the right orientation by two Langmuir–Blodgett monolayer depositions as shown in Fig. 13.

It is then possible to make surface forces experiments in which the bases face each other from one surface to another as shown in Fig. 14. This allows

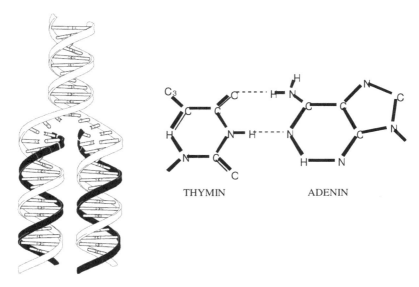

Figure 10. (Left) Double stranded DNA; (right): Watson–crick pair of adenin–thymin.

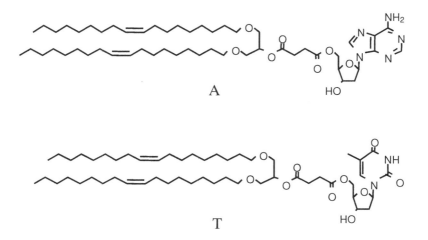

Figure 11. Di-oleoyl succinyl-thymidine (DOST) and -adenosine (DOSA).

to make surface forces measurements in which each of the surfaces can be covered with A or T.[16]

It then becomes possible to compare A/T interaction with A/A and T/T. As both surfaces are deformable, when the surfaces are in contact,

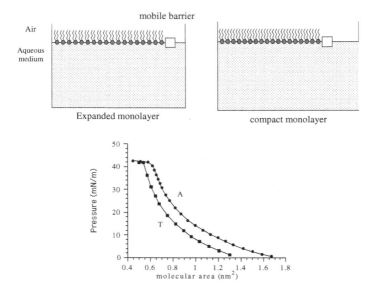

Figure 12. (up)Langmuir film balance in which a monolayer is spread (left) and compressed (right) at the air/water interface. (Bottom) Monolayer compression isotherms giving the surface pressure as a function of the area per molecule.

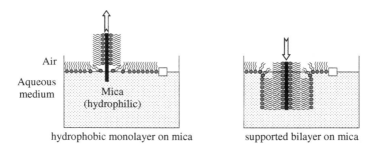

Figure 13. Langmuir–Blodgett deposition of a lipid bilayer on a solid substrate.

they flatten because of the adhesion and form a finite contact area. Upon pulling on one of them in order to separate them, the contact area diminishes but remains finite at separation. The force necessary to separate them is measured (see Fig. 15) with the SFA (the surface force apparatus described in Fig. 7).

Before any analysis is done, these forces already reveal the preferential A/T interaction relative to A/A and T/T. In the case of deformable surfaces

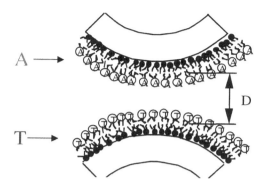

Figure 14.  Two curved mica surfaces coated with a bilayer of lipids bearing adenin and thymin.

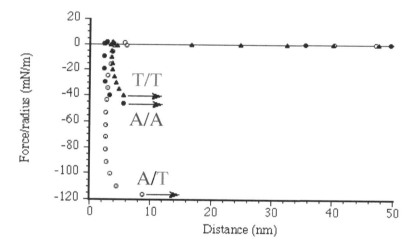

Figure 15.  Forces between two surfaces bearing DOST lipids (filled triangles) (see Fig. 11), DOSA lipids (filled circles) and between one surface with DOSA and one surface with DOST (open circles).

and strong adhesion which form a flat contact, Eq. 8 applies. It relates the pull off force $f_{sep}$, indicated by an arrow in the Fig. 15, to the adhesion free energy and the radius of curvature of the surfaces R:

$$W_a = 2f_{sep}/3\pi R$$

Table 1 gives the adhesion energies deduced from the pull off forces, the surface densities of sites, and the binding energies deduced from Eq. 16.

Table 1.  Adhesion energies, surface densities of sites, binding energies of the DNA bases.

|                       | A/T  | A/A  | T/T  |
| --------------------- | ---- | ---- | ---- |
| $W_a(mJ.m^{-2})$      | 23.3 | 10.6 | 9.1  |
| Nb molecules/$nm^2$   | 1.66 | 1.66 | 1.85 |
| $e(k_BT)$             | 3.6  | 1.9  | 1.5  |

Once the binding energies calculated, it is interesting to compare them to the ones known from the literature for the adenine/thymine (A/T) binding energy in water (or between adenin and uracyl because uracyl is very similar to adenin). This binding energy was measured by thermodynamical methods between adenin and uracyl[17]: $2.2 kcal/Mole$ or $3.6 k_BT$. Uracyl is almost identical to adenin and comprises a methyl group in a place which does not interfere with the formation of hydrogen bonds as can be seen in Fig. 16.

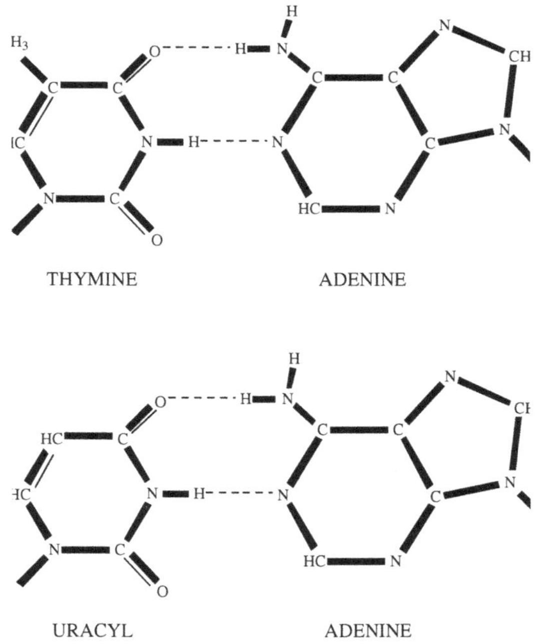

Figure 16.  Watson–crick pairs of adenine-thymine and adenine-uracyl.

One can nevertheless wonder why this approach allows to obtain binding energies in good agreement with literature with molecules which are immobilized on a solid surface instead of being in bulk. This may be because the monolayers can form a very compact layer of sites while maintaining the translational freedom of the nucleosides and their orientational freedom provided by a flexible spacer which makes them available for pairing.

### 2.3. *Electrostatic nanotitration of weak biochemical bonds*

Another approach to measure the binding energy of a weak biochemical link consists in balancing the adhesion it produces by a controlled electrostatic repulsion. The specific adhesion produced by biochemical bonds is titrated against a known repulsion: one can introduce electrostatic charges on the two surfaces bearing A and T in aqueous medium. Under a strong screening of the charges by salt, the surfaces can be brought in molecular contact. Then, if the screening is progressively reduced, there will be a stage at which the attractive force related to the A/T interaction will be balanced by a repulsive double-layer force. At this point, the adhering surfaces will separate spontaneously. This type of experiment can be achieved with giant lipid vesicles. Classically, vesicles are not functionalized and made of SOPC, a neutral lipid. By incorporating 10% of lipid bearing a DNA base, the vesicle is functionalized and will have these bases at its membrane. If one makes vesicles by adding both nucleoside lipids and negatively charged lipids to SOPC, and pushes two vesicles bearing the complementary bases in contact, the charged lipids will be segregated out of the contact region because they can move along the membrane and they will be pushed out by the repulsions between the charges. This will make the balancing of attractive and repulsive forces impossible to achieve. But if one uses one such vesicle and brings it into contact with an SOPC vesicle which includes nucleoside lipids with a negative charge attached to it, for instance an Adenine lipid with a $PO_4^-$ group, the charge will not be driven out of the contact region because of the attraction between the nucleosides. The measurement is possible in these conditions.[18] The adjustment of the negative charges in the former vesicle and the salt concentration in the solution allow a control of the double-layer interaction.

This double-layer repulsion scaled to one nucleoside depends on the density of charges $\rho_e$ of the other membrane (number of charges per $nm^2$) and the ionic strength of the solution[19] $c_i(mol/l)$:

$$\text{Sinh}(E_{dl}/2k_BT) \sim 1.36\rho_e/\sqrt{c_i} \qquad (17)$$

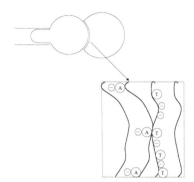

Figure 17.    One vesicle bearing DOSPA lipids with a negatively charged group (left) in contact with a vesicle bearing DOST lipid and negatively charged SOPS lipids (right).

In practice, this experiment can be made with two vesicles, one including a neutral classical lipid (SOPC) (90%) and a lipid bearing adenin and a negative charge (a phosphate group) (DOSPA) (10%); the other vesicle contains SOPC (90%), the lipid bearing thymin (DOST) (5%) and a classical lipid negatively charged (SOPS) (5%). By playing on the salt concentration, one can adjust the double-layer repulsion and therefore balance the adhesion due to the A/T bonds. The experiment consists therefore in forcing the two vesicles in contact. For this purpose, a mechanical action is not sufficient. Depletion forces[21,21] are applied with a non-adsorbing polymer (here, dextran). The depletion forces are attractive and, at high ionic strength, are sufficiently large to overcome the double-layer repulsion. Once the vesicles are in contact, the depletion forces are released without changing the salt concentration. The vesicles remain adherent. They are then transferred to a chamber with a weaker ionic strength. If they remain in contact, they are transferred to a chamber with an even weaker ionic strength, and so on until the vesicles detach spontaneously.

At this point, one deduces the upper and lower bounds to the binding energy from the salt concentrations respectively before and after the separation. One can also vary the proportion of lipids in the vesicles to refine the measurements.

Table 2 gives the proportions of the lipids used in the measurements, the salt concentrations and the upper and lower bounds of the binding energies obtained.

These values are in good agreement with the ones obtained with the SFA:

Table 2. Lipids used in the measurements, salt concentrations, upper and lower bonds of the binding energies obtained.

| Vesicle #1 | Vesicle #2 | $C*$ $(mM)$ | binding energy ($k_B T$) |
|---|---|---|---|
| nb DOSPA/SOPC (5 : 95) | DOST/SOPS/SOPC (5 : 5 : 90) | $1 < C* < 10$ | $1.83 < E_b < 3.83$ |
| DOSPA/SOPC (5 : 95) | DOST/SOPS/SOPC (5 : 10 : 85) | $1 < C* < 10$ | $2.97 < E_b < 5.18$ |
| DOSPA/SOPC (5 : 95) | DOSA/SOPS/SOPC (5 : 5 : 90) | $5 < C* < 50$ | $0.90 < E_b < 2.37$ |

Table 3. Comparison of experimental titration values with SFA values for binding energies.

| binding energy ) $k_B T$ | Experimental values titration | SFA values |
|---|---|---|
| A/T | $2.97 < E_b < 3.83$ | 3.6 |
| A/A | $0.90 < E_b < 2.37$ | 1.7 |

This new titration technique provides a simple approach to evaluate binding energies between 2 and $25 k_B T$ which is the typical range for biochemical bonds found in biological systems.

## 2.4. *Surfaces bearing mobile adhesion sites*

Consider two surfaces in contact bearing adhesion sites. If the sites are mobile on one of the surfaces, they will tend to migrate towards the contact zone (see Fig. 18), and this migration is only limited by the 2D osmotic pressure corresponding to the increase of density of sites.

In the mid-eighties, theory had been made of such systems and the outcome was a relationship between the adhesion energy and the enrichment of the contact region in sites[21,22]:

$$W_{spe} = \Pi_{osm2d} = k_B T(\rho_{contact} - \rho_{ext})$$ (18)

obtained in the following manner. The variations of the thermodynamic potential $G$ and the contact area $A_c$ are related to the adhesion free energy $W_{spe}$ by:

$$\delta G = -W_{spe} \delta A_c$$ (19)

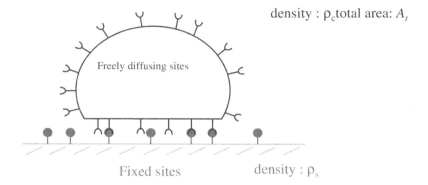

Figure 18.    Vesicle possessing mobile adhesion sites in contact with a solid surface bearing fixed sites.

$$G = \sum \mu_i N_i = \mu_c N_c + \mu_{nc} N_{nc} \qquad (20)$$

where the subscripts $c$ and $nc$ refer respectively to the contact and the non-contact regions of the vesicle. $N$, the total number of sites is constant:

$$N = \rho_c A_c + \rho_{nc}(A - A_c) \qquad (21)$$

the chemical potential in the contact region is:

$$\mu_c = \mu_0 - e_b k_B + k_B T \ln(\rho_c/\rho_0). \qquad (22)$$

Out of the contact region, the chemical potential is:

$$\mu_{nc} = \mu_0 + k_B T \ln(\rho_{nc}/\rho_0). \qquad (23)$$

Differentiating G gives:

$$\delta G = (\mu_{nc} - \mu_c)\delta N_{nc} - N_c k_B T \delta e_m + k_B T A_c \delta\rho_c + (A - A_c)k_B T \delta\rho_{nc} \quad (24)$$

Differentiating Eq. (21) gives:

$$A_c \delta\rho_c + (A - A_c)\delta\rho_{nc} = (\rho_{nc} - \rho_c)\delta A_c. \qquad (25)$$

At equilibrium, $\rho_{nc} = \rho_c$. From Eq. (20) to (25), one deduces:

$$\delta G = -N_c \delta e_m + k_B T \Delta \rho \delta A_c. \tag{26}$$

Which leads to the adhesion free energy:

$$W_{spe} = (\Delta \rho) k_B T.$$

However, no relation was established between the adhesion energy and the binding energy. Such a relation can be established by using a microcanonical model.[23] One bead has immobile adhesion sites with a density $1/A_0$. This bead interacts with a vesicle of area $A_v$ displaying N mobile sites. Each site has a field of attraction with an area $\alpha$. A vesicle site can be bound to one bead site if it is located within an area $\alpha$ around the bead site. $A_f$ is the part of the bead surface in which a site of the vesicle is bound: $A_f = A_c \alpha / A_0$. In practical situations, $A_f$ is very small compared to $A_v$. The partition function Z is:

$$Z = \sum_{\text{all states}} e^{-\beta e_b}. \tag{27}$$

The number of configurations in which n sites are bound (in $A_f$) is:

$$\binom{N}{n} (A_v - A_f)^{N-n} A_f^n e^{ne_b}. \tag{28}$$

The partition function is then:

$$Z = K \sum_{n=0}^{N} \binom{N}{n} (A_v - A_f)^{N-n} A_f^n e^{ne_b} \tag{29}$$

in which $K$ is a constant. This partition function can be directly calculated:

$$Z = K((A_v - A_f) + A_f e^{e_b})^N. \tag{30}$$

The adhesion energy is formally equal to:

$$W_{spe} = -\frac{\partial F}{\partial A_c} T |_{V,T} = k_B T \frac{\partial \ln Z}{\partial A_c} |_{V,T} \tag{31}$$

where F is the free energy and $A_c$ is the contact area. One gets finally:

$$W_{spe} \approx \frac{\alpha \rho_v}{A_0} (e^{e_b} - 1) k_B T \tag{32}$$

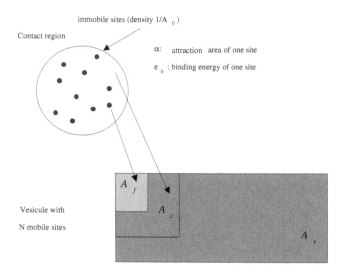

Figure 19.   Total area of the vesicle $A_v$, contact area $A_c$ and area $A_f$ in which the sites are bound.

where $\rho_v = N/A_v$.

Experiments can be made to test this theory by measuring the adhesion energy of a vesicle bearing mobile adhesion sites with a bead bearing the complementary sites which are in a fixed position (see Fig. 19). This adhesion energy can be measured by micropipette micromanipulation.

Lipids carrying nucleosides were incorporated (10%) in a vesicle membrane. Beads covered with streptavidin were coated by heterobifunctionalized polymers having at one end a biotin, and at the other end a nucleoside (A or T). The biotin anchored to the streptavidin therefore allows the attachment of nucleosides to the bead as shown in Fig. 20. The bead/vesicle adhesion energies were measured in the case A/T, T/A, A/A, T/T and SOPC/T as a control (in order to evaluate the contribution of forces which are not involved in the nucleoside recognition) shown in Fig. 21.

The adhesion energy values $W_{adh}$ deduced from the data of Fig. 21 and the specific adhesion $W_a$ due to the bonds between nucleosides are displayed in Table 4.

The binding energy values can be deduced from the adhesion energies $W_a$ by using Eq. 32 and using known values for $A_0$ and $\alpha$. These latter values are not known with great accuracy. However, it is possible as a test of the model to check that for realistic values of these parameters, one obtains

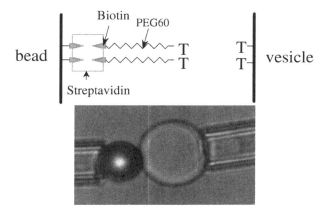

Figure 20.   (Top) Functionalized bead with an end-grafted polyethylene glycol bearing biotin at one end and one nucleoside on the other (bottom) micromanipulated vesicle in contact with a bead held by a micropipette. The bar represents $2\mu$m.

Table 4.   Adhesion energy $W_{adh}$ and specific adhesion $W_a$ due to bonds between nucleosides. $W_a$ is equal to $W_{adh} - W_{nspe}$ where $W_{nspe}$ is the T/PC adhesion energy.

|  | A/T | T/A | A/A | T/T | T/PC |
|---|---|---|---|---|---|
| $W_{adh}(10^{-6}J/m^2)$ | 12.0 | 12.9 | 9.9 | 9.3 | 8.6 |
| $W_a(10^{-6}J/m^2)$ | 3.4 | 4.3 | 1.3 | 0.7 | |

Table 5.   Binding energy values $e_b$ in $k_BT$ units as deduced from Eq. 32 for different values of $A_0/\alpha$.

| $A_0/\alpha$ | 800 | 1050 | 1300 | Literature values |
|---|---|---|---|---|
| Bead/vesicle | | | | |
| A/T | 2.9 | 3.2 | 3.4 | 3.6 |
| T/A | 3.1 | 3.4 | 3.6 | 3.6 |
| A/A | 2.0 | 2.3 | 2.5 | 1.9 |

the known values of the binding energies of these nucleosides. This is done in the Table 5 which shows that the agreement with the theoretical model is fair for $A_0/\alpha$ close to 1050. The microcanonical description relating adhesion energy measurements to binding energies is coherent with the experiments and can be applied to other systems.

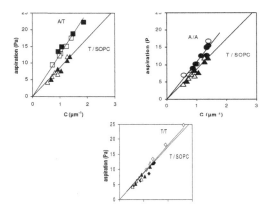

Figure 21.   Aspiration pressure as a function of C (cf. Eq. 15) for a specific interaction and comparison with the non-specific case T/SOPC (control). The slopes are equal to $W_a$. The solid lines are fits forced through zero.

## 3. Rupture force of a single biochemical bond

### 3.1. *What is a single bond?*

Usually, in all processes linked to cell adhesion, a single bond represents a complex made of two molecules: a ligand and a receptor. Often, the specific single bond is imaged as a " key-lock " bond where the ligand is the key and the receptor the lock. A bond frequently used by the scientific community is the streptavidin–biotin pair that forms one of the "strongest" non covalent bond. Here, we will call single bond any connection between a single molecule and another entity. This entity may be a receptor, as mentioned above, but also a surface on which the molecule could be grafted or anchored into.

### 3.2. *What is the strength of a single bond?*

A simple preconceived idea is to think it is possible to measure the force of a single bond by pulling on it and detecting the force at which it breaks. This approach is incorrect because these bonds, with very short life spans ranging from nanoseconds to several days, are weak. Therefore thermal fluctuations will have a huge impact on their breakage. To be convinced of it, it is easy to imagine that the pulling force is applied very slowly on the bond. In this case, after a finite time, typically of the order of the life span, the bond will break and the measured force will be close to zero. Thermal fluctuations will have provided the work required for bond breakage. On the other hand, if the pulling force is applied very fast, a finite force will be measured. Of course, experiments conducted twice in similar conditions will not give the same measured forces, again because of the thermal fluctuations. It is therefore incorrect to talk about the force of a bond. The only measurement that can be reported is a distribution of the rupture force under given pulling conditions.

### 3.3. *Connection between the energy landscape and the distribution of the rupture force*

When a single bond breaks, the ligand follows a most likely path. The interaction potential of the two entities making the bond (for instance, the ligand and a membrane), hereinafter called energy landscape, can therefore be considered unidimensional. For example, it can look like the one presented in Fig. 22. The problem of the rupture of a bond is therefore similar to the escape of a particle trapped in the potential and under a random force (due to thermal fluctuations).[24] This is an old problem that was popularized by Kramers in 1940.[25,26] The evolution of the position $\mathbf{r}$ and the speed $\mathbf{v}$ of the particle are given by the Langevin equation:

$$\begin{cases} \frac{d\mathbf{r}}{dt} = \mathbf{v} \\ M\frac{d\mathbf{v}}{dt} = -grad(U(\mathbf{r})) - \zeta\,\mathbf{v} + \xi(t) \end{cases} \tag{33}$$

where $\zeta$ represents the viscosity and $\xi$ verifies:

$$\begin{cases} \langle\xi(t)\rangle = 0 \\ \langle\xi(t)\xi(t')\rangle = 2\zeta k_B T \delta(t - t') \end{cases} \tag{34}$$

To simplify, we shall first assume that the energy landscape has only one barrier. There will then be only one well. The features of this energy landscape are given in Fig. 22. The barrier in $x_b$ often exists for biochemical bonds because of conformational changes during the association of the two entities forming the bond. Let's call $\rho(x, v, t)$, the probability of the states for a bond already formed at t=0. We shall study the temporal evolution of $\rho(x, v, t)$. It will also be assumed that the particle, once it has escaped the well, cannot go back in it. This means that once the bond is broken, it will not reform (hypothesis of an absorbing well). At the initial time, the particle is trapped in the well, therefore:

$$\int_0^{x_b} \int \rho(x, v, 0) dv dx = 1 \tag{35}$$

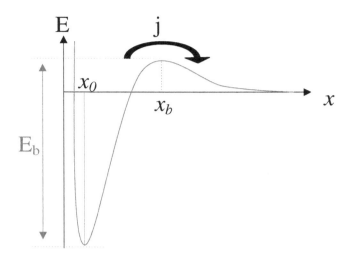

Figure 22.   Energy landscape with a single barrier. j is the initial flow of states from bound to unbound (cf. text).

The system is self similar at any time, normalized by the probability that the particle is still trapped at the time t, $P(t) = \int_0^{x_b} \int \rho(x, v, t) dv dx$. The escape flow from the well will therefore be proportional to $P(t)$. Assuming a constant "rethermalization" of the trapped states, j, the proportionality coefficient appears to be the initial escape flow. $P$ verifies:

$$P(t) = 1 - \int_0^t jP dt. \tag{36}$$

And, finally:

$$P(t) = \exp(-jt). \tag{37}$$

$j$ remains to be determined to fully characterize $P$. $j$ can also be written:

$$j = \int_0^{+\infty} v\rho(x_b, v, 0) dv \tag{38}$$

since the negative $v$ values can be ignored because of the absorbing well hypothesis. The knowledge of $\rho$ at the initial time will thus allow determining $j$. Close to the minimum of the well, the system is thermalized. We can therefore write, in this region:

$$\rho(x, v) = Z^{-1} \exp \left\{ - \left[ \frac{1}{2} Mv^2 + U(x) \right] \middle/ k_B T \right\}. \tag{39}$$

Most of the states are in this region, meaning that:

$$\iint Z^{-1} \exp\left\{-\left[\frac{1}{2}Mv^2 + U(x)\right]/k_BT\right\} dvdx = 1. \qquad (40)$$

Equation 40 allows calculating $Z$. Contrarily, at the level of the barrier $(x \sim x_b)$, the system is not thermalized anymore. It is then possible to use the Langevin Eq. in order to show that $\rho$ verifies the stationary Fokker–Planck equation:

$$\left[-\frac{\partial}{\partial x}v + \frac{\partial}{\partial v}\left[U'(x) + \zeta v\right] + \zeta k_BT\frac{\partial^2}{\partial v^2}\right]\rho(x,v) = 0. \qquad (41)$$

Taking $u = (x - x_b) + av$, one can find a solution in the form: $\rho(x,v) = \lambda(u)\exp\left\{-\left[\frac{1}{2}Mv^2 + U(x)\right]/k_BT\right\}$. With harmonic potentials close to the minimum and to the barrier:

$$\begin{cases} U(x) = \frac{1}{2}\omega_0^2(x - x_0)^2 \\ U(x) = E_b - \frac{1}{2}\omega_b^2(x - x_b)^2 \end{cases} \qquad (42)$$

and, imposing that the solution in $x_b$ has the asymptotic behavior the solution in $x_0$, we find:

$$j = \frac{\left[\frac{\zeta^2}{4} + \omega_b^2\right]^{1/2} - \frac{\zeta}{2}}{\omega_b}\frac{\omega_0}{2\pi}\exp\left(-\frac{E_b}{k_BT}\right). \qquad (43)$$

In the over damped regime ($\omega_b \ll \zeta$) where the characteristic time of the observation is larger than the viscous damping time, this expression simplifies to:

$$j = \frac{\omega_b\omega_0}{2\pi\zeta}\exp\left(-\frac{E_b}{k_BT}\right). \qquad (44)$$

A similar result can be derived from the Smoluchowski equation. This is the classical expression of the escape from a metastable state, used for instance in the kinetics of chemical reactions:

$$\frac{dP}{dt} = -v_0\exp\left(-\frac{E_b}{k_BT}\right)P \qquad (45)$$

with

$$v_0 = \frac{\omega_b\omega_0}{2\pi\zeta} = \frac{1}{\tau_0} \qquad (46)$$

$\tau_0\exp\left(\frac{E_b}{k_BT}\right)$ is the average time spent in the metastable state (called life span for a biochemical bond). It is worth noting that $\tau_0$ depends on the potential and thus depends on the considered bond. The probability p(t)dt that the bond breaks between time $t$ and time $t+dt$ can therefore be written:

$$p(t) = \upsilon_0 \exp\left(-\frac{E_b}{k_B T}\right) \exp\left(-\upsilon_0 \exp\left(-\frac{E_b}{k_B T}\right) t\right) dt. \qquad (47)$$

When a force $f$ is applied to the bond, the potential is tilted by the quantity $-fx$ (cf. Fig. 233.2). In first approximation, it seems reasonable to consider that the relative positions of $x_0$ and $x_b$ do not change, which leads to the life span of the bond under a given force $f$:

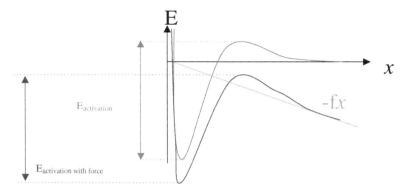

Figure 23. the energy landscape under force is tilted compared to the initial one. The activation energy decreases with the force.

$$\tau(f) = \tau_0 \; exp(E_b/k_B T).exp(-fx_b/k_B T). \qquad (48)$$

In the case of the harmonic approximation mentioned above (Eq. (42)), $\tau_0$ does not change when $f$ varies. As it could have been intuitively expected, the life span of the bond decreases when the force increases. The probability $p(t)$ can then be written:

$$p(t) = \upsilon_0 \exp\left(-\frac{E_b - fx_b}{k_B T}\right) \exp\left(-\upsilon_0 \exp\left(-\frac{E_b - fx_b}{k_B T}\right) t\right). \qquad (49)$$

### 3.4. *Quantification of the rupture of a bond with one barrier under a ramp of force*

Experimentally, it is often difficult to pull on a bond with a constant force. The measurement usually consists in applying an pulling force linearly increasing with time: $f = rt$ ; $r$ will be called the loading rate. Keeping the assumption that the relative position of $x_0$ and $x_b$ do not vary with the force, it is possible to analytically derive the probability distribution of the rupture force as a function of the energy landscape.[24] Indeed, at any time, $P$ verifies an evolution Eq. similar to (45):

$$\frac{dP(t)}{dt} = -\frac{P(t)}{\tau(f)} = -\frac{P(t)}{\tau_0 \exp\left(\frac{E_b}{k_B T}\right)} \exp\left(\frac{f\, x_b}{k_B T}\right) = -\nu\, P(t) \exp\left(\frac{f\, x_b}{k_B T}\right)$$

(50)

that integrates into:

$$\tilde{P}(f) = \exp\left(-\frac{\nu\, k_B\, T}{r\, x_b}\left(\exp\left[\frac{f\, x_b}{k_B\, T}\right] - 1\right)\right).$$

(51)

Thus, the probability density of rupture forces is:

$$\tilde{p}(f) = -\frac{d\tilde{P}}{df} = \frac{\upsilon}{r} \exp\left(\frac{x_b f}{k_B T}\right) \tilde{P}(f)$$

(52)

The knowledge of $x_b$, $E_b$, $\tau_0$ and $r$ allows calculating $\tilde{p}(f)$. The force distribution will vary with $r$. An example is given in Fig. 24. The most likely force and the width of the distribution increase with $r$. The experimental problem is not to find a distribution from a given energy landscape, but to conversely deduce an energy landscape from the rupture force distribution. In the example presented here, the measurement of the most likely force $f_m$ is enough. $f_m$ is obtained by: $\frac{dp}{df}(f_m) = 0$, which gives:

$$f_m = \frac{k_B T}{x_b}\left(\ln\left(\frac{x_b}{k_B T \upsilon}\right) + \ln(r)\right).$$

(53)

From this equation, it is clear that the curve giving $f_m$ as a function of $\ln(r)$ must be linear. The slope will allow determining $x_b$ while the intersection with the y-axis will give $\upsilon$, which represents $E_b$ provided that $\tau_0$ is known. This description has already been successfully used on several systems.

## 3.5. *Quantification of the rupture of a bond with several barriers under a ramp of force*

The previous approach can be generalized in the case of several barriers, in which there are several wells in the energy landscape of the bond. The problem becomes rapidly extremely complex. In this case, taking the example

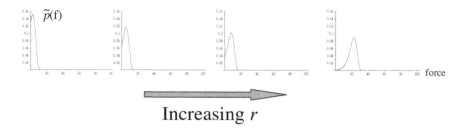

Increasing $r$

Figure 24.   Example of variation of the probability distribution with the loading rate. An arbitrary energy landscape was chosen.

of three wells, Eq. (50) becomes:

$$
\begin{cases}
\dfrac{dP_1}{dt} = -\upsilon_{12}\,(f)\,P_1(t) + \upsilon_{21}\,(f)\,P_2(t) \\[2mm]
\dfrac{dP_2}{dt} = -\,(\upsilon_{21}\,(f) + \upsilon_{23}\,(f))\,P_2(t) + \upsilon_{12}\,(f)\,P_1(t) + \upsilon_{32}\,(f)\,P_3(t) \\[2mm]
\dfrac{dP_3}{dt} = -\,(\upsilon_{32}\,(f) + \upsilon_{34}\,(f))\,P_3(t) + \upsilon_{23}\,(f)\,P_2(t)
\end{cases}
\tag{54}
$$

The subscripts correspond to the different states (cf. Fig. 25), state 4 being the unbound state. To determine the evolution of the system, it is also required to know the initial distribution of states, $P_1(0)$, $P_2(0)$ et $P_3(0)$.

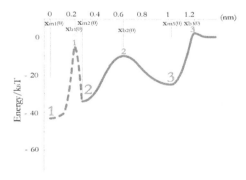

Figure 25.   Example of energy landscape with three barriers.

The transition rates from state i to the state j=i+1 or j=i-1 can be written (cf Eqs. 45 and 46):

$$
\begin{cases}
\nu_{ii+1}\left(f\right) = \left(\dfrac{(\kappa_{mi}\kappa_{bi})^{1/2}}{2\pi\zeta}\right) \exp\left(-\dfrac{E_{bi}(f)-E_{mi}(f)}{k_B T}\right) \\[4mm]
\nu_{ii-1}\left(f\right) = \left(\dfrac{(\kappa_{mi+1}\kappa_{bi})^{1/2}}{2\pi\zeta}\right) \exp\left(-\dfrac{E_{bi}(f) - E_{mi+1}(f)}{k_B T}\right)
\end{cases}
\tag{55}
$$

where $E_{bi}$ is the energy of barrier $i$ and $E_{mi}$, the one of the minimum $i$. $\kappa_m$ and $\kappa_b$ are the curvatures close to, respectively, the minima and the barrier. The parameters necessary to describe the energy landscape are numerous:
   - relative positions of wells and barriers
   - relative heights of wells and barriers
   - curvatures of the potential close to the wells (in $\omega_0$) and barriers (in $\omega_b$).

Therefore, each extra barrier brings six new parameters, meaning that to there are too many parameters to fit and it becomes impossible to deduce energy landscape from rupture force measurements. A way to proceed consists in considering that the highest barrier at any given time will dominate the kinetics (this barrier will depend on the pulling force). In this case, the results presented for a single barrier remain valid and it is expected that, different regimes will be found with the loading rate, each regime corresponding to one of the main barriers of the energy landscape. This approach may work in some cases. It is also possible to use additional data in order to reduce the number of parameters. Molecular dynamics simulations can be useful in this regard. It allows obtaining the positions and heights of the wells and the positions of the barriers with an accuracy sufficient to try fitting the experimental rupture force distribution with the remaining parameters. This approach has been validated on the streptavidin – biotin bond.[12]

One of the important conclusion of this part is that on "human" time scales, say from the order of a second, a bond is not necessarily in its most stable state. This raises the problem of the association and dissociation constants ($k_{on}$ and $k_{off}$) in chemistry that should be defined for each metastable state. Of course, in most bonds, the most stable state will be dominant.

### 3.6. *Breakage of several bonds under a ramp of force*

In this part will be briefly described the case of a adhesive junction containing several bonds rather than a single bond having many energy barriers.[24] A ramp of force is applied to this junction. The same approach as that presented for the single bond will be used. It will be assumed that the $n$ bonds are identical and that once the bond is broken, it cannot reform. As a last assumption, we will only consider the case of bonds with a single barrier in the energy landscape. In the same manner as for the single bond, the Eq. of evolution of the probability to have $n$ bonds, $P_n(t)$, can be written as:

$$dP_n(t)/dt = -\nu_{n->n-1}(t)P_n(t) + \nu_{n+1->n}(t)P_{n+1}(t) \qquad (56)$$

where $\nu$ represents the transition rate from one state to the other. The system of equations is similar to the Eq. (54). Three limiting cases can appear:

i. $n$ strongly correlated bonds in parallel

In this case, it will be assumed that the n bonds are always in exactly the same state and that the force is uniformly distributed between them. This system is therefore equivalent to a single bond of energy $nE_b$ and under a loading rate $r/n$.

Thus, the most likely force is (cf. Eq. 53):

$$f_m = \frac{k_B T}{x_b} \left( \ln\left(\frac{x_b}{k_B T \nu_0}\right) + \ln(r) + \ln(n) + \frac{nE_b}{k_B T} \right) \qquad (57)$$

ii. $n$ independent bonds in parallel.

Let's assume that the force is still uniformly distributed between the $n$ bonds formed at a given time t, but that the bond will break one after the other. Since, to pass from $n$ to $n-1$ bonds, each of the $n$ bonds has the same probability to break, the transition rate can be directly deduced from Eq. (45):

$$\nu_{n->n-1}(f) = n\nu_0 \exp\left(-\frac{E_b}{k_B T}\right) \exp\left(\frac{f x_b}{n k_B T}\right). \qquad (58)$$

It is then possible to show that the most likely force verifies:

$$\frac{r x_b}{\nu_0 k_B T} \exp\left(\frac{E_b}{k_B T}\right) = \left[\sum_{n=1}^{n} \frac{1}{n^2} \exp\left(-\frac{f_m x_b}{n k_B T}\right)\right]^{-1}. \qquad (59)$$

When $n$ is large, it gives:

$$f_m \approx \frac{nk_BT}{x_b}\left(\ln\frac{rx_b}{\nu_0 k_B T} - \ln\frac{f_m x_b}{k_B T} + \frac{E_b}{k_B T}\right) \tag{60}$$

and, if $n$ is small:

$$f_m \approx \frac{nk_BT}{x_b}\left(\ln\frac{rx_b}{\nu_0 k_B T} - 2\ln n + \frac{E_b}{k_B T}\right) \tag{61}$$

iii. $n$ independent bonds in series

In this case, the first bond to rupture will break the structure. The probability of rupture of the one of the $n$ bonds is therefore equal to $n$ times the probability of rupture of one bond:

$$\nu_0(n) = n\nu_0 \exp\left(-\frac{E_b}{k_B T}\right)\exp\left(\frac{fx_b}{k_B T}\right) \tag{62}$$

which gives the most likely rupture force:

$$f_m(n) = \frac{k_B T}{x_b}\left[\ln\left(\frac{rx_b}{\nu_0 k_B T}\right) - \ln(n) + \frac{E_b}{k_b T}\right]. \tag{63}$$

Equations (57), (60), (61) and (63) show that the study of $f_m$ as a function of $n$ allows determining the nature the adhesive junction (bonds in series or in parallel, correlated or uncorrelated).

### 3.7. *Example of technique to measure the rupture force of a single bond: the Biomembrane Force probe*

A well-known technique to quantify the breakage of a single bond is the Atomic Force microscope. Because it is widely spread, it will not be described here. Another technique has recently been developed: the Biomembrane Force Probe.

Its principle is described in Fig. 26. Briefly, the idea is to use a red blood cell as a spring. A bead coated with the studied ligand is chemically attached to the red cell which is micromanipulated with a micropipette. The bead is brought in contact with a substrate containing the counterpart of the ligand (a receptor, a membrane...). Once the bond is formed, a ramp of force is applied until the rupture of the bond. The probability distributions of the rupture force can then be obtained for several loading rates.

The main advantage of this technique over the AFM is that is allows changing the spring constant very easily by varying the aspiration in the micropipette.

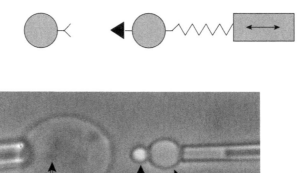

cell      functionalized bead
                              Red blood cell

Figure 26. Principle of the Biomembrane Force Probe. A red blood cell is held by a micropipette (right hand side). A bead functionalized with the ligand is attached to the red cell and approach to a substrate (the cell on the left hand side).

# References

1. B. Alberts (1998), The cell as a collection of protein machines, Cell 92, 291-294

2. "Studying Cell Adhesion", ed. P. Bongrand, P. Claesson and A. Curtis, Springer Verlag, Heidelberg, 1994

3. Molecular Biology of the cell, $4^{th}$ ed. By Alberts, Johnson, Lewis, Raff, Roberts, Walter, 2002, Garland Science Publishing

4. F. Brochard-Wyart, and P.G. de Gennes, *C.R. Physique*, **4**, 281 (2003)

5. K.L. Johnson, K. Kendall, and A.D. Roberts. *Proc. R. Soc. Lon. Ser. A.* **324**, 301 (1971).

6. D. Maugis, J. Colloid Interf.Sci. **150**, 243 (1992). B.D. Hughes, L.R. White, J. Chem. Soc. Farad. Trans. 1 **76**, 963 (1980).

7. P-G. de Gennes, in *Scaling Concept in polymer physics*, Cornell University Press, Ithaca and London, second edition (1985).

8. E. Evans, and D. Needham, Macromolecules **21**, 1822 (1988).

9. Y-S. Chu, S. Dufour, J-P. Thiery, E. Perez, F. Pincet. *Phys. Rev. Lett.* **94**, 028102 (2005).

10. M. S. Turner, and P. Sens. *Biophysical Journal* **76**, 564 (1999).

11. M. Daoudi, E. Lavergne, A. Garin, N. Tarantino, P. Debré, F. Pincet, C. Combadière, P. Deterre, *J. Biol. Chem.* **279**, 19649 (2004).

12. J.N. Israelachvili, D. Tabor, Proc. Roy. Soc. London Series A, **331**, 1584 (1972)

13. Israelachvili J.N., Adams G. E., *J. Chem. Soc. Faraday Trans*. *1* **1978,** 74, 975

14. Derjaguin B.V. Kolloid-Z **69** 155 (1934)

15. E. Evans, Colloid Surfaces **43**, 327-347 (1990)

16. F. Pincet, E. Perez, G. Bryant, L. Lebeau et C. Mioskowski, Physical Review Letters, **73**, 2780-2784 (1994)

17. I. Jr. Tinoco, P.N. Borer, B. Dengler, M.D. Levine, O.C. Uhlenbeck, D.M. Crothers, J. Gralla, Nature (London) **246**, 40-41 (1973)

18. F. Pincet, W. Rawicz, E. Perez, L. Lebeau, C. Mioskowski and E. Evans, Physical Review Letters, **79**, 1949-1952 (1997)

19. R.M. Peitzsch, S. Mc Laughlin, Biochemistry **32**, 10436 (1993)

20. J.F. Joanny, L. Leibler, P.G. de Gennes, J. Polym. Sci. **17**, 1073 (1979)

21. E. Evans, Biophys. J. **48**, 175-183 (1985)

22. G.I. Bell, M. Dembo, P. Bongrand, Biophys. J. **45**, 1051-1064 (1984)

23. F. Pincet, E. Perez, J.C. Loudet, L. Lebeau, Physical Review Letters, **87**, 178101-1-4 (2001)

24. E. Evans, P. Williams. Dynamic force spectroscopy. *In* Physics of Bio-Molecules and Cells. (2002). F. Julicher, P. Ormos, F. David, and H. Flyvb-jerg, editors. Springer Verlag, Berlin, Germany. 145 – 204.

25. P. Hänggi, P. Talkner, M. Borkovec. *Rev. Mod. Phys.* **62**, 251 (1990)

26. H.A.Kramers,. *Physica* (Utrecht) **7**, 284 (1940)

27. F. Pincet, J. Husson. *Biophys. J.* **89**, 4374–4381 (2005)

# Chapter 7

# Interfacial Growth Phenomena

Alain Pocheau

*IRPHE, CNRS & Universités Aix-Marseille I & II,*
*49 rue Joliot-Curie, B.P. 146, Technopôle de Château-Gombert,*
*F-13384 Marseille, Cedex 13, France*
*E-mail: alain.pocheau@irphe.univ-mrs.fr*

Growth by interface motion takes place in many different systems ranging from mesoscopic scales (solidification) to geophysical scales (earth crust formation). Despite this diversity, it displays a number of similar features that refer to a common understanding. They include the segregation of compositions by fractional crystallization, the generation of forms by instabilities, their selection by singular perturbations, the emission of sidebranches on finger-like forms and the occurrence of self-organized states. They then provide interesting implications in geophysics (gulf stream intensity, datation tools, rock segregation), metallurgy (macrosegregation), materials (microsegregation) and biophysics (algae developments, rhizoïd forms). These lecture notes are intended to review them from fundamentals to recent advances.

## Contents

## 1. Introduction

We live in a world where things assemble, develop and stagnate, before dying and recomposing. This is so not only for living entities but also for physical systems, as they are all prone to ultimately return to thermodynamic equilibrium. Even the earth itself is still solidifying from a melt with important consequences regarding the currents that brass its liquid core and the occurrence of earth quakes. On a smaller scale, all the solid metals we use in everyday life have grown out from a melt,[1] the only exception being those created since the last century by vapor deposition. Accordingly, almost all the objects, living or inert, natural or man-made, that surround us are the end result of growth phenomena.

Usually, growth phenomena involve interface motion. This is true for solidification processes since a definite interface separates the liquid phase from the solid phase. This also so, on a larger ground, for all systems undergoing phase transition (magnetic systems, liquid crystals), reaction (combustion), non-miscible fluid motion (viscous digitation, wetting/dewetting) or deposition (electrodeposition, sedimentation), since all show an interface between media whose motion determines the evolution of the whole. In addition, the concept of interfacial growth also extends to living organisms from unicellular algae and neurons to meristems or bacterial colonies. It will thus be the central issue of this course.

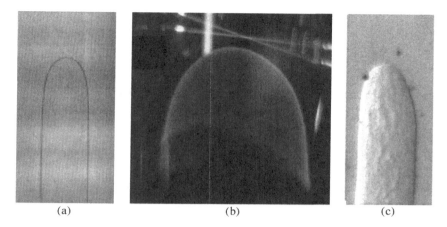

(a)                                    (b)                                    (c)

Figure 1. Digit growing shapes in viscous digitation (a), combustion (b) and on fungi (c). (a) Fluid interface between air pushed in oil in an Hele-Shaw cell. (b) Premixed flame propagating upward. Courtesy of J. Quinard and G. Searby. (c) Rhizoïd shape displayed during the development of the Rhizoctonia Solani fungus[89]. Reprinted figure with permission from S. Bartnicki-Garcia, C.E. Bracker, G. Gierz, R. López-Franco and H. Lu, *Biophysical Journal* **79**, 2382-2390 (2000). Copyright (2000) by the Biophysical Journal.

Interestingly, in all the above systems, the growth stage imprints the final state of the system regarding its composition, its structure or its shape. It thus stands as a dynamical mean for selecting the final steady state, its main features and its main functional capacities. Beyond a fundamental or applied interest for the growth stage, getting insight to its proper mechanisms will thus prove to be essential for understanding or controlling the nature of the resulting states.

Current observations of the end-result of growth reveals a number of striking analogies between radically different systems. Interfaces show similar digit shapes in viscous digitation, in combustion or on fungi (Fig. 1). The emission of sidebranches on a digit or needle forms looks surprisingly similar in solidification, electrodeposition, dielectric breakdown, neuronal arborization or meristem formation and may even give rise to multi-scale branched forms in vegetals (Fig. 2). The puzzling persistence of compositional heterogeneities in stirred systems is encountered at the mesoscopic scale in the microsegregation of solids, at the macroscopic scale in the macrosegregation of cast ingots, at the geological scale in the variety of rocks found in solidified volcano lava and even at the planetary scale in the picked distribution of minerals on the earth crust (Fig. 3). All this points

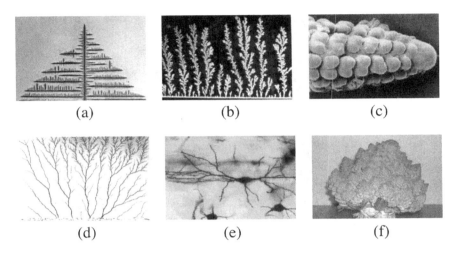

Figure 2. Branched growing forms in solidification (a), electrodeposition (b), dielectric breakdown (d), neurones (e) and vegetal formation (c,f). (a) Solidification dendrites formed in an ammonium-bromide mixture[93]. Reprinted figure with permission from Y. Couder, J. Maurer, R. González-Cinca and A. Hernández-Machado, *Phys. Rev. E* **71**, 031602 (2005). Copyright (2005) by the American Physical Society. (b) Electrodeposition pattern of copper involving dendrites[94]. Reprinted figure with permission from P.P. Trigueros, F. Sagués and J. Claret, *Phys.Rev.E* **49**, 4328-4335 (1994). Copyright (1994) by the American Physical Society. (c) Maize tassel displaying iterative meristem formations. (d) Dielectric breakdown. Courtesy of Professor K. Brecher, Boston University, Massachusetts, USA. (e) Neurone of a rat showing dendrites and axons. Courtesy of Dr. C. Isgor, Florida Atlantic University, Florida, USA. (f) Romanesco cauliflower showing multi-scale similar protuberances.

to universal mechanisms acting similarly on growth interfaces and whose study could help us understand in a unified framework the generation of forms and of inhomogeneities in so largely different systems.

The objective of this course is to report the current modeling of interfacial growth phenomena and to stress its implications on issues ranging from materials to geophysics and biophysics. The method applied will consist first in pointing out fundamental mechanisms on simple physical systems whose modeling is firmly set, and then in extrapolating them to more global or more complex systems so as to gain relevant insights on some of their main features. Two physical systems will serve us as a guide in that: viscous digitation[2-4] and, mainly, solidification.[5-7] Insights in geophysical issues and biophysical issues will be given regarding rock datation, magma segregation, thermohaline circulation, algae lobe formation and rhizoïd shape.

(a)          (b)          (c)

Figure 3. Segregation of solid composition at a mesoscopic scale (a), at a macroscopic scale (b) and at a geological scale (c). (a) Longitudinal section of a directionally solidified $Al - 3wt\%Cu$ alloy[8]. Reprinted figure from *Materials Science and Engineering* **A327**, M. Gündünz and R. Çadirli, Directional solidification of aluminium-copper alloys, 167-185, Copyright (2002), with permission from Elsevier. (b) Cut of an Aluminium ingot, 5cm wide. Courtesy of Professor H. Badeshia, University of Cambridge, UK. (c) Solidified lava lake of the Halemaunau crater located within the Kilauea Caldera in the Hawaii Volcanoes National Park.

Section 2 reports fundamentals on the modeling and the thermodynamics of growth interfaces. Growth without shape implications is detailed in section 3 on planar interfaces and geophysical applications are drawn on section 4. This is followed in section 5 by a review of the generation of forms by planar growth instability. Steady growth forms are then studied in section 6 from a theoretical and experimental point of view, with emphasis on selection issues. The phenomenon of sidebranching is then addressed in section 7 together with the fundamental alternative regarding its nature: noise amplifier or non-linear oscillator? This opens the way for reporting in section 8 two examples of non-linear growth dynamics in solidification: a limit cycle induced by a Hopf bifurcation and a self-organized dynamics on a large scale curved interface. The next section then addresses geometrical models of growth with applications to biophysical interfaces: the formation of lobes on the alga Micrasterias and the tubular form of steadily growing rhizoïds. A conclusion on this guided tour ends the course.

## 2. Fundamentals

### 2.1. *Types of interfaces and resulting modeling*

Interfaces separate two media that differ by at least the value of an essential variable. Their thickness is usually small enough compared to the characteristic scales of both media for allowing them to be considered as infinitely thin. Doing that, an important distinction has to be made regarding the physics that governs the interface and its surrounding media. As shown below, it yields interfaces to be classified in two main types, a transition sheet and a separation zone, with important implications regarding their modeling.

Transition sheets correspond to the kind of interfaces that appear in systems that are homogeneous regarding modeling. They reduce to a sharp zone of variation of variables or of order parameters with no need of additional specific modeling (Fig. 4-a). In particular, no specific frame needs to be specified to model their field evolution and the resulting jump of variables across them. These kinds of interfaces are mainly encountered in physics, for instance in media referring to different thermodynamic phases (solidification, combustion) or to non-miscible fluids (digitation). They will stand as the main concern of this course (Sect. 3 to 8).

However, some interfaces may refer to a material of a different nature than those involved in the surrounding media (Fig. 4-b). This is so for instance for solid or elastic interfaces separating fluid media or for membranes or cell walls displayed in biophysical or biological systems (Sect. 9). Then, a specific modeling of the interfacial domain is required together with an explicit specification of the frame in which it has to be applied. We call these kinds of interfaces separation zones.

As shown below (Sect. 2.2), solving for the interface dynamics will prove to reduce to determining the interface normal growing velocity. For this, the dynamics of each medium surrounding the interface will be required. In the limit of infinitely thin interfaces, they will be complemented, for transition sheets, by jump relations across the interface and by boundary value relations on one of its side (Sect. 2.3) and, for separation zones, by specific dynamical laws in an explicit frame (Sect. 9).

## 2.2. *Interface kinematics and normal velocity*

Consider a dynamical curve $\mathcal{C}(t)$ representing an interface in a plane within a given parametrization: $\mathcal{C}(t) = \{M(t,p), p \in \Re\}$. Change of parametrization $p \to p' = p'(p)$ consists in making the curve glide on itself while keeping the interface unchanged $M(t, p = s) \to M'(t, p' = s)$, i.e. $p = s \to p = p'^{-1}(s)$ (Fig. 5-a). This therefore corresponds to an internal symmetry of the interfacial system with respect to which all the relevant features of the interface must be invariant.

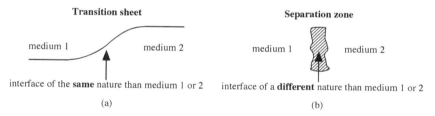

**Transition sheet**  **Separation zone**

medium 1    medium 2    medium 1    medium 2

interface of the **same** nature than medium 1 or 2    interface of a **different** nature than medium 1 or 2

(a)    (b)

Figure 4. Types of interfaces. (a) Transition sheets: the interface corresponds to a sharp change of variables or of order parameters but is of the same nature than the surrounding media. The curve symbolizes the change of variable amplitude from medium 1 to medium 2. (b) Separation zone: the interface corresponds to a medium of a different nature than the surrounding media. Whereas the usual modeling of continuous media can apply to transition sheets, a specific modeling is required for separation zones.

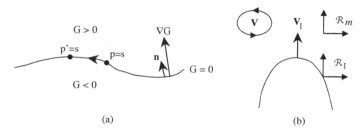

(a)    (b)

Figure 5. Interface kinematics. (a) Change of parametrization corresponds to making the curve glide on itself $M \rightarrow M'$ for $p = s \rightarrow p' = s$. Modeling must be invariant with respect to this symmetry mode. The interface may be viewed as a separation line between a $G > 0$ and a $G < 0$ domain. It involves a kinematics that only depends on its normal velocity $\mathbf{V}.\mathbf{n}$ where $\mathbf{n} = \nabla G/|\nabla G|$. (b) Frames used to model the interface dynamics: the barycentric frame $\mathcal{R}_m$ of the medium in which the interface advances and the reference frame $\mathcal{R}_I$ which co-moves with the interface at its mean growing velocity $V_I$. The velocity field $\mathbf{V}$ denotes the internal flows of the medium.

To express the curve dynamics, it is convenient to introduce a field $G(M,t)$ with respect to which $\mathcal{C}$ stands as an iso-G: $\mathcal{C}(t) = \{M; G(M,t) = 0\}$. The dynamics of $\mathcal{C}$ may then be deduced from that of $G$ by:

$$\frac{dG}{dt} = \frac{\partial G}{\partial t} + \mathbf{V}.\nabla G = 0 \ ; \ \nabla G // \mathbf{n} \tag{1}$$

where $\mathbf{V} = dM/dt$ and $\mathbf{n} = \nabla G/|\nabla G|$ denotes the normal to the curve.

Relation (1) shows that the dynamics of the curve, or equivalently that of $G$, only specifies the normal velocity $\mathbf{V}_n = (\mathbf{V}.\mathbf{n})\mathbf{n}$. In particular, the remaining tangential velocity $\mathbf{V}_t = \mathbf{V} - \mathbf{V}_n$ only makes the curve glide onto itself with no implication for the curve dynamics. This is confirmed by the fact that gliding is simply equivalent to a reparametrization of the curve (Fig. 5-a). By contrast, the normal velocity determines the rate of change of $G$ and thus the curve dynamics. It is thus independent of the parametrization or, in other words, gauge invariant.

The objective of modelings will thus be to determine interface normal velocities. This will be obtained on transition sheets by expressing the field dynamics (Sect. 2.3.1), the jump relations at the front (Sect. 2.3.2) and the specific conditions on the interface (Sect. 2.3.3). On separation zones, however, this will require specific modelings (Sect. 9).

## 2.3. Modeling interface dynamics

### 2.3.1. Field dynamics

In a reference frame $\mathcal{R}$, the diffusion equation of a scalar $x$ writes:

$$\frac{Dx}{Dt} = \frac{\partial x}{\partial t} + \mathbf{V}_{\mathcal{R}}.\nabla x = D\nabla^2 x \tag{2}$$

where $\mathbf{V}_{\mathcal{R}}$ denotes the velocity of the medium with respect to the frame $\mathcal{R}$ and $D$ the diffusivity of the scalar $x$ in this medium. In particular, if the frame corresponds to the barycentric frame $\mathcal{R}_m$ of the medium, $\mathbf{V}_{\mathcal{R}}$ simply corresponds to the velocity field $\mathbf{V}$ of its internal flows. However, if it is taken as the frame $\mathcal{R}_I$ co-moving with an interface at its mean growing velocity, $\mathbf{V}_{\mathcal{R}}$ then writes $\mathbf{V} - \mathbf{V}_I$ where $\mathbf{V}_I$ denotes the mean interface velocity in $\mathcal{R}_m$ (Fig. 5-b). One or the other frame $\mathcal{R}_m$ or $\mathcal{R}_I$ will be used in the following depending on convenience.

Application of this to specific interfaces yields:

• Solidification interface

In a pure medium, solidification is simply mediated by the temperature field. In frame $\mathcal{R}_m$, its diffusion equation writes, in a liquid at rest:

$$\frac{\partial T}{\partial t} = D_T\nabla^2 T \tag{3}$$

where $D_T$ denotes the thermal diffusivity. The same kind of dynamics also applies in the solid phase with a different diffusivity.

In a binary mixture, solidification depends on two scalar fields, the thermal field $x_1 \equiv T$ and the solute concentration field $x_2 = c$. Their dynamics write in a medium at rest:

$$\frac{\partial T}{\partial t} = D_T\nabla^2 T \ ; \ \frac{\partial c}{\partial t} = D_c\nabla^2 T \tag{4}$$

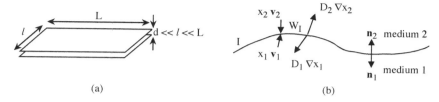

Figure 6. (a) Sketch of a Hele-Shaw cell in which a viscous fluid interface is formed between non-miscible fluids. (b) Sketch of the volumic advective, $x_i \mathbf{v}_i$, or diffusive, $-D_i \nabla x_i$, fluxes of a scalar $X$ and of its surfacic production rate $W_I$ at an interface. Conservation of scalar $X$ yields the jump relation (5) at the interface.

where $D_T$ and $D_c$ denote the thermal or solute diffusivities of the liquid phase or of the solid phase that is considered. It appears however that the thermal diffusivity is seemingly the same in both phases and that it is far larger than the solute diffusivity in any phase. Expressing diffusivities in $\text{cm}^2\,\text{s}^{-1}$, one notices for instance that, in aluminium, $D_T$ amounts to $3.7 \times 10^{-1}$ (resp. $7 \times 10^{-1}$) in the liquid (resp. solid) phase whereas the corresponding values of solute diffusivities are typically of order $3 \times 10^{-5}$ (resp. $3 \times 10^{-9}$). Also, in a plastic crystal, the succinonitrile SCN, thermal diffusivity is $1.1 \times 10^{-3}$ in both phases whereas solute diffusivities are only of order $1.3 \times 10^{-5}$ in the liquid phase and $1 \times 10^{-9}$ in the solid phase. Accordingly, the thermal field will reach equilibrium far more rapidly than the concentration field, so that solidification dynamics will be controlled by the solute field. In particular, the thermal field could be considered as quasi-steady in the remainder. In addition, solute diffusivity in the solid phase could be neglected in many instances.

- Viscous fluid interface

We consider the interface separating two viscous non-miscible fluids in a cell whose depth $d$ is small compared to its width and length (Fig. 6-a). In this Hele-Shaw cell, the dissipation scale is set by the small dimension $d$. Then, in the creeping flow approximation, the velocity field of either fluid satisfies $\nabla p \approx \mu \partial^2 \mathbf{V}/\partial z^2$ where $\mu$ denotes the fluid dynamical viscosity, $p$ the pressure field and $z$ the depth direction. The averaged flow velocity $< \mathbf{V} >$ over the channel depth then follows a Darcy law, $< \mathbf{V} >= -K\nabla p$ where $K = d^2/(12\mu)$. In particular, as the flow is divergent-free, the pressure field follows a Laplace equation: $\nabla^2 p = 0$. This still corresponds to a diffusion equation for a scalar field but in an infinite diffusivity limit.

### 2.3.2. *Jump relations at the front*

Jump relations are obtained by expressing the conservation of a scalar $X$ across the interface. For this, it is convenient to index by $I$ the interface and by $i = 1, 2$ the variables of the media standing on each side of $I$. Calling $x_i$ the volumic density of scalar $X$ and $\mathbf{v}_i$ the relative velocity of the medium with respect to $I$, the volumic flux $\Phi_i$ of scalar $X$ reads $\Phi_i = x_i \mathbf{v}_i - D_i \nabla x_i$. Conservation of scalar $X$ in a control volume co-moving with the interface then writes $W_I = [\Phi_1 \mathbf{n}_1 + \Phi_2 \mathbf{n}_2]_I$ where $\mathbf{n}_i$ denotes the interface normal pointing toward medium $i$ and $W_I$ the surfacic production rate of scalar $X$ at the interface (Fig. 6-b). As $\mathbf{n}_1 = -\mathbf{n}_2$, this yields the jump relation at the interface:

$$W_I = [x_2 \mathbf{v}_2 - x_1 \mathbf{v}_1]_I.\mathbf{n}_2 - [D_2 \nabla x_2 - D_1 \nabla x_1]_I.\mathbf{n}_2. \tag{5}$$

Interestingly, this relation provides a way of determining the normal velocity $\mathbf{v}_2.\mathbf{n}_2$ of the interface with respect to medium 2 from the values of the surface production rate, of the scalar densities $x_i$ and of their gradients on both parts of the interface. Expressing it in solidification or viscous fingering yields:

• Solidification interface

Usually, the difference between the volumic masses of the media is weak enough for being neglected: $\rho_2 \approx \rho_1$. One then obtains:

Mass conservation: $X = m$, $x_i = \rho_i$, $\nabla x_i = \mathbf{0}$, $W = 0$, so that $\rho_2 \mathbf{v}_2.\mathbf{n}_2 = \rho_1 \mathbf{v}_1.\mathbf{n}_1$ and, with $\rho_2 = \rho_1$, $\mathbf{v}_2.\mathbf{n}_2 = \mathbf{v}_1.\mathbf{n}_1$.

Conservation of enthalpy $H$: $X = H$, $x_i = h_i$ where $h_i = c_{p,i} T_i$ denotes the volumic enthalpy and $c_{p,i}$ the volumic specific heats. Calling $Q$ the volumic latent heat $(h_2 - h_1)_I$, $\mathbf{V}_I = -\mathbf{v}_2$ the interface velocity with respect to medium 2 and $\mathbf{n} = \mathbf{n}_2$ the interface normal, relation (5) reads, with $W_I = Q\mathbf{V}_I.\mathbf{n}$:

$$Q\mathbf{V}_I.\mathbf{n} = -[c_{p,2} D_{T,2} \nabla T_2 - c_{p,1} D_{T,1} \nabla T_1]_I.\mathbf{n}. \tag{6}$$

• Viscous fluid interface

Impenetrability of the interface between non miscible fluids forbids any advective flux across it in its proper reference frame: $(< \mathbf{V}_i >_I - \mathbf{V}_I).\mathbf{n} = 0$. This, together with the Darcy law (Sect. 2.3.1) determines its normal velocity:

$$\mathbf{V}_I.\mathbf{n} = < \mathbf{V}_i >_I .\mathbf{n} = -(K_i \nabla p_i)_I. \tag{7}$$

### 2.3.3. *Thermodynamic conditions at the front*

Consider, for an isotherm system, the thermodynamic potential $G_i$ of each phase: $G_i = U_i + p_i V_i - T S_i$ where $U_i$ denotes the internal energy, $V$ the volume and $S_i$ the entropy of the medium. The net free enthalpy of the system then writes $G = G_1 + G_2 + \gamma S$ where $S$ denotes the area of the interface between the two media and $\gamma$ the surface tension.

The variation of $G$ for elementary exchanges of volume $\delta V_1$ and of particle number $\delta N_1$ and for an elementary change of interface area $\delta A$ writes:

$$\delta G = (p_2 - p_1)\delta V_1 + (\mu_2 - \mu_1)\delta N_1 + \gamma \delta A \tag{8}$$

where $\mu$ denotes the particle chemical potential, i.e. the free enthalpy per particle.

Here, the whole system is in an out-of-equilibrium state, since it involves a moving interface. However, in many instance, the interface velocity is low enough for making the kinetic time of transfer of molecules from one phase to the other very short compared to the relevant diffusion times. This means that thermalization of thermodynamic quantities is completed at the interface, despite its motion and the existence of gradients. Then, the interface can be assumed to be in thermodynamic equilibrium, despite the whole system is not, so that the net free enthalpy $G$ must be at a minimum. The intensive variables $p_i$, $\mu_i$ and $\gamma$ must then be such that the variation $\delta G$ is of higher order than the primary variations $\delta V_1$, $\delta N_1$, $\delta A$. We shall express this by considering first an elementary displacement $\delta \xi$ of the interface and then an elementary exchange $\delta N_1$ of constituents.

• Interface displacement $\delta \xi$

To determine the resulting change $\delta dS$ of an elementary surface $dS$, we first consider the change $\delta dl$ of arc length $dl$ induced on a circle by a change $\delta R$ of its radius $R$. By similarity, we have: $\delta dl/dl = \delta R/R$. The relative change $\delta dS/dS$ of the elementary surface $dS$ then simply expresses as the sum of the relative changes of length $dl_1$, $dl_2$, on its principal axes: $\delta dS/dS = \delta dl_1/dl_1 + \delta dl_2/dl_2$ (Fig. 7-a). Turning back to the two-dimensional case of a circle with $\delta R \equiv \delta \xi$, one finally obtains $\delta dS/dS = \kappa \delta \xi$ where $\kappa$ denotes the total curvature $\kappa = (1/R_1 + 1/R_2)$ and $R_1$, $R_2$ the principal radii of curvature of the surface. As $\delta V_1 = \delta \xi dS$ for $\delta \xi$ defined positive when increasing the medium 1, one finally obtains $\delta dS = \kappa \delta V_1$.

This, together with (8), yields $\delta G = (p_2 - p_1 + \gamma \kappa)\delta V_1$ and, finally, at

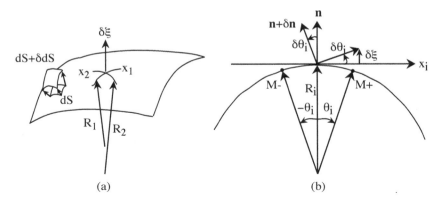

(a)                                                      (b)

Figure 7. Determination of the variations of free enthalpy implied by surface tension (a) and anisotropy (b) upon an interface displacement $\delta\xi$. (a) The variation $\delta S$ of elementary surface $dS$ is proportional to $\delta\xi dS$ and to the mean curvature $\kappa = R_1^{-1} + R_2^{-1}$. (b) A rotation of angle $\delta\theta_i$ of the normal $\mathbf{n}$ to the interface corresponds to a linear shift $\delta\xi(x_i)$ in the normal direction: $\delta\theta_i = \partial\delta\xi/\partial x_i$. This, together with $x_i \approx R_i\theta_i$, enables an integration by part of the additional variation brought about by the anisotropy of surface tension to the free energy. The net variation involves the second derivative $\partial^2\gamma/\partial\theta_i^2$ of surface tension as a result of the antisymmetry of the first order corrections: $\gamma(M\pm) = \gamma \pm \theta_i\partial\gamma/\partial\theta_i + (\theta_i^2/2)\,\partial^2\gamma/\partial\theta_i^2 + o(\theta_i^2)$.

equilibrium $\delta G = 0$, the Laplace law:

$$p_1 - p_2 = \gamma\kappa \tag{9}$$

• Exchange of constituents between the thermodynamic phases of a pure substance

At thermodynamic equilibrium, minimization of $G$ with respect to an exchange of constituents $\delta N_1$ between phases (8) yields equality of their chemical potential: $\mu_1(p_1, T) = \mu_2(p_2, T)$. It is convenient to develop this relationship around a common isobaric and isotherm equilibrium state of reference. Calling $(p_M, T_M)$ its pressure and temperature and denoting $v = \partial\mu/\partial p$ and $s = -\partial\mu/\partial T$ the specific volume and specific entropy of each phase, we obtain:

$$(p_1 - p_M)v_1 - (T - T_M)s_1 = (p_2 - p_M)v_2 - (T - T_M)s_2. \tag{10}$$

Neglecting the inhomogeneities of specific volume, $v_1 = v_2$, relation (10) provides the following change of equilibrium temperature:

$$T - T_M = v_2 \frac{p_2 - p_1}{s_2 - s_1} \tag{11}$$

• Gibbs–Thomson relationship; critical seed

Relation (11), together with the Laplace law (9), yields the relative reduction of equilibrium temperature by curvature. It is called the Gibbs–Thomson relationship:

$$\frac{T - T_M}{T_M} = -\frac{\gamma}{Q}\kappa \qquad (12)$$

where $Q = T_M(s_2 - s_1)/v_2$ denotes the volumic latent heat.

An important implication of the Gibbs–Thomson relationship addresses the early stages of nucleation in an undercooled medium of temperature $T_\infty < T_M$. It actually states that the energetic cost required to create highly curved interfaces is prohibitive since their equilibrium temperature $T$ may be reduced even below the medium temperature $T_\infty$ if their curvature radius $R$ is below a critical value $R_c = (2\gamma/Q)\, T_M/(T_M - T_\infty)$: $R < R_c \Longrightarrow T < T_\infty < T_M$. The corresponding germs are then superheated and recede. On the opposite, large enough germs involving a curvature radius larger than $R_c$ will keep being undercooled, $T_\infty < T < T_M$, and will therefore expand.

Interestingly, these surface tension considerations introduce a characteristic length scale in the system: the capillary length $d_0 = \gamma/Q\, (c_p T_M/Q)$. In particular, calling $\Delta$ the non-dimensional undercooling $\Delta = c_p(T_M - T_\infty)/Q$, the critical radius $R_c$ simply reads: $R_c = 2d_0/\Delta$.

Typical values of capillary lengths are of order of a few Ångströms: $d_0 = 2.5$ Å in aluminium, 3.7 Å in copper and 28 Å in succinonitrile. Their smallness indicates that no phenomenological interpretation has to be looked for. Instead, $d_0$ simply serves as a scale for evaluating in terms of critical radius $R_c$ whether an undercooling $\Delta$ is large or small.

• Exchange of constituents between thermodynamic phases in a binary mixture

In a binary mixture involving an interface between two thermodynamic phases, exchanges of constituents cannot occur between solvent and solute but, for each of them, between their respective thermodynamic phases. Indexing the solvent by superscript S and the solute by superscript s, one therefore has to minimize $G$ with respect to variations $\delta N_1^S$ and $\delta N_1^s$ of the number of constituents of each species in phase 1. One then obtains the equality of chemical potentials of solvent and solute in each phases, $\mu_1^S = \mu_2^S$, $\mu_1^s = \mu_2^s$, each of them being a function of pressure $p_i$, temperature $T$ and solute concentration $c_i$. In particular, for ideal solutions or for weak concentrations, they express as:

$$\mu_i^S(p_i, T, c_i) = \mu_i^S(p_i, T) - kTc_i \; ; \; \mu_i^s(p_i, T, c_i) = \mu_i^s(p_i, T) + kT\ln(c_i). \tag{13}$$

Accordingly, the additional dependence on solute concentration shifts the equilibrium conditions by an amount $kT\ln(c_1/c_2)$ for the solute and $kT(c_2 - c_1)$ for the solvent.

The former shift implies that the ratio of solute concentration in phases 1 and 2 is a constant, up to negligible surface tension effects:

$$c_1 \approx Kc_2 \; ; \; K = \exp[\frac{\mu_2^s - \mu_1^s}{kT_M}]. \tag{14}$$

Here, the chemical potentials are taken in the reference state $(p_M, T_M)$ and $K$ is called either the partition ratio or the segregation coefficient.

The shift of the solvent equilibrium condition brings about an additional, solute dependent, term in the Gibbs–Thomson relation (12): $(kT_M^2/Qv)(c_1 - c_2)$. Using the equilibrium condition for the solute, $c_1 \approx Kc_2$, one obtains a correction proportional to the solute concentration $c_2$:

$$T - T_M = -mc - \frac{Q}{c_p}d_0\kappa \tag{15}$$

where $c \equiv c_2$ denotes from now on the solute concentration in the liquid phase and where $m = (kT_M^2/Qv)(1 - K)$ stands for the opposite of the liquidus slope. We note that it is positive for $K < 1$, as usually satisfied in dilute mixtures. The simplicity of the corresponding phase diagram (Fig. 8-a) contrasts with that displayed for large concentration (Fig. 9-a) or for ternary mixtures (Fig. 9-b).

• Anisotropy of surface tension

As a vestige of molecular structures, the surface tension of an interface, even rough, slightly varies with its orientation: $\gamma \equiv \gamma(\mathbf{n})$ where $\mathbf{n}$ denotes the interface normal. For instance, for a cubic crystal, anisotropy can be modelled by $\gamma = \gamma_0[1 + \epsilon\cos(4\theta)]$ where the angle $\theta$ states the interface orientation and where $\epsilon$ is of order of a few percents. This anisotropy implies taking into account the variations of $G$ induced by elementary rotations of angle $\delta\theta$ of the interface. This brings about an additional term $\int \delta\gamma dS$ in (8) where the integral is performed over the interface.

Denoting $(x_1, x_2)$ the principal axis of the surface at a point $M$, $(\theta_1, \theta_2)$ the angles of the surface normal at this point and $(R_1, R_2)$ the principal curvature radii (Fig. 7-a), the anisotropy contribution to the variation of $G$ reads: $\int[\partial\gamma/\partial\theta_1 \, \delta\theta_1 + \partial\gamma/\partial\theta_2 \, \delta\theta_2]dS$. Here $\delta\theta_i$ results from additional

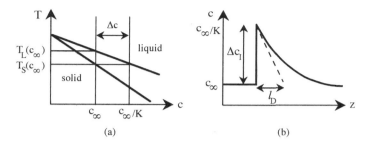

(a)                                    (b)

Figure 8. Thermodynamic phase diagram of binary mixtures at low solute concentration. (a) Binary mixture with small solute concentration $c$. The concentration $c_\infty$ denotes the concentration of the homogeneous mixture. The concentration step $\Delta c$ is called the miscibility gap. The temperature of the liquidus and of the solidus are denoted $T_L$ and $T_S$ respectively. (b) Solute concentration field for a steadily growing planar front (Sect. 3.2.2). A step $\Delta c_I$ is involved at the interface. It is equal here to the miscibility gap $\Delta c$ of the mixture. It is followed by an exponential relaxation in the liquid phase toward the concentration $c_\infty$ on a relaxation length equal to the diffusion length $l_D = D/V$ (27).

interface distortion: $\delta\theta_i = \partial\delta\xi/\partial x_i$ where $\xi$ denotes the coordinate of the normal axis to the surface (Fig. 7-b). This allows integrating by part on the surface since $\delta\gamma = \partial/\partial x_i \, [\delta\xi_i\partial\gamma/\partial\theta_i] - \delta\xi_i \, \partial^2\gamma/\partial x_i\partial\theta_i$. The bracket term only gives boundary contributions that are dominated by the remaining bulk term. The bulk term provides a contribution proportional to $\delta\xi dS$, i.e. $\delta V$, which writes $1/R_i\partial^2\gamma/\partial\theta_i^2$ since, by definition, $\partial/\partial x_i = 1/R_i\partial/\partial\theta_i$. This additional pressure term complements the contribution of surface tension to the relative shift of equilibrium temperature:

$$T - T_M = -mc - \frac{T_M}{Q}[\frac{1}{R_1}(\gamma + \frac{\partial^2\gamma}{\partial\theta_1^2}) + \frac{1}{R_2}(\gamma + \frac{\partial^2\gamma}{\partial\theta_2^2})]. \qquad (16)$$

The absence of first derivative of $\gamma$ in (16) and the presence of a second derivative may be simply recovered by considering a spherical interface (Fig. 7-b). Then, on any curved arc passing through a point $M$, first order corrections to $\gamma$ are, as the deviations $\delta\theta$, antisymmetric with respect to point $M$. They thus yield no total contribution. On the opposite, second order corrections are symmetric. They then provide a net contribution when integrated.

It should be noticed that, in crystals, the contribution of the second derivative of surface tension is actually the dominant contribution regarding anisotropy. This follows from the fact that, for a surface tension varying periodically with the orientation of the interface, e.g. $\gamma = \gamma_0[1+\epsilon\cos(4\theta)]$, the net surface tension coefficient in the Gibbs–Thomson relation (16) writes

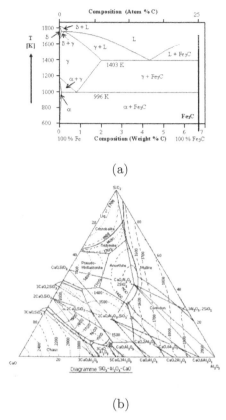

(a)

(b)

Figure 9.   Thermodynamic phase diagram of mixtures. (a) Binary mixture. Alloy Fe-C showing an eutectic point and different solid phases. (b) Ternary mixture. The various phases of the $SiO_2 - Al_2O_3 - CaO$ system are represented in a three axes diagram. Temperatures are given in Celsius degrees.

$\gamma + \partial^2 \gamma / \partial \theta_i^2 = \gamma_0 [1 - 15\epsilon \cos(4\theta_i)]$. In particular, the net relative anisotropy strength, $15\epsilon$, can then be large, even if $\epsilon$ is not.

## 3.  Growth of planar interfaces

We restrict attention here to the growth of interfaces that remain planar. This specificity allows us to avoid the implications that will be driven in the next sections by the deformations of interface geometry. In turn, it enables the fundamental structure of fields around growing interfaces to be highlighted.

## 3.1. *Solidification of a pure liquid*

The solidification of a pure liquid is thermally controlled by the evacuation by diffusion of the latent heat generated by the phase transition.

### 3.1.1. *Steady planar growth*

Consider a planar solidification interface propagating in an infinite undercooled medium. We wish to identify the conditions for which it can propagate at constant velocity and then determine the expression of the resulting velocity. Without loss of generality, we may reduce this study to a one-dimensional problem on the $z$-axis normal to the interface. Taking, for simplicity, the same physical constants in either phases, the dynamical equation for the thermal field writes in the liquid frame:

$$\frac{\partial T}{\partial t} = D_T \frac{\partial^2 T}{\partial z^2} + \frac{Q}{c_p} V_I \, \delta(z - z_I) \tag{17}$$

with boundary conditions: $T(-\infty) = T_M$, $T(+\infty) = T_L < T_M$. Here $\delta(.)$ denotes the delta-function, $T_L$ the melt temperature and $T_M$ the melting temperature.

For a front propagating at velocity $V_I$, $T(z,t) = T(z - V_I t) = T(\xi)$ with $\xi = z - V_I t$. Equation (17) then reduces to:

$$-V_I \frac{\partial T}{\partial \xi} = D_T \frac{\partial^2 T}{\partial \xi^2} + \frac{Q}{c_p} V_I \, \delta(\xi - \xi_I) \tag{18}$$

where $\xi_I = z_I - V_I t$. As temperature gradients vanish at infinity, integration from $-\infty$ to $+\infty$ yields: $V_I(T_M - T_L) = V_I Q / c_p$. Steady planar growth thus requires $\Delta = c_p(T_M - T_L)/Q = 1$ (resp. $S = \Delta^{-1} = 1$) where $\Delta$ (resp. $S$) denotes the mixture undercooling (resp. the mixture Stefan number). However, this condition being fulfilled, the interface velocity $V_I$ remains indeterminate.

The criterion $\Delta = 1$, $S = 1$, simply corresponds to energy conservation since the volumic latent heat $Q$ is fully used to warm the medium from the liquid temperature $T_L$ to the melting temperature $T_M$. Unusually, the solidification process is thus not limited here by diffusion since none is required to regulate heat exchange with outer systems. Thermal diffusivity $D_T$ then only serves to determine the temperature profile without any dynamical implication. This results in a lack of length and time scales to set the interface velocity and hence, to its indeterminacy.

### 3.1.2. *Solidification at a cold boundary*

To remove the above indeterminacy, we introduce a boundary that is maintained at a fixed temperature $T_B$ by absorption of thermal fluxes. Two situations are considered regarding the far liquid temperature $T_L$ and the boundary temperature $T_B$: an undercooled boundary $T_B < T_L = T_M$ or an undercooled liquid $T_L < T_B = T_M$.

- Undercooled boundary: $T_B < T_L = T_M$

As the interface temperature $T_M$ is the liquid temperature here, $T_L = T_M$, the liquid phase is isothermal and drives no thermal gradient (Fig. 10-a). The dynamics of solidification is thus fully controlled by the rate at which the released latent heat is diffused in the solid phase and evacuated at the cold boundary.

Diffusion in the solid phase satisfies the diffusion equation (3) with boundary conditions $T(0) = T_B$, $T(z_I) = T_M$ on temperature and $QV_I = c_p D_T (\partial T/\partial z)_I$ on the temperature flux at the interface (6) with $V_I = dz_I/dt$. The invariance of the diffusion equation (3) by the rescaling $z \to \alpha z$, $t \to \alpha^2 t$ allows a self-similar solution $T(u)$ to be proposed with $u = z(4D_T t)^{-1/2}$:

$$T - T_B = (T_M - T_B)\frac{\mathrm{erf}(u)}{\mathrm{erf}(u_I)} \tag{19}$$

where $u_I = z_I(4D_T t)^{-1/2}$ and where the error function writes: $\mathrm{erf}(u) = 2\,\pi^{-1/2} \int_0^u \exp(-x^2)dx$.

The boundary condition on the temperature flux at the interface then fixes the velocity. Interestingly, the time variable cancels out on both sides of this relationship, a fact which corroborates the relevance of the self-similar solution (19) to this issue. It then yields:

$$G(u_I) = \sqrt{\pi}\, u_I\, \exp(u_I^2)\, \mathrm{erf}(u_I) = \Delta_B = S_B^{-1} \tag{20}$$

where $\Delta_B$ (resp. $S_B$) denotes the undercooling (resp. the Stefan number) based on the boundary temperature $T_B$: $\Delta_B = c_p(T_M - T_B)/Q$.

A solution for criterion (20) exists for *any* positive Stefan number $S_B$, as displayed in Fig. 10-b. It provides the constant $u_I(S_B)$ from which the dynamics of the interface follows as $z_I^2 = 4u_I^2 D_T t$. This diffusive dynamics confirms that the solidification process is limited by diffusion and can proceed for any positive undercooling $\Delta_B = S_B^{-1}$.

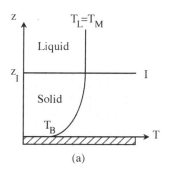
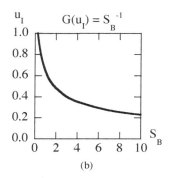

(a)                                                    (b)

Figure 10.   Planar solidification from an undercooled boundary $T_B < T_M$ of a pure liquid at the melting temperature $T_L = T_M$. A self-similar solution parametrized by $u_I(S_B)$ exists for any positive Stefan number $S_B$ or undercooling $\Delta_B = S_B^{-1} = c_p(T_M - T_B)/Q$. It corresponds to a diffusive dynamics for the interface: $z_I = 4u_I^2 D_T t$. (a) Sketch of the system and of the thermal field. (b) Graph of the relationship $u_I(S_B)$.

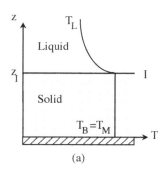
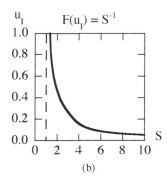

(a)                                                    (b)

Figure 11.   Planar solidification of an undercooled pure liquid $T_L < T_M$ from a boundary at the melting temperature $T_B = T_M$. A self-similar solution parametrized by $u_I(S)$ *only* exists for Stefan number $S$ larger than unity, i.e. undercooling $\Delta = S^{-1} = c_p(T_M - T_L)/Q$ smaller than unity. This corresponds to a released heat in excess compared to the thermal energy required to warm the liquid from $T_L$ to $T_M$. The interface dynamics is then diffusive: $z_I = 4u_I^2 D_T t$. (a) Sketch of the system and of the thermal field. (b) Graph of the relationship $u_I(S)$.

• Undercooled liquid: $T_L < T_B = T_M$

As the interface temperature $T_M$ is the boundary temperature here, $T_B = T_M$, the solid phase is isothermal and drives no thermal gradient (Fig. 11-a). The dynamics of solidification is thus fully controlled by the rate at which the released latent heat is diffused in the liquid phase.

The solution of the dynamics is similar to that reported above, the only change being in the boundary conditions for temperature. The self-similar solution writes:

$$T - T_L = (T_M - T_L)\frac{\text{erfc}(u)}{\text{erfc}(u_I)} \tag{21}$$

where the complementary error function $\text{erfc}(u)$ is $\text{erfc}(u) = 1 - \text{erf}(u)$.

The flux boundary condition at the interface then yields:

$$F(u_I) = \sqrt{\pi}\, u_I\, \exp(u_I^2)\, \text{erfc}(u_I) = \Delta = S^{-1}. \tag{22}$$

The resulting graph for the function $u_I(S)$, displayed in Fig. 11-b, reveals that no solution exists for $S < 1$ (resp. $\Delta > 1$) and that $u_I$ diverges for $S$ (resp. $\Delta$) approaching 1.

These features may be understood from energy and entropy considerations. For $S < 1$, $\Delta > 1$, the released volumic latent heat is too weak for warming the unit volume of liquid from temperature $T_L$ to the melting temperature $T_M$. Accordingly, the temperature profile cannot be stationary. In particular, as the energy budget is negative, the liquid phase must cool down as growth proceeds so as to provide the required energy to transform the liquid at temperature $T_L$ in a solid at temperature $T_M$: $(\partial T/\partial t)_{z-z_I} < 0$. However, this spontaneous *cooling* of a yet coldest domain is forbidden by the second principle of thermodynamics. There is then no surprise not to find it in the solutions provided by the diffusion equation.

On the opposite, for $S > 1$, $\Delta < 1$, latent heat is released in excess compared to the energy required to warm the liquid to the melting temperature. This energy then spreads by diffusion and raises the temperature profile: $(\partial T/\partial t)_{z-z_I} > 0$. However, the larger $S$, the larger the excess energy released by unit time and the resulting temperature rise in the profile. As this rise is achieved by diffusion, more time must be left to achieve it so that the interface velocity must be comparatively smaller. This explains the decrease of $u_I$ with $S$.

Finally, as far as solidification approaches the energy balance condition $S = 1$, the required dynamical change of the temperature profile reduces. This allows an ever larger interface velocity.

### 3.1.3. *Local thermodynamic out-of-equilibrium*

Despite the above analysis, experiment reveals that a planar growth can nevertheless operate at $S < 1$, $\Delta > 1$. This, of course, must not be interpreted as a violation of the second principle of thermodynamics but, instead,

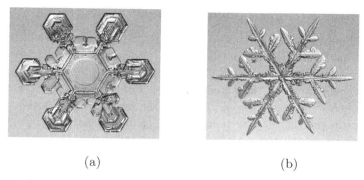

(a)                                    (b)

Figure 12. Snow-flakes evidencing a three-fold anisotropy and a planar growth. Folds result from the anisotropy of surface tension. Planar growth results from the anisotropy of the kinematic coefficient. Following it, growth is controlled by diffusion in the growing plane and by kinematics on its normal direction. As a result, the growth rate is much slower on this direction, thereby yielding an apparent two-dimensional growth. In the growing plane, the balance between anisotropy and diffusion, i.e. between faceting (a) and branching (b), results in a large variety of possible forms. Courtesy of Professor K.G. Libbrecht, California Institute of Technology, California, USA.

of the local thermodynamic equilibrium of growing interfaces: $T_I \neq T_M$. In particular, we recall that the temperature $T_I$ of a solidification interface is strictly the melting point $T_M$ when it is steady but that, when in motion, a kinetic undercooling $T_M - T_I$ may appear in proportion to the interface velocity $V_I$, at first order approximation: $T_I = T_M - \alpha V_I$.

Following this kinetic correction, a steadily growing interface may thus be achieved at $S < 1$, $\Delta > 1$ provided that the effective Stefan number $\tilde{S} = Q/[c_p(T_I - T_L)]$ based on the actual interface temperature $T_I$ is unity: $\tilde{S} = 1$. The interface then moves at velocity $V_I = \alpha^{-1}(T_M - T_L)(1 - S)$ with a steady co-moving temperature profile and at a rate controlled by the kinetics of molecular attachment.

A spectacular manifestation of the difference between solidification rates depending on whether they are controlled by diffusion or by kinetics is provided by the planar shape of a growing snow-flake (Fig. 12). This two-dimensional growth evidences a large anisotropy of solidification rates which traces back to that of the coefficient $\alpha$. In particular, whereas growth is limited by diffusion in the growth plane, it is limited by kinetics on its normal, owing to a large value of $\alpha$ on this direction. This result in a much weaker solidification rate on this axis and, thus, to an apparent planar growth.

## 3.2. Solidification of a binary mixture

We call liquidus (resp. solidus) the thermodynamic states for which an infinitely small solid (resp. liquid) germ may appear and persist when surface tension is ignored. Following (15), their temperature depends on the solute concentration $c$ in the liquid mixture: $T_L(c) = T_M - mc$ (resp. $T_S = T_M - mc/K$) where $L$ (resp. $S$) specifies the liquidus (resp. solidus).

Accordingly, the melting temperature $T_I$ of a planar interface now depends on its solute concentration $c_I$ in the liquid phase at the interface: $c_I(T_I) = c_L(T_I)$. On the other hand, following (14), the solid formed by solidification involves a different solute concentration: $c_S(T_I) = Kc_L(T_I)$. The difference $\Delta c = c_L(T_I) - c_S(T_I)$, called the miscibility gap (Fig. 8-a), then corresponds to a net solute rejection, $\Delta c > 0, K < 1$, or absorption, $\Delta c < 0, K > 1$, by the growing interface.

### 3.2.1. Isothermal solidification: free growth

We consider the isothermal growth of a binary mixture of mean solute concentration $c_\infty$. This condition implicitly assumes that no significant latent heat is released by the solidification process. Analysis then simply reduces to the dynamics of a single field: the solute concentration $c$ in a uniform thermal field $T = T_I$.

The mixture is taken undercooled, $c_\infty < c_L(T_I)$, so that solidification can naturally proceed, and metastable, $c_S(T_I) < c_\infty$, so that considering a thermodynamically stable liquid phase is relevant. We then denote $\Delta$ the relative position of the mixture concentration in the miscibility gap: $\Delta = [c_L(T_I) - c_\infty]/[c_L(T_I) - c_S(T_I)]$. By definition, it is thus positive and smaller than unity: $0 < \Delta < 1$.

The concentration field of the medium is constant in the solid phase and evolves by diffusion in the liquid phase with boundary conditions $c(+\infty) = c_\infty$, $c_I = c_L(T_I)$. Its dynamics is thus similar to that given by the solidification of a pure undercooled melt from a boundary at the melting temperature (Sect. 3.1.2). In particular, neglecting solute diffusion in the solid phase, solute conservation writes $V_I \Delta c = -Dc(\partial c/\partial z)_I$, so that the miscibility gap $\Delta c = c_L(T_I) - c_S(T_I)$ plays the role of the temperature raise $Q/c_p$ brought about by latent heat in a pure melt. In this analogy, the Stefan number writes $S = [c_L(T_I) - c_S(T_I)]/[c_L(T_I) - c_\infty] = 1/\Delta$ where $\Delta$ plays the role of the undercooling.

The fact that $S$ is larger than unity here means that the self-similar solution for diffusion-controlled growth always provides an actual front velocity $V_I$ (Fig. 11-b). According to it, the front advances in a diffusive way, $z_i(t) = 2u_I(D_c t)^{1/2}$, the constant of the motion $u_I$ being solution of $F(u_I) = S^{-1}$ (22). The interface velocity therefore decreases in time as $V_I(t) = u_I(D_c/t)^{1/2}$ so that no steadily growing interface is displayed except for $S = 1$ (Sect. 3.1.1).

### 3.2.2. *Solidification in a thermal gradient: directional growth*

We now turn to the solidification of a binary mixture of solute concentration $c_\infty$ in a uniform thermal gradient. Release of latent heat keeps being neglected.

The planar interface localizes itself at a point where its concentration lies on the liquidus: $c_I = c_L(T_I)$. The melt is then either moved at an imposed velocity $\mathbf{V}$ in the thermal field (Czochralsky type growth) (Fig. 22-b) or, equivalently, the thermal field is moved with respect to it at velocity $-\mathbf{V}$ (Bridgman type growth). Both methods give rise to a dynamics of the concentration field in the interface frame $\mathcal{R}_I$ (Fig. 5-b) which corresponds to an advection-diffusion equation (23) with boundary conditions (26) and, at the interface, the flux condition (24) and the jump conditions (25):

$$\frac{\partial c}{\partial t} - \mathbf{V}.\nabla c = D\nabla^2 c \tag{23}$$

$$[c_L(T_I) - c_S(T_I)]V_I = -D\frac{\partial c}{\partial z}|_I \tag{24}$$

$$c_S(T_I) = Kc_L(T_I) = Kc_I \tag{25}$$

$$c(+\infty) = c_\infty \quad ; \quad c(-\infty) = c_S(T_I) \tag{26}$$

where $D \equiv D_c$ denotes the solute diffusion in the liquid phase.

Steady state of growth implies that the resulting solid has the same solute concentration than the mixture: $c_S(T_I) = c_\infty$. This localizes the front at a definite temperature $T_I$ and implies, for the usual case $K < 1$, that the liquid solute concentration at the interface is larger than the mixture concentration: $c_I = c_L(T_I) = c_S(T_I)/K = c_\infty/K$. An impure layer is thus formed ahead of the interface (Fig. 8-b). Its structure is given by the advection-diffusion equation:

$$c(z) = c_\infty + \Delta c_I \exp(-z/l_D) \tag{27}$$

where $\Delta c_I = (c_L - c_S)_I$ denotes the miscibility gap at the interface temperature and where the diffusion length $l_D = D/V$ provides the characteristic scale of the advection-diffusion layer in the liquid phase. Interestingly, $l_D$ is no longer a free variable here since it is imposed by the translation motion with respect to the thermal field, as a result of the introduction of a thermal gradient.

Note that, as $c_{L,I} = c_\infty/K$ and $c_{S,I} = c_\infty$ (Fig. 8-b), the miscibility gap at the interface is equal here to the miscibility gap $\Delta c = c_\infty(1-K)/K$ of the mixture. On the other hand, the net excess of solute $\mathcal{C}$ in the whole diffusive layer writes: $\mathcal{C} = \int_0^\infty (c - c_\infty)dz = \Delta c_I\, l_D$.

### 3.2.3. *Zone melting*

Solute conservation in a closed system implies that the excess of solute $\mathcal{C}$ present in the diffusive layer is compensated by a solute depletion in the solid. Purification of a material can then be achieved by pushing the impure layer at one of its boundaries by solidification. This may be achieved by melting a part of the material and translating the molten zone from the beginning of the sample to its end. This procedure, called zone melting, is widely employed in industry in particular to purify ingots.

To further increase purification, the zone melting process can be made iterative after having re-homogenized the sample. This process, introduced by W.G. Pfann[9] and called zone refining, is applied as a last step of purification of a material. The expression of $\mathcal{C}$ shows that is all the more efficient that the velocity $V$ is low and the mixture is impure. In particular, removing the impure diffusive layer decreases the solute concentration of the medium by $\delta c_\infty = -\mathcal{C}/(L - l_D) \approx (1-K)/K\,(l_D/L)\,c_\infty$ where $L >> l_D$ denotes the mixture length. The relative efficiency $\delta c_\infty/c_\infty$ of the purification is thus a constant $\mu = (1-K)/K\,(l_D/L)$. Neglecting the reduction of sample length, it yields, after $n$ cycles of zone refining, an exponential decrease of the mixture concentration: $c_\infty(n) \approx c_\infty(0)\exp(-\mu n)$.

## 4. Geophysical implications of fractional crystallization

Before addressing the development of forms on growing interfaces, we find it interesting to consider here some geological implications of the compositional difference between liquid and solid phases, so-called fractional crystallization. They refer to a motor of the gulf stream, to rock datation and to the origin of heterogeneity of rock composition in the earth crust.

### 4.1. *Bank solidification and gulf stream motors*

Sea water is salty. However, the ice that it forms when freezing contains less salt than the initial water. The excess is then rejected in the surrounding medium yielding a salter water. This phenomenon occurs in particular on the artic bank every winter.

Interestingly, as salter water is also denser, solidification of the artic bank yields localized currents called chimneys, about 1 km wide, that go from the ocean surface to deeper levels. Following this coupling between solidification and hydrodynamics, a source of thermohaline convection is periodically activated in the artic zone. It stands as a heart for the large scale circulation of ocean currents and especially here, for the gulf stream.

During the last half million years, the thermohaline circulation has slowed down and even stopped several times. The origin of this was a weakness of the chimney currents that resulted either from a too thick bank preventing freezing from the atmosphere or from a melting of icebergs yielding a massive load of weakly salt water in the ocean. Memory of this has been kept in fossils deposited on the ocean floor. Implication of a weak thermohaline circulation has ever been a cooling of western europe, the last event of this type having occurred 17000 years ago. Nowadays, concerns refer to the expected consequences of the melting of the artic ice field, which has yet induced measurable decrease of chimney activities.[10]

### 4.2. *Segregation at the geophysical scale*

Interestingly, solidification of magma chambers in the earth mantle or of lava lakes in a volcano crater gives rise to issues similar to those found at a much smaller scale in metallurgy (Fig. 21). This is because both the scales of geophysical or man-made solidifying systems are far larger than the micro scale of solidification structures. In particular, examination of solidified lava lakes shows a textural division between columnar zones close to the crater boundaries and an equiaxed zone formed in the interior of the lake, that are both strikingly reminiscent of the macrosegregation structures of an ingot (Sect. 5.4, Fig. 21).

In magma chambers, scales are so large that solidification issues mainly concern their bulk. The large diversity of components then yield a fractional crystallization whose complexity is similar to that of fractional distillation in the industry of perfumes (Fig. 9-b). In addition, intense compositional convection forbids restricting analysis to thermodynamics and calls for a dynamical approach of the solidifying process. This results in a complex system from which one may fear not being able to reach a satisfactory level of understanding or of prediction. However, comparison of the stratigraphy of different magma chambers, as those of the Bushveld[11] and Stillwa-

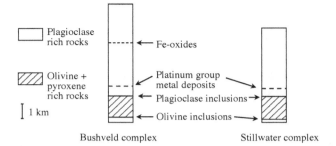

Figure 13. Compositional structure of the Bushveld and Stillwater complexes revealed by stratigraphy[11,12,13]. Their similarities reveal a striking determinism in the global solidifying process that imprints the resulting macrosegregation, beyond the specific initial differences and the involved dynamical phenomena.

ter[12] complexes, surprisingly reveals striking similarities.[13] In particular, all show olivine and pyroxane first formed, followed by platinum group metals and, finally, plagioclaste-rich rocks (Fig. 13).

A deterministic history of solidification in large chambers or in vast complex thus remains to be understood from modeling. Interestingly, it should answer a fundamental question regarding the origin of the production or largely inhomogeneous magmas from relatively homogeneous mantle source. Beyond its academic interest, this question refers to the possibility of easily extracting resources from the earth's crust. For instance, most of the world platinum resources come from the Bushveld complex of South Africa in which solidification has differentiated, from a homogeneously mixed magma, a shell that is particularly rich in this rare element (Fig. 13).

## 4.3. Rock datation

A method of datation in geophysics relies on radioactive desintegration of elements. The objective consists in determining the age of rocks so as to rebuild the scenario of formation of the earth or of the moon since its first stages. As the age of the earth is estimated to 4.5 Gy, the method has to be extremely accurate. Interestingly, we shall see that the phenomenon of fractional crystallization greatly helps in improving the accuracy of datation and the identification of the history of melting and solidification of rocks and magma. It relies on the fact that, in a multi-species inhomogeneous mixture, the differences between melting temperatures and segregation coefficients drive variations of the compositions of a given element in the resulting solid.

The datation procedure from radioactive desintegration consists in de-ducing, from the actual concentration ratio of a couple of isotopes embed-ded in rocks, the time at which these rocks have solidified. It thus relies on the assumption that no modification of composition of the corresponding solid has occurred except from spontaneous desintegration. This is actually true up to solid diffusion which homogenizes compositions on scales of or-der $L \approx (Dt)^{1/2}$ where $D$ denotes the species diffusivity in solids, of order $10^{-9} \text{cm}^2 \text{s}^{-1}$ and $t$ the time on which diffusion has proceeded. This yields, for the largest time of 4.5 Gy, a length of about 200 meters. This scale is for-tunately small enough compared to the size of geophysical inhomogeneities to make the method still available to date.

The method relies on the spontaneous desintegration of a "father" $F$ into a "son" $S$ where $F$ and $S$ refer to different elements since they involve a different number of protons. On the other hand, the son $S$ involves a stable isotope $s$ that has the specificity of not being produced by any other element including $F$ and $S$. When embedded in rocks, these elements $F$, $S$ and $s$ then allow a budget to be made with, following desintegration, a decrease of the solid concentration of $F$, an increase of that of $S$ and a stability of that of $s$. It is then convenient to consider the concentration ratios $\phi(t) = [F]/[s]$, $\sigma(t) = [S]/[s]$ whose time evolution is given by the exponential radioactive decay on a scale $\tau = T/\ln(2)$ where $T$ denotes the half-time of the father element:

$$\phi(t) = \phi(0) \exp(-t/\tau) \; ; \; \sigma(t) = \sigma(0) + \phi(0)[1 - \exp(-t/\tau)]. \qquad (28)$$

Here, the time scale $\tau$ and the concentrations $[\phi(t), \sigma(t)]$ at the time of the measurement are known, but the initial conditions $[\phi(0), \sigma(0)]$ and the time-delay $t$ are not. One therefore have one more unknown than the number of equations so that the system is undetermined. We thus face a situation similar to that occurring when finding a sandglass loosing sand at a measurable rate: we can trace the evolution backward but are unable to determine the time at which the sandglass was opened as far as we ignore its initial level.

In this instance, one may either assume an initial ratio $\sigma(0)$ as in data-tion methods based on $^{14}C$ or use several sets of similar equations (28) with some common variables. The latter method relies on fractional cristalliza-tion following which the father element $F$ will be incorporated in rocks at *different* concentrations depending on the nature of their dominant el-ements and on the evolution of the melt composition with solute rejection and convection as magma solidification proceeds. For a set of *different* rocks, $R_i, 1 \leq i \leq n$, having solidified at the same time from the *same* inhomogeneous mixture, this therefore provides a set of *different* initial solid concentrations $\phi_i(0)$ referring to about the same initial time $t = 0$ in (28). However, the important thing is that, because cristallization makes no difference between an element and its isotope, the initial ratios $\sigma_i(0)$ of the concentration of the element $S$ over that of $s$ will be uniform and *independent* of the nature of rocks: $\sigma_i(0) = \sigma_j(0), j \neq i$. Accordingly, the different sets of equations referring to the different rocks involve the *same* variable $\sigma(0)$: $\sigma_i(0) = \sigma(0), \forall i$. This provides an overdetermined system from which a surprisingly large datation accuracy may result (Sect. 4.3.1, 4.3.2).

### 4.3.1. *Moon formation*

An interesting example of the above situation is provided by the datation[14] of the moon rocks brought back by Apollo 11 (Fig. 14-a). The father element was 87 rubidium $^{87}Rb$, the son element 87 strontium $^{87}Sr$ and the stable element 86 strontium $^{86}Sr$. Initial compositions were unknown but many minerals formed at the same time were available: cristobalite, ilmenite, pyroxene, plagioclase ... All provided a specific data point $[\phi_i(t), \sigma_i(t)]$ indexed by the mineral number $i$. Interestingly, relation (28) yields an affine relationship between $\phi_i(t)$ and $\sigma_i(t)$ which applies to all: $\sigma_i(t) = A + B(t) \, \phi_i(t)$ with $A = \sigma(0)$ and $B(t) = \exp(t/\tau) - 1$. In particular, $A$ is independent of the minerals as discussed above and $B$ is the same for all since they solidified at the same time. Plotting the data points for the different minerals provides the expected affine relationship, thereby confirming the absence of other factors of composition change (Fig. 14-a). This gives access to $B(t)$ and thus to the age $t$ of the rock formation, i.e. of the moon formation: $B(t) = 0.0529$, $T = 48.8 \, Gy$, $\tau = 70.4 \, Gy$, $t = \ln(1 + B)\tau = 3.63 \, Gy$.

Notice the essential role of fractional cristallization in this datation: without segregation, rock composition would not keep memory of solidification; without a variation of segregation coefficients $K_i$ with constituents, data could not be made accurate since all the initial concentrations $\phi_i(0)$ would be the same. Both $K_i \neq 1$ and $K_i \neq K_j$ are thus required to recover rock history from radioactive desintegration.

### 4.3.2. *Earth crust history*

Another fascinating example of application of fractional cristallization to geological datation refers to the history of the formation of the earth crust.

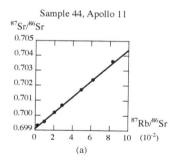

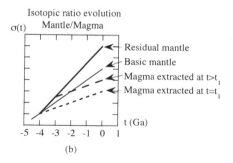

Figure 14. Datation of rock formation from radioactive desintegration. (a) Moon rocks brought back by Apollo 11 (sample 44) dated[14] from the radiochronometer $^{86}Sr/^{87}Rb$. Ordinates correspond to the relative concentration $\sigma_i(t) = [^{87}Sr]/[^{86}Sr]$ of strontium isotopes and abscissa to the relative concentration $\phi_i(t) = [^{87}Rb]/[^{86}Sr]$ of rubidium $^{87}Rb$ with respect to the stable strontium isotope $^{86}Sr$. The index $i$ refers to different kind of rocks formed at the same time from the same initial mixture. Points on the graph correspond, for increasing abscissa, to plagioclase, total sample, pyroxene-B, pyroxene-A, ilmenite, cristobalite-A and cristobalite-B. The slope $B(t) = \exp(t/\tau) - 1$ enables the determination of the delay $t = 3.63\,Gy$ between rock solidification and actual observation, given the radioactive decay time $\tau = 70.4Gy$ of rubidium $^{87}Rb$ into strontium $^{87}Sr$. (b) Isua rocks dated[15,16] from the radiochronometer $^{147}Sm/^{143}Nd$. Ordinates correspond to the relative concentration $\sigma(t)$ of neodyme isotopes $^{143}Nd$ and $^{144}Nd$ and abscissa to time. When a liquid magma is extracted by fusion from the mantle, it gets enriched in neodyme so that the relative concentration of Samarium appears comparatively smaller. As it drives the slope of the further evolution of concentration $\sigma(t)$, the time of this event is imprinted in the rock (30). Conversely, the residual mantle displays a slope increase that is caused by neodyme depletion and which enables the datation of a next extraction of magma.

The issue consists in recovering the different times at which rocks have solidified on the surface of our planet to form a crust. Here, the segregation resulting from fractional cristallization may also be very useful in giving rise, at the very instant of solidification, to composition steps that the rocks has kept in memory.

Consider the samarium/neodyme ($^{147}Sm,^{143}Nd$) radio-chronometer that has been used to trace back the history of the Isua rocks, one of the oldest rocks on earth, located on the south-west of Greenland, beneath the ice calot. These rocks originate from the mantle in which their composition was evolving according to relation (28), the stable element $s$ being 144 neodyme $^{144}Nd$. However, if, at an instant $t_1$, the mantle has liquefied following a temperature rise or a pressure decrease, segregation has enriched the liquid magma with neodyme and depleted the remaining solid accordingly: $[Nd]_m(t_1^+) = K[Nd]_M(t_1^+)$ where $m$ denotes the solid mantle, $M$ the liquid magma and where $K < 1$ (29). The crust formed from the magma ascending to the surface and the residual mantle have thus displayed at this instant $t_1^+$ different compositions of neodyme and thus different ratios $\phi(t_1^+) = [^{147}Sm]/[^{144}Nd]$: a larger one in the poor residual solid mantle than in the rich liquid magma: $\phi_M(t_1^-) = K\phi_m(t_1^-) < \phi_m(t_1^-)$ (30). Meanwhile, the ratio of neodyme isotopes $\sigma(t_1) = [^{143}Nd]/[^{144}Nd]$ has kept unchanged since the nucleus composition is indifferent to the solidification process (29). However, the amplitude of the further raise of $\sigma(t) \equiv \sigma_M(t)$ in the liquid magma (resp. $\sigma(t) \equiv \sigma_m(t)$ in the solid mantle) by samarium desintegration is proportional to $\phi_M(t_1^+)$ (resp. $\phi_m(t_1^+)$), taken as a new initial condition for the concentration evolution (28). Accordingly, the evolution of the ratio $\sigma(t)$ undergoes a change of slope at the instant $t_1$ of magma formation from the mantle: a slope decrease in the liquid magma and a slope raise in the residual solid mantle (30) (Fig. 14-b).

$$t = t_1 \; : \; \phi_M(t_1^-) = K\phi_m(t_1^-) < \phi_m(t_1^-) \; ; \; \sigma(t_1^+) = \sigma(t_1^-) \qquad (29)$$

$$t > t_1 \; : \; \phi_{M,m}(t) = \phi_{M,m}(t_1^-)\exp(\frac{t_1 - t}{\tau}) \qquad (30)$$

$$\sigma_{M,m}(t) = \sigma(t_1) + \phi_{M,m}(t_1^-)[1 - \exp(\frac{t_1 - t}{\tau})].$$

Notice that, as the half-time $T = \tau \ln(2)$ is dramatically large, $T = 106Ga$, the exponentials trends reduce to linear trends (Fig. 14-b).

Similar conclusions will be achieved on further extractions of magma from the mantle (Fig. 14-b). In particular, a second crust formation from the residual solid mantle at a later time $t_2 > t_1$ will occur for a ratio $\sigma_m(t_2)$ larger than would have been expected without the first crust formation. One therefore conceives that both the existence and the times of successive crust formations may be recovered this way from the sole measurement of the actual rock compositions (Fig. 14-b). In particular, measurements performed in the Isua rocks have revealed an anomalously large ratio $\sigma(t)$ which points to the formation of a primary crust *before* the formation of the main actual earth crust.[15,16]

## 5. Generation of forms

We address the primary and secondary instabilities that are responsible for the generation of forms in interfacial growing systems and we report their main implications regarding segregation at the micro- and macro-scales and in metallurgy.

## 5.1. *Issue and procedure*

We consider a planar interface that advances at a normal velocity $V_I$ in a medium and we question the existence of an instability breaking its planar shape. It is convenient for this to identify the formal mechanism of instability that might develop on the interface and the formal way to determine the resulting main instability features.

In section 2.3 the dynamics of the system has been identified as involving a field dynamics in its bulk, a thermodynamic condition at the interface and a conservation equation specifying the normal interface velocity. They formally write:

$$\frac{D\varphi}{Dt} = D\nabla^2\varphi \tag{31}$$

$$\mathcal{F}(\varphi, \psi, \nabla)|_I = 0 \tag{32}$$

$$\mathbf{V}_I.\mathbf{n} \propto \nabla\varphi|_I.\mathbf{n} \tag{33}$$

where $\varphi$ denotes the scalar field corresponding to the order parameter of the two phases in contact, $D$ its diffusivity and $\psi$ an imposed field as the temperature field in directional growth. The index $I$ labels variables taken at the interface and the gradient operator in (32) allows the implications of interface curvature to be formally handled.

Relation (33) characterizes the interface normal velocity which drives the evolution of form. It depends on the normal gradient of $\varphi$ on the interface which, in turn, depends on the interface form by (32). In particular, as the interface is an iso-$\mathcal{F}$, any of its distortion will reduce or enlarge the field tubes of $\mathcal{F}$, i.e. its gradient, and thus, possibly the gradient of $\varphi$. One obtains this way a looped system: an interface deformation changes the gradient on which its own evolution depends. If a positive counter-reaction sets in, a growing deformation will follow.

A canonical method for determining a criterion for instability and the resulting main instability features consists in performing a linear analysis of the system based on the normal modes:

$$\tilde{\varphi} = \hat{\varphi}\exp(\sigma t - qz + ikx) \;\; ; \;\; \tilde{\xi} = \hat{\xi}\exp(\sigma t + ikx) \tag{34}$$

where the $x$-axis is along the planar interface and the $z$-axis along its normal (Fig. 15-a) and where $\tilde{\varphi}$ and $\tilde{\xi}$ denote the modulations of the field $\varphi(x,t)$ and of the positions $z_I(x,t)$ of the interface.

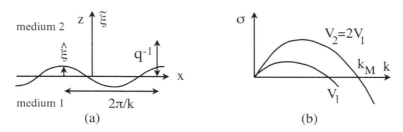

Figure 15. Planar front instability. (a) Sketch of the normal modes of deformation $\tilde{\xi}$ of the interface. Here, $2\pi/k$ denotes the period of the modulation and $q^{-1}$ the decay length of the field modulation $\hat{\phi}$ ahead of the interface. (b) Dispersion relation of the Saffman–Taylor instability for a velocity $V_2$ and its half $V_1 = V_2/2$.

Relation (31) then gives the dispersion relation $\sigma \equiv \sigma_1(q, k)$. Relations (32)(33) provide, at the first order in the normal mode amplitudes, a linear homogeneous system $\mathcal{M}(\hat{\varphi}, \hat{\xi}) = 0$ whose solvability condition, $\det(\mathcal{M}) = 0$, provides a second relationship $\sigma \equiv \sigma_2(q, k)$. Equating them yields the determination of both the growth rate $\sigma(k)$ and the decay rate $q(k)$ of perturbations of wavenumber $k$. The existence of a positive growth rate provides a criterion for the planar instability.

### 5.2. Instability of planar interfaces

• Saffman–Taylor instability[2]

We consider two non-miscible fluids enclosed in an horizontal Hele-Shaw cell of depth $d$ (Fig. 6-a). Their mutual frontier forms an interface which advances together with the fluids (Fig. 1-a). We index the media with $i = 1, 2$ and we denote $p_i$ their pressure, $\mathbf{V}_i = -K_i \nabla p_i$ their velocity fields (7) and $\mu_i$ their viscosity with $K_i = d^2/(12\mu_i)$.

The pressure fields satisfy the Laplace equation $\nabla^2 p_i = 0$ and undergo a jump $p_1 - p_2 = \gamma(\kappa_1 + \kappa_2)$ at the interface where $\kappa_j$, $j = 1, 2$, denote the principal curvatures of the interface (12). On the other hand, the impenetrability of the interface implies the equality of the normal fluid velocities on it: $\mathbf{V}_1.\mathbf{n}|_I = \mathbf{V}_2.\mathbf{n}|_I = V_I$.

The normal modes correspond to (34) with $\varphi \equiv p_i$. The Laplace equation $\nabla^2 p_i = 0$ provides the dispersion relations $q_i^2 = k^2$. Avoiding divergence, they reduce to $q_i = (-1)^i k$, the medium 2 being located in the $z > 0$ domain (Fig. 15-a). The remaining equations provide the following homogeneous linear system, care being taken of the change of position of the interface:

$$\sigma\hat{\xi} = q_1 K_1 \hat{p}_1 = q_2 K_2 \hat{p}_2 \tag{35}$$

$$\hat{p}_2 - \hat{p}_1 = [V(K_2^{-1} - K_1^{-1}) - \gamma k^2]\hat{\xi}. \tag{36}$$

Elimination of the mode amplitudes $\hat{p}_1$, $\hat{p}_2$, $\hat{\xi}$ then gives, with $q_i = (-1)^i k$, the dispersion relation:

$$\sigma = \frac{\mu_2 - \mu_1}{\mu_2 + \mu_1} kV - \gamma \frac{d^2}{12} \frac{k^3}{\mu_1 + \mu_2}. \tag{37}$$

Instability requires $(\mu_2 - \mu_1)V > 0$, i.e. an interface advancing into the most viscous medium. In this case and in absence of surface tension, i.e. $\gamma \equiv 0$, all length scales are instable. This is because, as there no longer exists any characteristic scale in the system, all scales must have the same fate. When surface tension is in order, the dispersion relation discriminates between unstable scales and stable scales, the unstable scales being above a critical scale $k_M^{-1}$ such that $\sigma(k_M) = 0$, i.e. $k_M^2 = (12\gamma/d^2)(\mu_2 - \mu_1)V$. Its graphs are reported in Fig. 15-b for a given velocity and its half.

• Directional solidification instability[17]

A mixture with solute concentration $c_\infty$ is pulled at velocity $\mathbf{V} = V\mathbf{z}$ in a thermal gradient $\mathbf{G} = G\mathbf{z}$. Solute diffusivity is labelled $D$ in the liquid phase and is neglected in the solid phase. From (16) (23) (24), the equations governing the system in the frame $\mathcal{R}$ of the temperature field (Fig. 5-b) are:

$$\frac{\partial c}{\partial t} - V\mathbf{z}.\nabla c = D\nabla^2 c \tag{38}$$

$$T_I = T_M + Gz_I = T_M - mc_I - T_M \frac{\bar{\gamma}}{Q}(\kappa_1 + \kappa_2) \tag{39}$$

$$c_I(1 - K)\mathbf{V}_I.\mathbf{n} = -D\nabla c|_I.\mathbf{n} \tag{40}$$

with boundary conditions (25) and steadily growing planar solution (27). Here, following (16), $\bar{\gamma}$ denotes $\gamma + \partial^2\gamma/\partial\theta^2$, the difference of variation of surface tension with the principal axis of curvature being overlooked.

The normal modes correspond to (34) with $\varphi \equiv c$. The field equation (38) provides the dispersion relation $\sigma + qV = D(q^2 - k^2)$. The expansion at the first order of equations (39) and (40) yields respectively:

$$m\,\hat{c} + \left[G - \frac{m\Delta c_I}{l_D} + T_M \frac{\bar{\gamma}}{Q}k^2\right]\hat{\xi} = 0 \tag{41}$$

$$D\left[\frac{1 - K}{l_D} - q\right]\hat{c} + \Delta c_I \left[\sigma + K\frac{D}{l_D^2}\right]\hat{\xi} = 0 \tag{42}$$

where $\Delta c_I$ denotes the miscibility gap at the planar interface and $l_D = D/V$ the diffusion length (Fig. 8-b). Here, care has been taken to handle the displacement $\tilde{\xi}$ of the interface in any value of variables taken *at* the interface, namely: $c_I$, $\mathbf{V}_I$, $\nabla c|_I$, $z_I$. In particular, following (27), $\delta(c_I) = \hat{c} + \tilde{\xi}\mathbf{z}.\nabla c|_I = \hat{c} - \Delta c_I \hat{\xi}/l_D$, $\delta(\nabla c|_I.\mathbf{n}) = -q\hat{c} + \Delta c_I \hat{\xi}/l_D^2$, and $\delta(\mathbf{V}_I.\mathbf{n}) = \sigma\hat{\xi}$.

We introduce the thermal length $l_T = m\Delta c/G$ and the analogous of the capillary length for solute growth $d_0 = (\bar{\gamma}/Q)\,(T_M/m\Delta c)$ where $\Delta c = c_\infty(1 - K)/K$ denotes the miscibility gap for the mixture. In particular, we notice that $\Delta c_I = \Delta c$ here (Sect. 3.2.2). We then adopt the diffusion length $l_D = D/V$ and the diffusion time $\tau_D = D/V^2$ as length and time units and we label $\tilde{x}$ the corresponding non-dimensional form of a variable $x$.

In non-dimensional form, the dispersion relation and that obtained from the solvability condition of the linear system (41) (42) write:

$$\tilde{\sigma} + \tilde{q} = \tilde{q}^2 - \tilde{k}^2 \tag{43}$$

$$\tilde{\sigma} + K = [\tilde{q} - (1 - K)][1 - \frac{\tilde{l}_D}{\tilde{l}_T} - \tilde{d}_0 \tilde{k}^2] \tag{44}$$

Eliminating $\tilde{k}$ and solving for the growth rate $\tilde{\sigma}$ yields $\tilde{\sigma} = f(\tilde{q})/g(\tilde{q})$ with:

$$\tilde{d}_0^{\ -1} f(\tilde{q}) = (\tilde{q} - a)\,(b - \tilde{q}^2 + \tilde{q})\ -\ c \tag{45}$$

$$g(\tilde{q}) = 1 - \tilde{d}_0\,(\tilde{q} - a) \tag{46}$$

and $a = 1 - K$, $b = \tilde{d}_0^{\ -1}(1 - \tilde{l}_D/\tilde{l}_T)$, $c = \tilde{d}_0^{\ -1} K$.

The onset of instability corresponds to the minimum of the marginal stability curve $\tilde{\sigma} = 0$. It thus satisfies both $\tilde{\sigma} = 0$ and $d\tilde{\sigma}/d\tilde{q} = 0$, i.e. $f = 0$ and $df/d\tilde{q} = 0$. This provides the following criterion:

$$(\tilde{q} - a)^2(1 - 2\tilde{q}) + c = 0. \tag{47}$$

It is convenient at this point to consider some orders of magnitude. Usually, capillary lengths are of the order of some Angströms, $d_0 \approx 10^{-9}$m, diffusion length of the order of micrometers to millimeters depending on the velocity, $l_D \approx 10^{-3}$m, and thermal length of the order of millimeters to centimeters, $l_T \approx 10^{-2}$m. Altogether, they provide a range of scales extending at least over 6 decades: $l_T/d_0 > 10^6$ . Such a large scale range would correspond in fluids to a Reynolds number of $(10^6)^{4/3} = 10^8$, i.e. to a fully developed turbulence. Needless to say, it allows and commands controlled approximations.

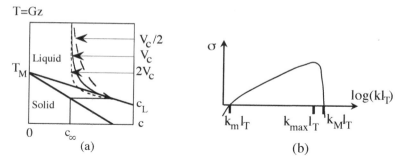

(a)                                                (b)

Figure 16.   Mullins–Sekerka instability. (a) Interpretation of the Mullins–Sekerka instability from constitutional supercooling. Three profiles of the evolution of concentration
ahead of a planar interface are shown. Above the critical velocity $V_c$, the liquid concentration $c$ ahead of the interface is lower than the liquidus concentration $c_L(T)$ at its
temperature $T$: the liquid mixture is then metastable. (b) Dispersion relation of the
Mullins–Sekerka instability (49). Logarithmic coordinates are taken on abscissa, owing
to the extremely large scale range. For $V \gg V_c$, one has $k_m \approx V/D$, $k_M \approx (V/Dd_0)^{1/2}$
and $k_{max} = k_M/\sqrt{3}$.

We first notice that, the unit length being $l_D$, the factor $c$ amounts to
$c = Kl_D/d_0 \approx 10^{-4}/10^{-9} = 10^5$ so that $c \gg 1$. As $a$ is at most of order
unity, criterion (47) reduces to $\tilde{q}^3 \approx c/2$ and implies $\tilde{q} \gg 1$. Turning
back to $df/d\tilde{q} = 0$, this yields at the dominant order in $\tilde{q}$, $b \approx 3\tilde{q}^2$ and,
finally, $b^3 \approx 27c^2/4$. One obtains this way the following criterion for planar
instability:

$$1 - \frac{l_D}{l_T} > 3\left(\frac{K^2}{4}\right)^{1/3}\left(\frac{d_0}{l_D}\right)^{1/3}. \tag{48}$$

A finely approximate expression of (48) may be obtained by noticing
that, as $K^2 d_0/l_D$ is of order $10^{-6}$ at most, the right hand side of criterion
(48) can be neglected, except in the very vicinity of the onset. As $l_D/l_T = m\Delta c/(DGV)$ this yields the approximate criterion of instability $V > V_c$
with the critical velocity $V_c = DG/(m\Delta c)$.

An heuristic interpretation of the Mullins–Sekerka instability has been
proposed by considering the thermodynamic stability of the liquid ahead
of the interface, so-called constitutional supercooling criterion. For this,
it is convenient to draw in the phase diagram the curve representing the
state of the mixture at each location (Fig. 16-a). As the thermal gradient
is a constant, $T \equiv Gz$, the ordinates can represent both variables $T$ and
$z$. The solution (27) for solute concentration of a steady planar interface is
then represented in Fig. 16-a by an exponential decay toward the $c = c_\infty$

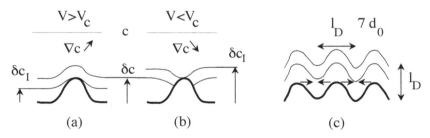

Figure 17. Destabilization/restabilization of a planar solidification front. (a) (b) Interpretation of the Mullins–Sekerka instability from the increase of concentration gradient $\nabla c$ ahead of a bulge. Heavy lines corresponds to the interface and light lines to isoconcentration lines. Here, $\delta c_I < 0$ represents the actual change of concentration at the bulge tip and $\delta c < 0$ the concentration drop displayed at the same distance on a planar front configuration. When $|\delta c| > |\delta c_I|$, the isoconcentration lines are bumped in a way that increases concentration gradient. This promotes diffusion and instability (a). On the opposite, for $|\delta c_I| > |\delta c|$, the decrease of concentration gradient at the bulge tip weakens diffusion and depletes instability (b). (c) Interpretation of the restabilization at large velocity. Here the diffusion length is so short that it is comparable to the capillary length that bounds by below the distortion scales. The whole boundary layer is then wrinkled so that concentration gradients (arrows) are mainly tangential. They thus tend to reduce the distortion amplitude and to restabilize the interface.

asymptot. However, the return to the asymptote is all the more short that the diffusion length $l_D$ is small. A transition then occurs when the corresponding curve enters the coexistence domain between the two phases since the melt gets undercooled in the vicinity of the front. This happens when the concentration of the liquid mixture $c(z)$ drops more rapidly than the liquidus line $c_L(z) \equiv c_L[T(z)]$, i.e. when $|dc/dz(z_I)| > |dc_L/dz(z_I)|$. As $dc_L/dz = -G/m = -\Delta c/l_T$ and, following (27), $dc/dz(z_I) = -\Delta c_I/l_D$ with $\Delta c_I = \Delta c$, this occurs for $l_D < l_T$ or $V > V_c$, i.e. at the same approximate onset than that determined above from stability analysis. In this instance, the liquid ahead of the front is simply metastable according to thermodynamics. Strictly speaking, this does not suffice for concluding about instability since thermodynamics refers to equilibrium whereas the planar growth instability deals with out-of-equilibrium.

Another, quantitatively equivalent, but physically more sound, phenomenological interpretation of the planar instability is provided by an hand-wavy estimation of the evolution of concentration gradients at the interface for an elementary displacement $\delta\mathbf{z}$ in the case where surface tension effects can be overlooked (Fig. 17-a). For a planar interface, the variation of concentration $\delta c$ on the distance $\delta\mathbf{z}$ is $\nabla c|_I\delta\mathbf{z}$ with, following (27),

$\nabla c|_I = -\Delta c_I/l_D \, \mathbf{e}_z$ with $\Delta c_I = \Delta c$ and, thus, $\delta c = -\Delta c/l_D \delta z$. However, if the interface moves to the location $z_I + \delta z$, its concentration will actually drop by $\delta c_I = -(G/m)\delta z = -\Delta c/l_T \delta z$ so as to satisfy the Gibbs–Thomson relation (15). If the amplitude $|\delta c_I|$ of this concentration shift is smaller than the shift $|\delta c|$ involved ahead of a planar interface, the new concentration at this location will be larger than before. On the other hand, as $\tilde{q} = q l_D \gg 1$, the iso-concentration lines that stand at a distance of a diffusion length $l_D$ are too far for being significantly perturbed. As a result, the iso-concentration lines in between them and the interface will get tightened and the concentration gradient will raise. Following relation (40), the interface velocity will locally increase since the evacuation by diffusion of the rejected solute will then be more efficient and the corresponding interface bulge will grow. This instability condition, $|\delta c_I| < |\delta c|$, writes as above $l_D < l_T$ or $V > V_c$. It is however more satisfactory than the previous heuristic argument since it explicitly refers to the interface velocity and thus, to out-of-equilibrium.

To further analyze the instability, it is useful to notice that relation (43) yields $0 \le \tilde{q} - \tilde{k} = (\tilde{\sigma} + \tilde{q})/(\tilde{q} + \tilde{k})$ and relation (44) $\tilde{\sigma} < \tilde{q} - 1$, so that $(\tilde{\sigma} + \tilde{q})/(\tilde{q} + \tilde{k}) < (2\tilde{q} - 1)/(\tilde{q} + \tilde{k}) < 2$ and $\tilde{q} - 2 < \tilde{k} < \tilde{q}$. The decay rate $\tilde{q}$ and the mode wavenumber $\tilde{k}$ are thus close if they are large.

At the onset of instability, $\tilde{q} \approx (c/2)^{1/3} \gg 1$, so that $\tilde{k} \approx \tilde{q}$. Calling $k_c$ and $\lambda_c$ the critical wavenumber and the critical wavelength, we then obtain, in dimensional form, $k_c l_D \approx (K l_D/2d_0)^{1/3}$ and $\lambda_c \approx 2\pi(2/K)^{1/3}(d_0 l_D^2)^{1/3}$ with, at onset, $l_D \approx l_T$ from (48). The critical wavelength $\lambda_c$ scales the typical size of disturbances that should develop on a planar front. It stands as a geometrical mean between the smallest characteristic scale, $d_0 \approx 10^{-9}$m, and the largest characteristic scale, $l_T$, usually of order of a millimeter. For moderate segregation coefficient, $K > 0.1$, it is about a hundred microns.

Farther from onset, the largest growth rate corresponds to $d\tilde{\sigma}/d\tilde{q} = 0$ with, as above, $\tilde{k} \approx \tilde{q} \gg 1$. Relations (45) (46) then straightforwardly yield, at the dominant order in $\tilde{k}$, $\tilde{k}^2 \approx \tilde{q}^2 \approx b/3$. In dimensional form, this provides the following expressions for the most amplified wavenumber $k_{max}$ and the most amplified wavelength $\lambda_{max}$: $k_{max} = (3d_0 l_D)^{-1/2}(1 - l_D/l_T)^{1/2}$, $\lambda_{max} = 2\pi[3Dd_0/(V - V_c)]^{1/2}$. In particular, the ratio between the most amplified wavelength $\lambda_{max}$ and the critical wavelength $\lambda_c$ writes: $(\lambda_{max}/\lambda_c)^2 = 3(K^2/4)^{1/3}(d_0/l_D)^{1/3}(1 - l_D/l_T)^{-1}$. Criterion (48) shows that, as expected, it is unity at onset.

## 5.3. Stability diagrams

### 5.3.1. Primary instability

With the approximations $k \approx q \gg 1$, the dispersion relation (44) writes in dimensional form:

$$\sigma = (V - V_c)k - Dd_0k^3[1 - (1 - K)\frac{V}{kD}] - \frac{V^2}{D}[1 - (1 - K)\frac{V_c}{V}] \quad (49)$$

For $V/V_c \gg (1-K)^{-1}$, it yields a band of unstable wavenumbers $(k_m, k_M)$ with $k_m \approx (V/D) [V/(V - V_c)]$, $k_M \approx [(V - V_c)/Dd_0]^{1/2}$ and a most amplified wavenumber $k_{max} \doteq k_M/\sqrt{3}$ (Fig. 16-b).

The relative extent $k_M/k_m$ of the unstable wavenumber band is $k_M/k_m \approx (D/d_0)^{1/2} (V - V_c)^{3/2}V^{-2}$. Interestingly, it thus collapses at large velocity. This is corroborated by the expression of the largest growth rate $\sigma(k_{max})$:

$$\sigma(k_{max}) = \frac{2}{3}\frac{(V - V_c)^{3/2}}{(3Dd_0)^{1/2}} - \frac{V^2}{D}[1 - (1 - K)\frac{V_c}{V}] \quad (50)$$

which, for $V \gg V_c$, vanishes for $V \approx 4/27\, D/d_0 \gg 1$: the interface then restabilizes.

A phenomenological interpretation of this restabilization may be given on the same lines as for the planar destabilization, by determining the change of concentration gradients on a deformed interface (Fig. 17-c). The main difference with the study performed at threshold is that the decay rate $q$ of concentration fluctuations ahead of the front is such that $\tilde{q} = ql_D$ is order unity here, instead of being large at threshold. This is because the wavenumber band reduces here to $k = k_{max} \approx (d_0l_D)^{-1/2}$, with $l_D \approx 27/4\, d_0$, so that $\tilde{k} \approx 2.6$ and $2.6 < \tilde{q} < 4.6$. This means that the whole diffusive layer is perturbed by the fluctuations, so that we can no longer assume its iso-concentration lines to be straight at some distance ahead of the front. Accordingly, it is then no longer relevant to deduce concentration gradients from the comparison between the concentration on the interface and that which should be displayed by a planar front at the same location. Instead, the actual picture corresponds to a diffusive layer involving a uniform thickness $l_D \approx 7d_0$ and a series of wrinkles at the same scale (Fig. 17-c). Additional concentration gradients are then no longer normal to the interface but tangential. They thus do not promote instability but, instead, the re-homogenization of concentration along the interface.

In addition, we notice that, as the wavenumber $k$ is close to $d_0^{-1}$, the curvature radius $R_I$ of an interface displaced by $\delta z \approx d_0$ is as small as $d_0$.

Then, the capillary term of the Gibbs–Thomson relation raises to $-m\Delta c$ so that the interface concentration $c_I \equiv (c_L)_I$ is shifted by $-(c_L - c_S)_I$. It thus reaches the bulk value $(c_S)_I = c_\infty$: $c_I = c_\infty$. It is clear that further decrease of concentration would inhibit growth since a depletion of concentration, $c < c_\infty$, would be generated at front bulges. The maximal possible distortion is thus bounded by $d_0$.

In practice, restabilization occurs at velocities $V_r$ of order $V_r = 4/27\ D/d_0$, i.e. $1.5\ \mathrm{cm}^2\mathrm{s}^{-1}$ for $d_0 \approx 10^{-8}\mathrm{m}$ and $D \approx 10^3\mathrm{mm}^2\mathrm{s}^{-1}$. Such macroscopic velocities may be achieved by solidification following a laser induced fusion.

A sketch of the instability domain of a planar front is plotted in Fig. 18. It reveals a closed domain due to restabilization at large velocity (Fig. 18-a) and an extremely flat marginal stability curve near the onset of instability (Fig. 18-b). In particular, the relative extent of the range of unstable wavenumbers soon gets dramatic values. For instance, at $V = 2V_c$, $k_M/k_m = (D/d_0 V_c)^{1/2}[2(1 + K)]^{-1} \approx 10^2$. This is because the system being dominated by diffusion, it is close to being Laplacian, in which case all its wavenumbers would have been unstable. This is not the case here, due to the presence of the stabilizing effects of thermal gradient and of capillarity. In contrast, however, unstable fronts show cells whose width extends over a factor 2 or 3 only (Fig. 19). In this sense, a dramatically huge selection of wavenumbers is in order in this growth system. It is mediated by secondary instabilities (Figs. 19, 20).

### 5.3.2. *Secondary instabilities*

The phenomena responsible for the above sharp selection of primary structures are the secondary instabilities of cells. Too thin cells are eliminated by recession as a result of the screening of their diffusion field by those of their larger neighbors (Fig. 19-a). Too large cells are eliminated by splitting at their tip or by overgrowth of a lateral branch (Fig. 19-b). These events forbid steady cells involving too large or too small width.

In a periodic array of cells, one may identify cell width with the wavelength $2\pi/k$. In a diagram $(V/V_c, kl_T)$, one then obtains a wavenumber band bounded by the instabilities of tip splitting or of elimination[18,19] (Fig. 20-b). Within this band, large cells may emit sidebranches and cells at moderate velocities may involve an oscillatory coupled dynamics of their width and position in the thermal gradient, in phase opposition with their neighbors. They will be addressed in more detail in sections 7 and 8.1 respectively.

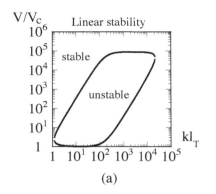
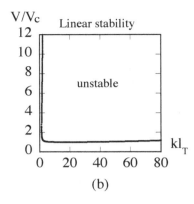

Figure 18.    Linear stability diagram of the Mullins–Sekerka instability. (a) The linearly unstable domain is closed due to restabilization at large velocity (Fig. 17-c). (b) Enlargement of the stability diagram close to onset. The marginal stability curve is then extremely flat, in contrast with the familiar parabolic-like curve relevant to usual pattern formation systems.

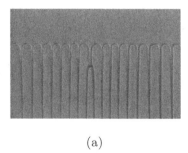
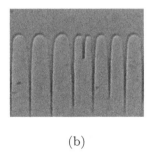

(a)                                              (b)

Figure 19.    Secondary instabilities bounding the stability domain of cells in a succinonitrile alloy[18,19]. (a) Cell elimination by screening of a cell that dangerously swept down behind the tips of its neighbors. (b) Cell creation by tip splitting of a cell.

Above $V/V_c > 3$, the secondary instabilities select a small stability channel (Fig. 20-b). However, closer to onset, the stability band drastically enlarges. In particular, increasing velocity, the channel reduction occurs sharply in the vicinity of $V/V_c \approx 3$ by the sudden elimination of a large range of small to moderately large cells[18–20] (Fig. 20-b).

Analysis of solidification patterns from a statistical viewpoint have reported a S-shape evolution[21,22] of the mean cell spacing $\bar{\Lambda}$ with the velocity (Fig. 20-a). As the slope inversion occurs while some cells emit sidebranches, it has been argued[23] that cells and dendrites refer to different

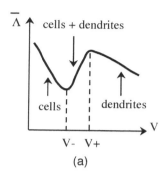
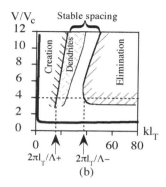

Figure 20. Secondary instabilities and mean cell spacing. (a) Evolution with the velocity of the averaged cell spacing over the interface. It shows a $S$-shape displaying a sudden raise in the velocity domain where dendrites occur[21,22]. (b) Stability diagram in a succinonitrile alloy[18,19] showing the marginal linear stability curve (heavy line), the domains referring to secondary instabilities and the channel of stable spacing for cells. Notice the dramatic reduction of this channel at $V \approx 3V_c$ due to large cell eliminations. The implications of this diagram on the evolution of the averaged cell spacing enables its $S$-shape (a) to be recovered (51).

branches of solutions for solidification, a small spacing branch for cells and a large spacing branch for dendrites, both branches involving a decreasing spacing with velocity. In a transition region, however, the stable solution for solidification would change branch, thereby increasing cell spacing and emitting sidebranches at the same time.

This point of view was stimulating in that it was able to propose a scheme of the mechanisms at work on a cell that was compatible with observations on mean cell spacing and on the occurrence of sidebranching. It is however in contradiction with the detailed observation of cells, since no transition with velocity is noticed regarding their density of occupation in the stable range of the stability diagram (Fig. 20-b). It nevertheless remains that the available spacing range suddenly decreases at about $V/V_c \approx 3$, as a result of the elimination of many cells. We evaluate below the implication of this event on the mean cell spacing.

Guided by observations, we assume that the probability density $p(\Lambda)$ of finding a cell spacing in the elementary range $[\Lambda, \Lambda+d\Lambda]$ only depends on the spacing bounds $(\Lambda_-, \Lambda_+)$ of the stable range $[\Lambda_-, \Lambda_+]$: $p \equiv p(\Lambda; \Lambda_-, \Lambda_+)$. This is supported by the end result of the instabilities of creation or of elimination, since both make wavenumbers return within the stable spacing range. In particular, these catastrophes provide the same kind of boundary conditions for the wavenumber dynamics at all velocities. Assuming the similarity of the resulting probability density of spacing, one is led to consider this probability as depending on the relative location of the spacing $\Lambda$ in the stable band $[\Lambda_-, \Lambda_+]$: $p \approx p(\xi)$ with $\xi = (\Lambda - \Lambda_-)/(\Lambda_+ - \Lambda_-)$. Normalization then implies that the probability density in the variable $\xi$ is $\Pi(\xi) = (\Lambda_+ - \Lambda_-)p(\xi)$.

This modeling of the probability density of spacing provides an appropriate tool for inferring the implications of the stability diagram on the mean cell spacing $\bar{\Lambda}$. In particular, $\bar{\Lambda}$ appears to solely depend on the spacing bounds $(\Lambda_-, \Lambda_+)$ since:

$$\bar{\Lambda} = \int_{\Lambda_-}^{\Lambda_+} \Lambda \, p(\Lambda; \Lambda_-, \Lambda_+)d\Lambda = (\Lambda_+ - \Lambda_-) \int_0^1 \xi \, \Pi(\xi) \, d\xi + \Lambda_- \qquad (51)$$

It will therefore be sensitive to the evolutions of the spacing bounds and, in particular, will suddenly increase around $V/V_c \approx 3$, following the large raise of $\Lambda_-$. Above and below this transition zone, it will follow the general trend of a weak decrease of both $\Lambda_+$ and $\Lambda_-$.

According to this analysis, the sudden raise of mean cell spacing simply follows from the sudden occurrence of the elimination instability on a large range of *small* cells. It is thus totally decoupled from the sidebranching instability undergone by *large* cells. That sidebranching occurs quasi-simultaneously to the increase of mean cell spacing therefore stands as a coïncidence: the occurrence in the *same* domain of velocities of *both* the sidebranching instability and the elimination instability (Fig. 20-b).

## 5.4. *Micro/Macro segregation*

### 5.4.1. *Microsegregation*

An important implication of instabilities is the breaking of homogeneity of solute concentration in the solid phase. This follows from the breaking of planarity of the interface following which temperature $T_I$ and thus, solute concentration $c_I = c_L(T_I)$, varies along it according to the Gibbs–Thomson relationship (15). As the concentration of the solid phase is proportional to that of the liquid phase at the interface, $c_S = Kc_I$, this inhomogeneity transfers into the solid where, following the extremely low concentration diffusivity, it remains frozen (Fig. 21-a). In particular, following (15) in the usual case $m > 0$, the colder the interface, the larger its solute concentration $c_I$ and thus, that of the resulting solid. Grooves then build a more impure solid than cell tip. As a result, a steady cellular interface generates a

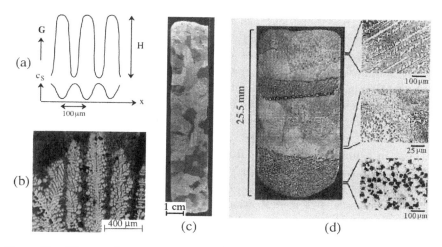

Figure 21. Micro- and macro-segregation. (a,b) Microsegregation at the cell scale[8]: (a) Sketch of a cellular array and of the resulting concentration profile in the solid phase; (b) Longitudinal section of a directionally solidified $Al - 3wt\%Cu$ alloy. Reprinted figure from *Materials Science and Engineering* **A327**, M. Gündüz and R. Çadirli, Directional solidification of aluminium-copper alloys, 167-185, Copyright (2002), with permission from Elsevier. (c,d) Macrosegregation at the ingot scale[24,25]: (c) Cross-sectional view of an ingot of a $Al - 1wt\%Cu - 0.67wt\%TiB_2$ alloy cooled by a side[24]. (d) Solidification of a Pb-Sn eutectic alloy showing the different kinds of microstructures brought about by macrosegregation[25]. Reprinted figures (c,d) from reference 24.

stratified solid involving impure lines behind grooves, purer lines behind cell tips, and a horizontal concentration gradient in between (Fig. 21-a).

An estimation of the level of heterogeneity can be obtained from the Gibbs–Thomson relation. Denoting $H$ the groove length, the increase of concentration at its ends writes $\delta c = HG/m = H\Delta c/l_T$ (Fig. 21-a). With a thermal length $l_T = D/V_c$ of about 1 mm and grooves length in between 3 and 10 mm, one arrives at a concentration increase as large as several miscibility gap $\Delta c$ in the grooves. Accordingly, the solid formed there involves a concentration $c_S$ several times larger than the mixture concentration $c_\infty$ from which it is formed: $\delta c_S = K\delta c = (1 - K)c_\infty H/l_T \gg c_\infty$. This corresponds to a noticeable segregation of solids (Fig. 3-a and 21-b). As it stands on the scale of the growth cells, it is called microsegregation.

### 5.4.2. *Macrosegregation*

Other mechanisms[1,6,26] can yield noticeable heterogeneities of solid composition or of solid structure at a scale large compared to the cell scale

(Fig. 21-c,d). They then yield a so-called macrosegregation that is responsible for the variation of quality of man-made solidified objects and for the picks of distribution of minerals on our planet crust.

The main origin of macrosegregation lies in pre-existing modulations of concentration or of thermal gradient. However, even in a uniform mixture, the transient phase required to build a diffusive layer ahead of a planar front at the beginning of growth gives rise to a gradient of solid concentration extended from a small initial value $Kc_\infty$ to the nominal concentration $c_\infty$. At last, in a finite volume, an enriched zone is formed when the impure diffusive layer reaches the end. Only in between can solidification generate a homogeneous solid.

The mechanisms of solidification show that the formation of solid at the interface only depends on a diffusion flux built on a diffusion length $l_D$ and on the configuration of isothermal lines in the interface vicinity. Accordingly, in an inhomogeneous mixture put in a non-constant thermal gradient, the composition of the resulting solid will only depend on the concentration field on a diffusion length ahead of the interface and on the thermal gradient in between the solidus and the liquidus line, i.e. over a thermal length around the interface. Their variations on scales larger than $l_D$ or $l_T$ will then imprint the solid, therefore resulting in macrosegregation.

Inhomogeneities of thermal gradient induce changes of morphologies of the solidifying interface and thus, large scale heterogeneities of microsegregation. In comparison, large scale variations of concentration are more important since they result in variations of the mean concentration of the resulting solid. Their main origin stands in convective flows induced by thermo-convection (i.e. thermally-induced density differences) or by compositional convection (i.e. composition-induced density differences). They are now recognized as the main cause of macrosegregation in metallurgy (Fig. 21-c,d) or in geophysics (Fig. 3-b,c and 13). Their modeling imposes to take into account the solidification microstructures that form at the solid-liquid interface on a scale that may even exceed the thermal length, due to solute enrichment in cell grooves. In particular, the dense array formed by dendrites, the so-called mushy zone, is so tight that it requires calling for the Darcy law of porous media to model the flow field across it.

A last cause of macrosegregation consists in the transport of solid detached from the interface on long distances. They then act as germs in supercooled regions which initiate solidification prior to the arrival of the main interface. This results in a different morphology, usually more isotropic than would had been obtained otherwise.

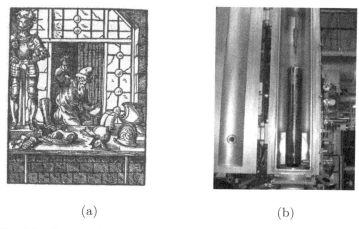

(a) (b)

Figure 22. Metallurgy. (a) In 1568, in Germany, from Jost Amman and Hans Sachs, Frankfurt am Main. (b) Nowadays: Czochralski growth of an aluminium ingot.

### 5.4.3. *Metallurgy*

Metallurgy aims at determining the relevant solidification procedures to reach a designed objective. This may address strength, concentration homogeneity, mono-crystal property or resistance to temperature for instance. It has long consisted of an art that made the difference between a blacksmith capable of forging swords as strong as Excalibur and others (Fig. 22-a). It has now become a science whose foundations rely on the analysis developed above (Fig. 22-b).

Usually, single crystals are sought for a good homogeneity of physical transport at the microscale, as now required in electronics. There are also requested for improving high temperature properties as for instance in turbine blades. A clever trick to obtain a single crystal consists in promoting a dynamical selection between dendrites in a long narrow channel prior to bulk solidification. Selection is based on the growth direction undertaken by dendrites at large growth velocities where, following the large value of the Péclet number, dendrite dynamics is independent of their environment. They then grow in a direction close to one of their principal crystalline axis. Those that move close to the channel axis will then be favored, the others being screened out when encountering the channel wall or the well aligned dendrites. As a result, only those dendrites whose principal cristalline axis is aligned with the channel axis succeed in entering the bulk. They then promote the solidification of a single crystal.

When a single crystal cannot be achieved, metallurgists often prefer to take on the opposite limit of numerous grains of smallest possible size. This is because grain boundaries yield preferred fracture paths when extended on a long range. Achieving a small grain distribution in the materials then allows this danger to be reduced.

A main concern of metallurgists refers to the macrosegregation that occurs in casting and ingots according to the following events (Fig. 21-c,d). First, nuclei spontaneously form in the liquid phase close to the mould wall due to a large thermal gradient amplitude there. They then give rise to dendrites growing in random directions to form a so-called outer equiaxed zone. Here "equiaxed" means that, in average, all cristalline directions are displayed and "outer" is added to make distinction between this equiaxed zone and a similar one that will appear in the bulk. Competitive growth between these dendrites favors those which have a cristalline axis aligned with the direction of the heat flow. This gives rise to a so-called "columnar zone" in which the cristalline axis are far more ordered than in equiaxed zones (Fig. 3-b). Several columnar zones may be distinguished depending on their cristalline orientation. Usually, all faces of the mould give rise to one columnar zone with grain boundaries in between. Lastly, some branches of dendrites may be detached by convective flows and transported in the liquid bulk where they initiate additional solidification zones. They may be accompanied by spontaneous germ nucleations too. This therefore gives rise to a new equiaxed zone called inner equiaxed zone. In addition to these variety of morphologies, convective flows mix the liquid medium prior to solidification and induce this way concentration variations at the scale of the casting (Fig. 3-b and 21-c,d).

Instead of a homogeneous solid in composition, one then gets an heterogeneous material in composition and in morphology (Fig. 3-b, 21). Important variations may be obtained depending on the nature of the material: in stainless steels, the structure is often fully columnar; in well-grain-refined metallurgy, the structure is fully equiaxed. Usually, in columnar zones, the thinner the dendrites, the better the material will be in strength and in transport capability. For this, large growth velocity and thus, large thermal gradient, are required.

When a larger control of solidification microstructures is required, a number of refined crystal growing methods may be used in which the direction of growth and of thermal gradient are monitored. For instance, in the Bridgman method, a crucible is translated within a furnace, yielding directional growth. In another, the Czochralski method, a seed crystal attached to a rod is placed in contact with a melt and pulled out slowly so as to continuously form a solidified rod (Fig. 22). This method is especially well suited to produce single crystals. One concern with these crystal-translating or crystal-pulling methods is the constancy of velocity since any variation of it makes the interface enter the diffusion layer and change both the solute concentration at the interface and in the resulting solid. The corresponding fluctuations of concentration are called zone banding.

## 6. Steady growth forms

In section 3, it has been shown that a steadily growing *planar* form could only be achieved at $\Delta = 1$ or, with kinetic effects, at $\Delta > 1$. In particular, in the main practical case of low undercooling, $\Delta < 1$, planar front velocity can only be unsteady and, in a self-similar regime, of a diffusive kind: $V \propto t^{-1/2}$. These statements call for seeking *non-planar* forms to achieve steady growth.

In the following, an important role will be devoted to surface tension. Forms which neglect it will loose a characteristic scale of the system, the capillary length, and will then display some indeterminacy. Those which take it into account will provide either a discrete set of solutions or none. In the latter case, anisotropy will offer an additional parameter to recover one or several branches of solutions. In all cases, stability considerations will select a definite branch: the quickest branch.

### 6.1. *Forms in absence of surface tension*

#### 6.1.1. *Ivantsov paraboloïds*

We address the form of the steadily growing curved interfaces displayed during the solidification of a pure melt.[27,28] We denote $\mathbf{V} = V\mathbf{e}_z$ their growth velocity with respect to the melt. The temperature field of these so-called dendrites thus simply undergoes a translation at velocity $\mathbf{V}$: $T(x, y, z, t) \equiv T(x, y, z - Vt)$. In absence of surface tension, it then satisfies the differential system:

$$\frac{\partial T}{\partial t} = D_T \nabla^2 T = -V \frac{\partial T}{\partial z} \tag{52}$$

$$Q V \mathbf{e}_z.\mathbf{n} = -D_T c_p \nabla T|_I.\mathbf{n} \tag{53}$$

$$T_I = T_M \quad ; \quad T(x, y, \infty) = T_\infty \tag{54}$$

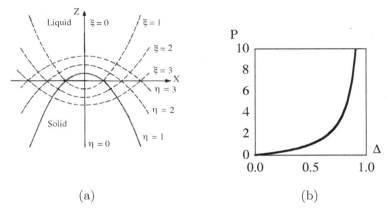

(a)                                              (b)

Figure 23.    Ivantsov paraboloïds. (a) Parabolic coordinates $(\xi, \eta)$. In absence of surface tension, the growing dendrite is an iso-$\eta$ surface, e.g. $\eta = 1$. (b) Variation of the Péclet number $P = \rho V / D_T$ with the undercooling $\Delta$. A growth solution exists for any $\Delta < 1$. It only prescribes the product $\rho V$ of the dendrite tip radius and the dendrite growing velocity.

where diffusion has been neglected in the solid phase. For the sake of simplicity, we restrict analysis to planar forms, the generalization to volumic forms being straightforward.

To infer the resulting form of the interface, it is convenient to consider the heuristic model of a point heat source moving at constant velocity $\mathbf{V}$. This source diffuse heat according to $\delta x \approx (D_T \delta t)^{1/2}$ while translating according to $\delta z \approx V \delta t$. As a result, iso-thermal lines correspond to $\delta z \approx \delta x^2$, i.e. to parabolas.

Following this remark, it is thus relevant to consider parabolic coordinates $(\xi, \eta)$ defined by:

$$x = \rho \, (\xi \eta)^{1/2} \ ; \ \ z = \rho \frac{\eta - \xi}{2} \tag{55}$$

where $\rho$ denotes a scaling length. The iso-$\eta$ lines satisfy $2\eta^{-1} z / \rho = 1 - \eta^{-2}(x/\rho)^2$. They are thus parabolas whose curvature radius at the tip is $\eta \rho$ (Fig. 23-a). In particular, $\rho$ appears as the curvature radius of the parabola $\eta = 1$. The iso-$\xi$ lines are parabolas normal to the iso-$\eta$ lines.

Without loss of generality, we seek a solution for which the interface is the parabola $\eta_I = 1$. As the interface is an iso-$T$, this means that, on the line $\eta = 1$, $T$ is independent of the coordinate $\xi$. This motivates us to investigate whether this property could extend to any $\eta$ line by seeking a self-similar solution only dependent on $\eta$: $T(\xi, \eta, t) \equiv T(\eta)$.

To express the differential system in parabolic coordinates, we recall that $\mathbf{V} = \rho V/2 \, (-h_1^{-1}, h_2^{-1})$, $\nabla = (h_1^{-1}\partial/\partial\xi, h_2^{-1}\partial/\partial\eta)$, $\nabla^2 = (h_1 h_2)^{-1}[\partial/\partial\xi(h_2/h_1\partial/\partial\xi) + \partial/\partial\eta(h_1/h_2\partial/\partial\eta)]$ with $h_1 = \rho/2 \, (1 + \eta/\xi)^{1/2}$, $h_2 = \rho/2 \, (1 + \xi/\eta)^{1/2}$. This yields for the self-similar solution $T(\xi, \eta, t) \equiv T(\eta)$ with $\eta_I = 1$ the ordinary differential equation:

$$-V\eta^{1/2}\frac{dT}{d\eta} = \frac{2D_T}{\rho}\frac{d}{d\eta}(\eta^{1/2}\frac{dT}{d\eta}) \; ; \; V\frac{Q}{c_p} = -\frac{2D_T}{\rho}\frac{dT}{d\eta}|_1 \; ; \; T(1) = T_M$$

(56)

where $\rho$ stands as the actual curvature radius of the interface since $\eta_I = 1$.

Interestingly, the coordinate $\xi$ has canceled out from the differential system, despite the dependence of the factors $h_1$, $h_2$ on it. This attests that a self-similar solution actually exists. Its expression is obtained by straightforward integration and reads, for an undercooling $\Delta = c_p(T_M - T_\infty)/Q$:

$$\Delta = \frac{P}{2}\exp(\frac{P}{2})\int_1^\infty \eta^{-1/2}\exp(-\frac{P\eta}{2})d\eta$$

(57)

where $P = \rho V/D_T$ denotes the Péclet number based on curvature radius $\rho$.

Following (57), a given undercooling $\Delta$ corresponds to a definite Péclet number $P(\Delta)$ and thus to a family of solutions satisfying $\rho V = D_T P(\Delta)$ (Fig. 23-b). In particular, slow dendrites with a large curvature radius or quick dendrites with a small curvature radius can be equally displayed. The solution for a parabolic dendrite solution with no surface tension is thus indeterminate.

Notice that this indeterminacy could have been expected since the beginning from the two following properties: the absence of surface tension denies prescribed characteristic length in the system; both relations (52)(53) are invariant by the rescaling $(x, z, V) \to (\alpha x, \alpha z, \alpha^{-1}V)$. Then, the scale-invariance of the system transposes itself as a velocity invariance, thereby resulting in the dynamical indeterminacy.

### 6.1.2. *Saffman–Taylor fingers*

We consider a fluid displacing a more viscous fluid in a Hele-Shaw cell (Fig. 6-a, 24-a). This yields a steady finger-like interface advancing at a constant velocity $\mathbf{U} = U\mathbf{e}_x$ (Fig. 1-a). Meanwhile the more viscous fluid is repelled at velocity $\mathbf{V} = V\mathbf{e}_x$ at the channel extremity. The issue[2] consists in determining the interface shape with respect to the ratio $V/U$.

The channel corresponds to the space $-1 \le y \le 1$, the $z$-axis being the depth axis (Fig. 24-a). We denote $\mathbf{v}_i$, $i = 1, 2$, the velocity of each medium

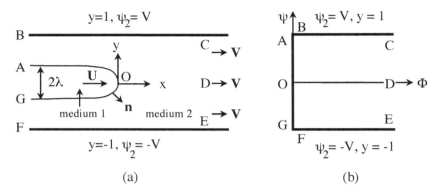

(a)                                          (b)

Figure 24.   Saffman–Taylor finger. (a) In the real space. The channel is bounded by
the $y = \pm 1$ planes. The finger velocity $U$ and its half-width $\lambda < 1$ are related to the
fluid velocity far from the finger by $U = \lambda V$. (b) In the $\Omega$-plane $(\Phi, \Psi)$. The system
boundaries are the $\Psi_2 = \pm V$ lines and the $\Phi_2 = 0$ line. The finger corresponds to the
segment $(\Phi_2, \Psi_2) = (0, \pm V)$. Letters show the correspondence between the $z$-plane $(x, y)$
(a) and the $\Omega$-plane $(\Phi, \Psi)$ (b) on definite points.

averaged over the cell depth, $p_i$ their pressure field, $\mu_i$ their viscosity, $d$ the
cell depth and $K_i = d^2/12\mu_i$. They satisfy the system (Sect. 2.3):

$$\mathbf{v}_i = -K_i \nabla p_i \; ; \; \nabla^2 p_i = 0 \; ; \; \mathbf{v}_i.\mathbf{n}|_I = U\frac{\mathbf{V}}{V}.\mathbf{n} \tag{58}$$

where $\mathbf{n}$ denotes the interface normal pointing towards the medium 2 in
which the interface advances (Fig. 24-a).

We denote $\lambda$ the half width of the finger far from its tip. Mass conservation implies that the space win by unit time by the finger $2\lambda U$ corresponds
to that expelled by unit time at the channel end $2V$, so that $V/U = \lambda < 1$.

As the fluids are two-dimensional, incompressible and potential, it is
convenient to introduce potentials $\Phi_i = -K_i p_i$ and stream functions $\Psi_i$ so
that:

$$\mathbf{v}_i = \nabla \Phi_i = \nabla \wedge \Psi_i.\mathbf{e}_z \tag{59}$$

The complex potential $\Omega_i = \Phi_i + i\Psi_i$ is then an analytical function of
$z = x + iy$ satisfying both $\nabla^2_{x,y} \Omega_i = 0$ and $\nabla^2_{\Phi_i, \Psi_i} z = 0$. This reveals a
duality between $\Omega$ and $z$ that allows the interface to be treated as a line in
either the $z$ plane or the $\Omega$ plane.

The clue then consists in noticing that if the issue involves a *free* boundary in the real space $(x, y)$, it involves a *fixed* boundary in coordinates $(\Phi, \Psi)$
(Fig. 24-b). The main difficulty of the problem can thus be overcome by

seeking to determine the digit boundary in term of $(x, y)$ as a function of $(\Phi, \Psi)$ instead of the opposite, i.e. as a line $z(\Omega)$ in the $\Omega$-plane instead of a line $\Omega(z)$ in the $z$ plane.

The boundary conditions of the system in the $\Omega$ plane are the following. As $\mathbf{n} = (-dy/dl|_I, dx/dl|_I)$ and $\mathbf{v}|_I = (\partial\Psi_i/\partial y|_I, -\partial\Psi_i/\partial x|_I)$ in the $z$-plane, the boundary condition on the interface (58) reduces to $d\Psi_i/dl|_I = U dy/dl|_I$. Taking $\Psi_i|_I = 0$ at the finger tip then yields $\Psi_i|_I = Uy|_I$ on the interface. In addition, without surface tension, the pressure involves no discontinuity at the interface. Neglecting the viscosity of the less viscous fluid then yields a constant pressure in medium 1, and finally, on the finger and in both media: $p_i|_I = p_0$. Without loss of generality, we may then take $\Phi_2|_I = 0$ on the interface. It remains to determine the boundary conditions on the channel side and at infinity. They straightforwardly write from (59): at $x = +\infty$, $\Phi_2 = Vx$ and $\Psi_2 = Vy$ since $\mathbf{v}_2 = V\mathbf{e}_x$; at $x = -\infty$, uniform $\Phi_2$ and $\Psi_2$ since the fluid is at rest; at $y = \pm 1$, $\partial\Phi_2/\partial y = 0$, $\partial\Psi_2/\partial x = 0$ since $\mathbf{v}_2.\mathbf{e}_y = 0$. This implies $\Psi_2(x, \pm 1) = \pm V$, $\Phi_2(-\infty, y) = 0$, $\Psi_2(-\infty, y) = Uy_I$ and $\Phi_2|_I = 0$. Accordingly, in the $\Omega$-plane, the boundaries of medium 2 simply appear as a fixed channel, $(\Phi_2 > 0, \Psi_2 = \pm V)$, $(\Phi_2 = 0, |\Psi_2| \leq V)$, the interface being located on the segment $\Psi_2 \leq V$ of the $\Psi$-axis (Fig. 24-b).

The normal modes of the Laplacian that are appropriate to the boundary conditions are $z = \exp(-in\pi\Psi_2/V)\exp(-n\pi\Phi_2/V)$, $z = \Phi_2$, $z = \Psi_2$. As the function $z(\Phi_2, \Psi_2)$ is Laplacian, i.e. $\nabla^2_{\Phi_i, \Psi_i} z = 0$, expanding it in terms of these modes reduces the issue to an algebraic problem whose solution is provided by the projection of the boundary conditions on the normal modes. One then obtains:

$$z = \frac{\Omega_2}{V} + \frac{2}{\pi}(1 - \frac{V}{U}) \ln\left[\frac{1 + \exp(-\pi\Omega_2/V)}{2}\right] \tag{60}$$

The finger shape now simply corresponds to the segment $\Omega_2 = (0, \pm V)$ with $\Omega_2|_I = (0, Uy|_I)$. It then reduces to the Saffman–Taylor profile:

$$x = \frac{1 - \lambda}{\pi} \ln\left[\frac{1 + \cos(\pi y/\lambda)}{2}\right] \tag{61}$$

As a solution is obtained this way for any half-finger width $\lambda = V/U$, the issue remains indeterminate. This is because, without surface tension, the system lacks a characteristic scale to fix a characteristic scale ratio. This scale-ratio invariance then transposes itself as a velocity-ratio invariance and thus to the above indetermination.

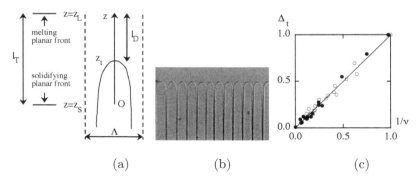

(a)                          (b)                          (c)

Figure 25.   Cell tip undercooling in directional solidification. (a) Sketch of cell position.
The solidus (liquidus) line $z = z_S$ ($z = z_L$) corresponds to the position of a solidifying
(melting) planar front. Cell tip stands in between them at a distance from the liquidus
which, following measurements (67), is the diffusion length $l_D$. (b) Images of an array
of cells showing different cell spacings $\Lambda$ but the same position of cell tips[32]. This shows
the independence of cell tip undercooling $\Delta_t$ with $\Lambda$. (c) Measurements[32] of cell tip
undercooling $\Delta_t$ with reduced pulling velocity $\nu = V/V_c$ from cell tip position (light
circles) and cell tip temperature (full circles). Both yield an affine dependence of $\Delta_t$ on
$1/V$ which reduces to the linear relationship $\Delta_t = 1/\nu$.

### 6.1.3. *Directional growth cells*

A stimulating analogy[29] can be drawn between Saffman–Taylor fingers and
the cell tips that are displayed in directional solidification at low Péclet
numbers $P = \Lambda V/D$ where $\Lambda$ denotes the cell spacing and $\mathbf{V} = V\mathbf{e}_z$ the
cell growing velocity. It relies on the possibility of neglecting concentration
advection in the tip region in this regime, and thus of reducing solidification
to a diffusive dynamics.

For simplicity, we reduce analysis to a plane. We place the origin of coordinates $(x, z)$ on the cell axis and at the location of a virtual steady planar
front, $z_S = 0$, and we denote $(0, z_t)$ the position of the cell tip (Fig. 25-a).
In the cell tip region, solute concentration varies from the cell tip concentration $c_t = c_I(0, z_t)$ to the boundary concentration $c(\pm\Lambda/2, z_t)$. Following
the Gibbs–Thomson relationship (15), the departure of the interface from a
straight line $z = z_t$ actually confirms the variation of solute concentration
on this line. This implies transverse solute concentration gradient on the
scale $\Lambda$ so that space derivatives are of order $\Lambda^{-1}$.

With this scaling, the relative magnitude of concentration advection
over concentration diffusion in the cell tip region writes: $|V\mathbf{e}_z.\nabla c/D\nabla^2 c| \approx$
$\Lambda V/D = Pe$. In the present low Péclet number regime, $Pe << 1$, advection
may therefore be neglected, thereby yielding, for steady cell, a Laplacian
dynamics: $\nabla^2 c \approx 0$ (23).

The corresponding scalar field, $c$, varies on the interface according to the Gibbs–Thomson relationhip (15), i.e. linearly with respect to the interface position $z_I$ in the thermal gradient: $c_I = c_\infty/K - z_I\,G/m$. It is therefore appealing to turn attention to another scalar field $\Phi = c - c_\infty/K + z\,G/m$ that provides the opportunity of still being Laplacian, $\nabla^2\Phi \approx 0$, while being constant on the interface: $\Phi_I = 0$. In addition, the conservation relation at the interface reads:

$$\nabla\Phi.\mathbf{n}|_I = [-(1 - K)c_I\frac{V}{D} + \frac{G}{m}]\frac{\mathbf{V}}{V}.\mathbf{n}. \tag{62}$$

Here the variation of $c_I$ on the cell tip region, $\delta c_I \approx \Lambda G/m$ can be neglected since it amounts to $\delta c_I \approx \Delta c\,\Lambda V_c/D << \Delta c$: $c_I \approx c_t$.

The system satisfied by the scalar $\Phi$ in the tip region is thus the same as in the digitation problem (58). Accordingly, the shape of steady cell tip is similar to that of a viscous digit. It is thus represented by a Saffman–Taylor profile with half width $\lambda$ (61).

The value of $\lambda$ is given by the analogous of the velocity ratio in the digitation problem, i.e. by the ratio of the flux at the interface (62) and at infinity. The latter is obtained by considering the intermediate asymptotic domain $\Lambda << z << l_D$ that is close enough to the cell tip for $\Lambda$ still being a characteristic scale and far enough from the cell tip for $c$ approaching the far field value $c(x, z) = (c_t - c_\infty)\exp(-Vz/D) + c_\infty$. Then $\nabla\Phi.\mathbf{n}|_{\Lambda << z << l_D} = -(c_t - c_\infty)V/D + G/m$. The velocities analogous to $\mathbf{U}$ and $\mathbf{V}$ in the Saffman–Taylor problem are then: $\mathbf{U} \equiv [-(1 - K)c_t V/D + G/m]\mathbf{e}_z$ and $\mathbf{V} \equiv [-(c_t - c_\infty)V/D + G/m]\mathbf{e}_z$, so that the parameter $\lambda \equiv V/U$ writes:

$$\lambda = \frac{(c_t - c_\infty)V/D - G/m}{(1 - K)c_t V/D - G/m}. \tag{63}$$

Here again, a family of forms is identified but the actual value of $\lambda$, or equivalently of tip concentration $c_t$ or of tip position $z_t$, remains undetermined.

This family of forms ceases to apply at a distance of order $\Lambda$ from the cell tip, since advection can then no longer be neglected (Fig. 25-a). Transverse diffusion being dominant over longitudinal diffusion, the advection-diffusion equation for solute concentration reduces to $-V\partial c/\partial z \approx D\partial^2 c/\partial y^2$. However, as the distance left between the cell groove and the cell boundary is small with respect to the diffusion length, we may neglect the variation of the concentration gradient $\partial c/\partial z \approx G/m$ on it. Integrating the above equation from $y = y_I$ to $y = \Lambda/2$ then yields: $V\partial c/\partial z|_I(\Lambda/2 - y_I) = D\partial c/\partial y|_I - D\partial c/\partial y|_B$ where the index $B$ denotes the cell boundary.

However, the transverse flux at the cell boundary vanishes by symmetry, $\partial c/\partial y|_B = 0$, and the transverse flux at the interface satisfies the mass conservation equation (24):

$$V \, \Delta c_I \frac{dy}{dz}|_I = -D \left[ \frac{dy}{dz} \frac{\partial c}{\partial z} - \frac{\partial c}{\partial y} \right]|_I. \tag{64}$$

where $\Delta c_I$ denotes the miscibility gap $(c_L - c_S)_I$. This finally yields the evolution equation on the interface:

$$V \frac{\partial c}{\partial z}|_I \, (\Lambda/2 - y_I) = \left[ V \Delta c_I + D \frac{\partial c}{\partial z}|_I \right] \frac{dy}{dz}|_I. \tag{65}$$

Following the Gibbs–Thomson equation (15), the term $(\partial c/\partial z)_I$ can be approximated by $-G/m$. Considering that the miscibility gap is a constant on the domain studied, i.e. $\Delta c_I = c_\infty (1 - K)/K$, then yields by integration an exponential profile[30] $y_I(z) - \Lambda/2 \propto \exp[Pe/(\nu - 1) \, z/\Lambda]$.

Another kind of approximation, introduced by Scheil,[31] consists in keeping the actual miscibility gap $\Delta c_I = (1 - K)c_I$ but in neglecting the term $D(\partial c/\partial z)_I$ in (65). This is legitimated by the fact that within the above approximation of the longitudinal gradient, it amounts to about $DG/m \approx V_c \, \Delta c$ and is thus negligible compared to $V \Delta c_I$ for $V \gg V_c$. Integration then yields a power law variation for the interface profile $y_I(z) - \Lambda/2 \propto B z^{-1/(1-K)}$.

The link between the two approximations follows from the fact that a constant miscibility gap formally corresponds to $K = 1$.

## 6.2. *Experimental determination of forms and undercooling in directional solidification*

### 6.2.1. *Tip undercooling*

Apart from the interface shape, another more integral variable specifies the solution for solidification: the undercooling $\Delta_t$ of cell tips. In a binary mixture, it corresponds to the relative position of the temperature $T_t$ of the cell tip within the range bounded by the liquidus temperature $T_L$ and the solidus temperature $T_S$ at the concentration $c_\infty$ of the mixture: $\Delta_t = (T_t - T_L)/(T_S - T_L)$. However, the uniformity of the thermal gradient and the linearity of the Gibbs–Thomson relationship (15) enable it to also express with respect to the tip position $z_t$ or the tip concentration $c_t$:

$$\Delta_t = \frac{T_t - T_L}{T_S - T_L} = \frac{z_t - z_L}{z_S - z_L} = \frac{c_t - c_L}{c_S - c_L} \tag{66}$$

Here the index L and S label the liquidus and the solidus respectively. In particular, $z_L$ and $z_S$ correspond to the positions where the liquidus temperature and the solidus temperature are reached (Fig. 25-a) and $c_L$ and $c_S$ to the concentrations at which the temperatures $T_L(c_\infty)$ and $T_S(c_\infty)$ are encountered on the liquidus line: $c_L = c_\infty$, $c_S = c_\infty/K$ (Fig. 8-a). In practice they correspond to the concentrations at a planar interface steadily growing at the liquidus position $c_I = c_\infty = c_L$ or on the solidus position $c_I = c_\infty/K = c_S$.

For a prescribed steady shape, the tip concentration $c_t$ sets an origin for the evolution of concentration along the interface. It thus directly affects the net solute concentration flux $\Phi$ absorbed by it. In particular, shifting $c_t$ by $\delta c_t$ while keeping the same shape changes $\Phi$ by $\delta\Phi = \delta c_t V \int_I (\mathbf{e}_z.\mathbf{n})dl$.

Shape and tip concentration thus appear intimately coupled in steady states: both monitor the concentration flux $\Phi$ absorbed by the whole interface which, in steady states, equals the advective flux coming from the far liquid: $\Phi = c_\infty V \Lambda$ where $\Lambda$ denotes the cell spacing. This means that $\Delta_t$ (resp. $\Delta_t(V)$) is liable to specify the solution (resp. the branch of solution). Determining $\Delta_t(V)$ thus stands as a simple but acute mean for characterizing the type of growth solution.

We have determined experimentally the evolution of cell tip undercooling in directional solidification of a succinonitrile mixture by combining two measurements methods.[32] Each is appropriate only in a given velocity regime but, taken together, they provide a complete determination over the whole available range. None of them revealed a noticeable variation of tip undercooling with the cell spacing.[33] This was confirmed with the naked eye by the alignment of tip cells on the same isothermal line, whatever the cell spacing: $\partial\Delta_t/\partial\Lambda \approx 0$ (Fig. 25-b).

The method appropriate at low velocities consists in determining the cell tip position $z_t$ in the thermal field. We observed a linear variation of $\Delta_t$ with the inverse pulling velocity $V^{-1}$ and no variation with the cell spacing $\Lambda$: $\Delta_t = a/V + b$ (Fig. 25-c). This method failed to be accurate at large velocities when the shift of the temperature field due to the advective transport by the moving sample became of order of the position shift $z_t - z_S$ to measure. We then turned to a direct measure of cell tip temperature $T_t$ by a thermocouple immersed in the mixture. Tip temperature was caught at the time the thermocouple crossed the isothermal line joining the cell tips. Here too, a linear relationship of tip undercooling with the inverse pulling velocity was obtained: $\Delta_t = a'/V + b'$ (Fig. 25-c).

In practice, the liquidus or the solidus variables correspond to a steadily advancing planar front either melting (L) or solidifying (S). The former provides $c_L$, $T_L$ or $z_L$ and the latter $c_S$, $T_S$ or $z_S$. However, as the solute diffusion time scales as $\tau = D/kV^2$, large velocities are required to accurately reach steady state in a moderate time. In practice, this can be achieved on melting fronts since they do not exhibit instabilities but not on solidifying front since planarity is limited to low velocities $V < V_c$. Only liquidus variables have then be directly measured. Fortunately, this did not prevent the whole relationship to be identified for the following reasons.

Using the measured values for $T_L$ or $z_L$, it appeared that the constants $b$ and $b'$ vanish. This means that cells solidifying at large velocities display a tip temperature close to the liquidus temperature and a tip concentration close to $c_\infty$. Interestingly, the domains available for each kind of measurements displayed a large overlap on which the slopes $a$ and $a'$ could be compared. They were identical, thereby denying a second order transition of the tip undercooling. This, together with the specific planar front condition $c_t(V_c) = c_S$, i.e. $\Delta_t(V_c) = 1$, prescribes the slopes, $a = a' = V_c$, and yields the simple relationship: $\nu\Delta_t = 1$.

This relation corresponds to that proposed by Bower, Brody and Flemings on phenomenological arguments.[34] It may be simply written $\Delta = l_T/l_D$ or:

$$z_L - z_t = l_D \;\; ; \;\; z_t - z_S = l_T - l_D. \tag{67}$$

Its simplicity worths being emphasized: among the characteristic length scales $(d_0, l_D, l_T, \Lambda)$ involved in solidification, only one of them is involved in the distance of cell tips to the liquidus line: the diffusion length $l_D = D/V$.

### 6.2.2. *Cell form*

Several models have been proposed to determine the form of definite cell parts in various regimes. They concluded to exponential or power law profiles in cell tails (Sect. 6.1.3), to a finger-like (resp. parabolic) shape of cell tips at low (resp. large) Péclet number (Sect. 6.1.3, resp. 6.1.1). In particular, at large Péclet numbers, it has been argued that the diffusive length is so short that the temperature difference $\delta T = GD/V$ across it is negligible compared to the temperature difference $(T_L - T_S) = m\Delta c$ between liquidus and solidus: $\delta T/(T_L - T_S) = (GD/m\Delta c)\, V^{-1} = V_c/V << 1$. Then, neglecting the implications of the thermal gradient on the solidification interface yields the same form as in free growth: a parabolic shape.

Although these studies provided interesting informations on the form of definite cell parts, they raise two main problems:

The first problem questions the status of the cut that is performed to define cell parts. This is because there usually stands a part of arbitrary in cutting shapes which may affect the measurements of variables in practice. An example of this is provided by the curvature radii of cell tips. Their unambiguous definition refers to the second derivative of the cell profile at cell tip. However, for accuracy reasons, curvature radii are never measured that way. Instead, they are determined from fits of cell tips to circles or parabolas in a definite domain, not too small to provide enough accuracy, and not too large to keep cell tips close to a circle or a parabola. However, the resulting data significantly varies in this range, so that even the relevance of curvature radius defined this way may be questionned.[35,36]

The second problem refers to finding a solution for a global cell shape by matching different kinds of geometry, each of them being valid in a specific cell region. A tentative for this was provided by Dombre and Hakim[30] but it failed in selecting a definite Saffman–Taylor branch of solution for the cell tip. A second tentative has been brought about by Spencer and Huppert[37,38] by taking into account cell interactions in a cellular array. It is however restricted to large velocities, $V/V_c >> 1/K$, where cells usually emit sidebranches.

These statements have motivated us to perform a global study of cell shape, both in the real space and in the control parameter space.[39] Global study in the real space means that the whole cell shape, from its tip to its groove, is determined in a single procedure; global study in the control parameter space means that we are not seeking for the shape of some definite cells but for the branch of solutions that are relevant to cells.

Experiments were performed in a succinonitrile mixture. The procedure has involved the visualization of a number of steady shapes to construct a cell library extended over a significant range of variables $(V, \Lambda, G)$ (Fig. 26-a). Each cell was skeletonized to provide a definite curve for the cell boundary. Applying a reflection symmetry to half of it then provided a function $x(z)$ where $z$ denotes the growth axis (Fig. 26-b).

Different families of functions were used to fit a definite cell accurately. However, as our goal was to identify the features of the cellular branch of solution, we took a special care to obtain fit parameters all displaying coherent variations with the state variables $(V, \Lambda, G)$ in their studied range. This, in particular, required restricting the number of fit parameters to a minimum so as to avoid sporadic compensations of the variations of one by the variations of another. We obtained this way the following complete characterization of cell shape, from their tip to their groove and from the

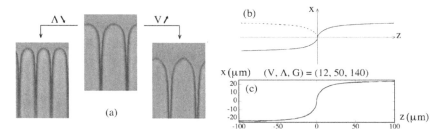

Figure 26.   Cell forms in directional solidification of a succinonitrile alloy[39]. (a) Evolution of cell form with cell spacing $\Lambda$ at fixed velocity $V$ and thermal gradient $G$, and with velocity at fixed spacing and gradient. From left to right, in $(\mu m s^{-1}, \mu m, K cm^{-1})$ units: $(V, \Lambda, G) = (12, 50, 140)$, $(V, \Lambda, G) = (12, 90, 140)$, $(V, \Lambda, G) = (24, 90, 140)$. Increasing velocity sharpens tip cell. Changing spacing nearly keeps the tip geometry unchanged. (b) Symmetry $z \to -z$ for $x > 0$ yielding a single-valued graph for cell boundary. (c) Cell form at $(V, \Lambda, G) = (12, 50, 140)(\mu m s^{-1}, \mu m, K cm^{-1})$ and its optimal fit (68) (69) (70). Form and fit are indistinguishable one from the other.

deep cell regime to the near dendritic regime (Fig. 26-c):

$$x = \frac{\Lambda}{2} \, c_x \, (\frac{l_D}{d_0})^{-1/10} \tanh^{1/2}[c_z \, (\frac{l_D}{d_0})^{9/20} \, (\frac{\Lambda}{d_0})^{-1/4} \, (\frac{l_T}{d_0})^{-1/2} \, \frac{z}{\Lambda}]. \qquad (68)$$

Here $c_x$ and $c_z$ denote the following non-dimensional prefactors:

$$c_x = c_L (\frac{D}{d_0})^{1/10} \qquad (69)$$

$$c_z = 8 \frac{c_\rho}{c_L^2} (T_L - T_S)^{1/2} d_0^{-3/10} D^{-9/20} \qquad (70)$$

where $c_L = 0.66 \ \mu m^{-1/10} \ s^{1/10}$ and $c_\rho = 0.095 \times 10^2 \mu m \ s^{-1/4} \ K^{-1/2}$.

The branch of solution for cells thus differs from the Saffman–Taylor shape whatever the cell part that is considered. Moreover, although the vicinity of cell tip is close to a parabola, the interface quickly departs from a parabola as the distance from the cell tip is increased. Accordingly, none of the kinds of forms proposed to partially fit cell forms prove to be convenient. It nevertheless remains that the cellular branch of solutions is now identified by a simple analytical representation which should be amenable to analytical understanding from modeling.

### 6.2.3. Synthesis

The determination of cell tip position (67) and cell shape (68) provides a complete characterization of the branch of solution for cells.[32,39] Recovering

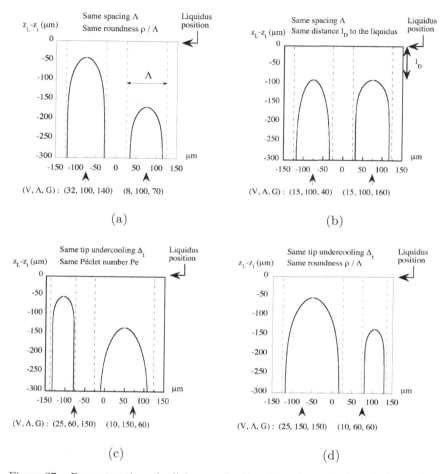

Figure 27. Reconstruction of cell form and cell position from relations (68) and (67). The units of $(V, \Lambda, G)$ are $(\mu m s^{-1}, \mu m, K cm^{-1})$. (a) Same spacing and same roundness but distinct positions. (b) Same spacing and same position but distinct shapes. (c) Same tip undercooling $\Delta_t$ and same Péclet number $Pe$ but distinct forms and spacings. (d) Same tip undercooling $\Delta_t$ and same roundness but different Péclet numbers $Pe$ and spacings.

it from modeling constitutes a theoretical challenge since no theory has succeeded to date in catching a global solution for cell geometry.

To illustrate the identified branch of solution, we report in Fig. 27 some forms and positions of cells for different curvature radius $\rho$, spacing $\Lambda$, roudness $\rho/\Lambda$, undercooling $\Delta_t$ (or $z_t$) and Péclet numbers $Pe$. Interestingly, as the number of variables necessary to specify a cell state is three (e.g. $V, \Lambda, G$), three variables among those listed above can be set independently one of the other. The remaining variables are then prescribed by them and may in particular take the same value for different triplets. This is illustrated in Fig. 27 by couples of cells sharing the same values of two variables among the five.

## 6.3. Selection of forms by surface tension and anisotropy

A large attention has been devoted to identify the mechanism of selection of a particular form among the continuous families available without surface tension (Sect. 6.1). A first obvious track has been to take into account surface tension, in particular for introducing a specific length in the system. Two other clues were however necessary to reach a more satisfactory picture: anisotropy and stability considerations.

A simple framework for fingering out how these considerations may yield selection is provided by geometrical models.[40–46] In their simplest expression, they aim at deducing forms from the normal velocity of the corresponding moving line (Sect. 2.2 and 9).

### 6.3.1. Geometrical model

We consider the geometrical model proposed by Brower, Kessler, Koplik and Levine,[40] following which the interface moves at the following normal velocity in suitable units:

$$U = \kappa + \gamma \frac{\partial^2 \kappa}{\partial l^2} \tag{71}$$

where $\kappa$ denotes the interface curvature and $\gamma$ the surface tension.

We restrict attention to a steady form advancing at velocity $\mathbf{V} = V\mathbf{e}_z$ (Fig. 28-a). In the parametrization where interface points $M$ move on the $z$-axis only, their velocity simply writes $\mathbf{V}_M = \mathbf{V}$, so that the normal velocity reads: $U = \mathbf{V}_M.\mathbf{n} = V\cos\theta$, where $\theta = (\mathbf{V}, \mathbf{n})$ denotes the angle of the normal with the growth axis (Fig. 28-a).

- No surface tension: $\gamma = 0$

  For $\gamma = 0$, the interface satisfies $\kappa = V\cos\theta = d\theta/dl$. However, as along the interface $dx = \cos\theta dl$ and $dz = -\sin\theta dl$, integration yields $x = \theta/V$ and $z = \ln|\cos\theta|/V$. This implies a curvature radius $\rho = 1/V$ following

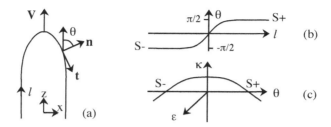

Figure 28. Geometrical model for growth interface with surface tension. (a) Sketch of cell geometry. (b) Evolution of normal orientation, i.e. angle $\theta$, with curvilinear abscissa $l$. The asymptotes correspond to the two fixed points $S\pm$ of the geometrical model. (c) Representation in the phase space $(\theta, \kappa, \epsilon)$ of the sought heteroclinic solution which connects the two fixed points $S\pm$. Whereas it usually involves a non-zero $\epsilon$, it should nevertheless satisfy by symmetry $\epsilon = 0$ at $l = 0$, i.e. at $\theta = 0$.

which the interface geometry writes:

$$z = \rho \ln |\cos(\frac{x}{\rho})| \tag{72}$$

We obtain this way a family of forms involving the indetermination $\rho V = 1$. The situation is therefore similar to that found on physical interfaces (Sect. 6.1) but in a much simpler framework.

Numerical studies[42] have shown that this form is unstable but can be stabilized by surface tension $\gamma$ for $V > V^* = (\tilde{\gamma}^*/\gamma)^{1/2}$ where $\tilde{\gamma}^* = 2.63$. Although interesting, this observation is only partially relevant to our problem since the stabilization by surface tension of a form derived at zero surface tension by no way ensures the existence of a steady state solution at non-zero surface tension.

- Non-zero surface tension: $\gamma \neq 0$

The evolution of form along the interface reduces from (71) to the following dynamical system in the variables $(\theta, \kappa, \epsilon)$:

$$\frac{\partial \theta}{\partial l} = \kappa \; ; \; \frac{\partial \kappa}{\partial l} = \epsilon \; ; \; \frac{\partial \epsilon}{\partial l} = \gamma^{-1}(V \cos \theta - \kappa) \tag{73}$$

This autonomous dynamical system involves the fixed points $S_\pm = (\pm \pi/2, 0, 0)$ which correspond to the asymptotes of the left and right grooves (Fig. 28-b). Its solution for the present problem represents an heteroclinic solution connecting the fixed point $S_-$ to the fixed point $S_+$ as $l$ increases from $-\infty$ to $+\infty$ (Fig. 28-c).

To analyse the conditions for such an heteroclinic solution to exist, it is useful determine the spectrum of the Jacobian of the system at both fixed points.[41] The Jacobian at $S_-$ involves the spectrum $\lambda(\lambda^2 + \gamma^{-1}) = \gamma^{-1}V$ whose solution for $\gamma << 1$ is $\lambda_0^- \approx V$, $\lambda_\mp^- \approx -V/2\pm i\gamma^{-1/2}$. In the same way, the spectrum at the other fixed point $S_+$ is $\lambda_0^+ \approx -V$, $\lambda_\mp^+ \approx V/2 \mp i\gamma^{-1/2}$. Accordingly, there exists a single repelled trajectory in the vicinity of $S_-$ corresponding to the eigenvalue $\lambda_0^-$ and a single attracted trajectory in the vicinity of $S_+$ corresponding to the eigenvalue $\lambda_0^+$. Both must connect somewhere in between the fixed points for an heteroclinic solution to exist. This condition is particularly hard to meet in a three-dimensional space $(\theta, \kappa, \epsilon)$. Would it be so, symmetry implies that, at the interface tip $l = 0$, the second derivative of $\theta$ would vanish $\epsilon(0) = 0$. However, numerical integration[41] reveals a transcendant mismatch: $\epsilon(0) \approx \exp(-c/\gamma)$ where $c$ is a constant. This corresponds to a second order non-differentiability at the tip that denies the possibility of connecting grooves in this geometrical differential system.

When anisotropy is considered, the normal velocity depends not only on curvature but also on the orientation $\theta$ of the interface:

$$U = [1 + \mu\cos(m\theta)]\kappa + \gamma\frac{\partial^2\kappa}{\partial l^2}$$

Here $\mu$ states the magnitude of anisotropy correction and $m$ its order of symmetry. Analytical study of the implications of anisotropy stands beyond actual analytical capabilities. However, numerical simulations[42] show that the matching condition $\epsilon(0) = 0$ is actually reached at two growing velocities. Whereas the smallest leads unstable trajectory, the largest provides a stable interface.[44,46,47] This therefore corresponds to a stable solution for the complete problem that removes the previous indeterminacy. Notice, however, that *both* surface tension and anisotropy have been required for reaching this unique stable solution.

### 6.3.2. *Viscous digitation*

When surface tension is involved, a curved interface between non-miscible fluids is no longer an isobar line. Assuming still that the pressure $p_1$ of the less viscous fluid is a constant, the Laplace relation (9) yields the potential $\Phi_2|_I$ of the most viscous fluid at the interface to depend on the digit curvature $\kappa$: $\Phi_2|_I = K_2\gamma\kappa$.

McLean and Saffman[48] have deduced an integral equation for this interface geometry. It expresses the requirement of a smooth differentiable form at the tip in term of a so-called solvability condition. It has been shown to involve a countable set of solutions[49] among which a single one is stable.[50]

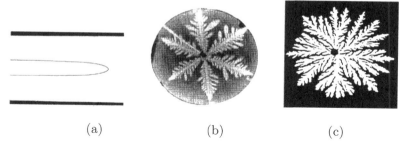

(a)          (b)          (c)

Figure 29. Anomalous or dendritic fingers. (a) A thin groove is engraved on both plates in the middle of the channel and along its axis. The digit form is then still a Saffman–Taylor-like shape but for an unusually small relative width[51]. Reprinted figure with permission from M. Rabaud, Y. Couder, N. Gerard, *Phys.Rev.A* **37**, 935-947 (1998). Copyright (1998) by the American Physical Society. (b) Bottom plate is engraved with a regular sixfold lattice of grooves. This results in the formation of sixfold dendritic interface emitting sidebranches[52]. Reprinted figure with permission from E. Ben-Jacob, R. Godbey, N.D. Goldenfeld, J. Koplik, H. Levine, T. Mueller, L.M. Sander, *Phys.Rev.Lett* **55**, 1315-1318 (1985). Copyright (1985) by the American Physical Society. (c) A viscous digitation experiment is performed with a nematic liquid crystal as the most viscous fluid and air as the less viscous fluid. This yields a dendritic interface emitting sidebranches[53]. Reprinted figure from Coherent Structures in Complex Systems, vol.567, 2001, P.298-318, Patterns in the bulk and at the interface of liquid crystals, Á Buka, T. Börzsönyi, N. Éber, T. Tóth-Katona, figure 6DD, with kind permission of Springer Science and Business Media.

Surface tension and stability consideration have thus succeeded in selecting a unique definite solution without additional considerations.

Despite fluids are isotropic, an extrinsic anisotropy can be introduced in the system by engraving plates,[51] by introducing a periodic lattice[52] or by using a liquid crystal.[53] In the former case, grooves etched on both plates in the middle of the channel width gave rise to so-called anomalous fingers[51] that still satisfy a Saffman–Taylor shape (61) but for unusually small relative widths $\lambda$ (Fig. 29-a). In the latter cases, interfaces were found to emit sidebranches in a way reminiscent of solidification dendrites (Fig. 29-b,c).[52,53]

### 6.3.3. *Cristalline dendrite*

When surface tension is involved, the interface of a dendrite solidifying in a pure melt is no longer an isothermal line. In particular, it is colder at the tip than on the sides so that a heat flow $\mathbf{J}_\gamma$ transports energy along the interface to its tip (Fig. 30-a). An estimation of this flux can be given using the curvature radius $\rho$ as a scale to estimate the temperature difference $\delta T \approx (d_0/\rho)(Q/c_p)$ and the resulting thermal flux: $\mathbf{J}_\gamma \approx d_0\rho^{-2}D_T Q$. In comparison, a diffusive heat flow $\mathbf{J}_d = Q\mathbf{V}_I.\mathbf{n}$ is in order at the tip to release heat in the surrounding medium. The ratio of both fluxes provides a non-dimensional parameter $\sigma/2 = l_D d_0/\rho^2$ where $l_D = D_T/V$ stands as the thermal diffusion length. This parameter may be rewritten $\sigma = (\lambda_M/2\pi\rho)^2$ where $\lambda_M = 2\pi(2l_D d_0)^{1/2}$ is the minimum instability wavelength on a paraboloïd (the corresponding wavenumber square $k_M^2$ is twice that reported in section 5.3 because two principal curvatures are taken into account here instead of one on a parabola).

Analytical[54–57] and numerical studies[45,58,59] have shown that the existence of a solution reduces to a criterion on $\sigma$. In particular, Langer and Müller-Krumbhaar[60] have argued that sidebranching makes curvature radius grow until the dendrite state is marginally stable with respect to tip splitting modes. This condition corresponds to $\rho \equiv \lambda_M$, i.e. to $\sigma = 1/4\pi^2 = 0.025$. Comparison with experiments shows a good agreement at low anisotropy[61] but a noticeable dispersion at larger anisotropies.[62] On the other hand, analytical studies similar to that performed in the geometrical model (71) have shown that the nature of this criterion consists in requiring a smooth differentiable dendrite tip in term of a solvability condition.

Without anisotropy, no steady solution with surface tension can be found and experiment[63] and simulations[64] show that dendrites exhibit tip splitting or the formation of a pair of asymmetric cells, the doublons. A finite anisotropy $d_0(\theta) = d_0(1 - 15\epsilon\cos\theta)$ yields a discrete set of solutions[47,55–57] with $\sigma(\epsilon) \propto \epsilon^{7/4}$. As in the geometrical model (Sect. 6.3.1), the quickest solution stands as the only stable one,[46] the other branches undergoing tip splitting.

Note that the solvability criterion on $\sigma$ corresponds to requiring an equilibrium between the strength of diffusive fluxes on a domain of width $\rho$. This needs a diffusive time scale to be satisfied. In comparison, the equilibrium of the whole dendrite and thus the relation $Pe(\Delta)$ (57) requires a much longer time to be satisfied, since the involved spatial scale is much larger.

## 6.4. Synthesis

A similar picture for the quest of stable steady forms thus emerges in any kind of system. No surface tension yields an absence of characteristic scale following which a family of self-similar solutions is obtained (Sect. 6.1). All are however unstable against perturbations. Introducing finite surface tension then not only sets a characteristic scale $d_0$ in the system; it also raises the differential order of the system and thus the difficulty in achieving an heteroclinic orbit representative of a form extended from one groove to the other. As a result, the continuum of solutions reduces to a discrete family in viscous fingering,[49] among which only the quickest is stable,[50] and to an absence of solution in solidification.[55] Hopefully, anisotropy provides an additional parameter while keeping the dimension of the dynamical system the same. It thus offers additional opportunities of finding solutions. Following them, a discrete family of solutions is finally recovered in solidification,[47,55–57] the quickest of all being here too the only stable.[46]

Accordingly, if considering surface tension was mandatory for breaking indetermination, it raised the difficulties inherent to parameters driving higher order derivatives in differential systems.[41] To cure them, stability considerations and use of an additional parameter, anisotropy, proved to be necessary.[41–46]

## 7. Branching dynamics

A crystal growth mode of paramount importance is dendritic sidebranching. It occurs till moderate velocities $V \approx 4V_c$ to much higher values and stands as the dominant growth mode in industrial processes and in nature.

Sidebranching consists in the repetitive emission of branches on the sides of tubular or parabolic growth forms (Fig. 30-b). As branches stay mainly at rest with respect to the medium in which the dendrite advances, they seem to move back to its grooves as growth proceeds. They thus leave room for the generation of a new bulge near the dendrite tip. Growth of the new branch yields the same scenario to resume, hence giving rise to a branched structure.

Interestingly, beyond solidification, a seemingly similar mode also occurs in other topics as viscous fingering, electrodeposition, dielectric breakdown, neuronal arborization and meristem development (Fig. 2). It therefore seems that its origin refers to general features common to them as geometry, diffusion and advection. On the other hand, a fundamental alternative sets in regarding the nature of this dynamics: do sidebranch emissions refer to a noise amplification or to non-linear oscillations? We report it below together with the theoretical and experimental investigations that may help it to be clarified.

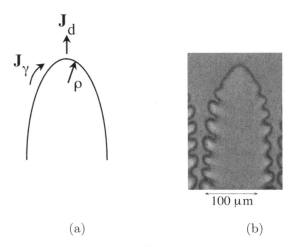

(a)                                                    (b)

Figure 30.  Dendrite and sidebranching. (a) When surface tension is anisotropic, the dendrite is no longer isothermal.  A thermal flux along the interface warms its tip. The ratio of this flux $\mathbf{J}_\gamma$ to the diffusive flux $\mathbf{J}_d$ at the tip provides a non-dimensional parameter $\sigma = 2J_\gamma / J_d$ on which solvability criterion is set. (b) Dendrite with developed sidebranching in a directionally solidified succinonitrile alloy[75].

### 7.1.  The phenomenon of branching: noise amplifier or non-linear oscillator?

Beyond the proper mechanism of branch emission, sidebranching gives rise to a pattern of branches whose structure raises the following alternative.

Branches may be uncorrelated one from the other in size, amplitude or occurrence. They then form a noisy wake as if they had grown out from noise independently. This scenario is put forward by theories relying on noise amplification in a linear regime (Sect. 7.2). Here, linearity allows each branch to develop independently of the others and forbids self-organization to emerge by non-linear interactions. Dendrites then behave like a noise amplifier.

Branches may display some correlations in size, amplitude and, more interestingly, in occurrence. They then correspond to a coherent wake reminiscent of the vortical wakes generated behind a bluff body. Here, non-linearity is essential for yielding a self-organized emission of branches or, equivalently, the emergence of a global mode. A number of mechanisms may be proposed for that, all consisting in bringing out a coupling between yet formed branches and the new branches to form. They yield dendrites to behave as a non-linear oscillator.

This kind of alternative is especially interesting in that it offers a number of different faces. On a fundamental ground, it points to explaining how spatial and temporal coherence may emerge in a growth system. This is reminiscent of the debates that have occurred in hydrodynamics about the convective/absolute transition in bluff body wakes.[65] On an applicative ground, it turns out addressing the possibility of monitoring growth from a macroscopic level or the impossibility of succeeding in that, except by a direct control at the mesoscopic scale. Finally, on a larger ground, it questions the origin of the regularity of living organisms, as revealed for instance by phyllotaxis,[66] and the possibility that it might refer to a widespread mechanism.

## 7.2. *Sidebranching as a convective instability*

We report here the noise amplification theory of sidebranching. This theory was brought about by Ya. B. Zel'dovitch[67] in the context of combustion to explain the surprisingly large stability of curved premixed flame fronts in comparison to planar fronts. It has then been transposed by P. Pelcé and P. Clavin[68] to solidification in order to infer sidebranch amplitudes.

We notice, however, that the context prevailing in solidification is somewhat different than in combustion since all growing cells survive to the development of sidebranches. Accordingly, the issue consists more in understanding the main sidebranching features, i.e. its onset, its amplitude, its regularity than in determining which cell is robust. Attention is thus put on longer time scales on which non-linear features should prevail. This has stimulated modelings in linear[56,69–72] and non-linear[43,73,74] directions.

The foundations of the noise-amplification theory are linear amplification and convection. Branches are modeled as bulges whose amplitude is linearly amplified while they drift away from the tip. The main objective of the modeling consists in determining their net growth factor $\Gamma$ as they arrive in the groove domain where they no longer amplify. For this, three assumptions are performed:

- Locality:
  The bulge dynamics is local, i.e. independent of the state of the interface on other parts than the bulge domain.

- WKB approximation:
  Despite their compact character and the heterogeneity brought about by the interface curvature, bulges are modeled as plane waves $A(t) \exp(ikx)$.

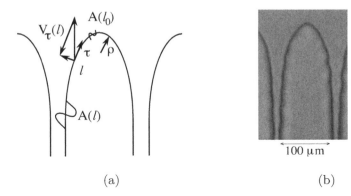

(a)                                                    (b)

Figure 31. Convective instability on a dendrite. (a) Sketch of a perturbation which develops from an initial, unnoticeable, amplitude $A(l_0)$ to the noticeable amplitude $A(l)$ as it drifts toward the groove. (b) Dendrite at the sidebranching onset in a directionally solidified succinonitrile alloy[75].

The wave amplitude follows a linear dynamics that is based on a growth rate $\sigma$ only dependent on the wavenumber $k$: $dA/dt \equiv \sigma(k)A$.

- Planar approximation:
    Despite noticeable curvature, it is assumed that $\sigma(k)$ refers to the dispersion relation (49) of the planar front tangent to the interface.

Altogether, they yield the following dynamics to integrate:

$$\frac{dA}{dt} = \sigma(k)A \tag{74}$$

$$\sigma(k) = \sigma = (V - V_c)(\mathbf{e}_z.\mathbf{n})k - Dd_0k^3 - \frac{V^2}{D}(\mathbf{e}_z.\mathbf{n})^2. \tag{75}$$

Here, two factors have been neglected in relation (49) in comparison to unity: $(1 - K)V/kD$ and $(1 - K)V_c/V$. The first involves $V/kD$ which, for the most amplified wavenumber, writes $[12/27 \; V/V_r \; V/(V - V_c)]^{1/2}$ where $V_r = 4/27 \; D/d_0$ denotes the onset of restabilization. As $V_c/V_r < 10^{-4}$, it is thus negligible except in the very vicinity of onset. The second factor displays a dependence in $V_c/V$ that is small in the regime where sidebranching occurs. In addition, the factor $(\mathbf{e}_z.\mathbf{n})$ has been introduced to handle the change of orientation of the interface with respect to the thermal gradient. This, in particular, changes the normal velocity, $V_n \to V(\mathbf{e}_z.\mathbf{n})$, the gradient normal to the interface $G \to G(\mathbf{e}_z.\mathbf{n})$ and thus the critical velocity $V_c \to V_c(\mathbf{e}_z.\mathbf{n})$.

To integrate the bulge dynamics, it is convenient to turn from a temporal integration to a spatial integration (Fig. 31-a). This is obtained by noticing that the bulges which grow normally to the interface drift at a tangential velocity $V_\tau = -\mathbf{V}.\tau$. This is because bulges should translate at the growing velocity $\mathbf{V}$ to remain at the same relative location on the interface. The difference $-V_\tau\tau$ between their actual velocity $V_n\mathbf{n}$ and the growing velocity $\mathbf{V} = V_n\mathbf{n} + V_\tau\tau$ corresponds to the velocity $V_\tau$ of their drift along the interface.

A correction to this kinematic drift is provided by the antagonist migration of bulges in the thermal gradient. This effect comes from the fact that, following the Gibbs–Thomson relationship, a distortion of a cell groove involves an inhomogeneous concentration and thus diffusion currents which prevent it to be steady. Conservation relation between solute rejection and solute diffusion (24) writes $-D\partial c/\partial z|I = V_m(1 - K)c_I$ where $V_m$ denotes the migration velocity. This, with $\partial c/\partial z|I = -G/m$ and $c_I \approx c_\infty/K$ gives $V_m = V_c$. Accordingly, interface distortion drifts back toward the tip at a velocity $V_c(\mathbf{e}_z.\tau)$ following which the net drift velocity writes $V_\tau = -(V - V_c)(\mathbf{e}_z.\tau)$.

As the interface is curved, the drift velocity changes on the interface from 0 at the tip to $-V$ in the grooves: $V_\tau \equiv V_\tau(l)$ where $l$ denotes the curvilinear abscissa. It thus displays a gradient which stretches bulges $\partial V_\tau/\partial l \neq 0$. Its implication on wavenumber $k$ may be derived by introducing a phase variable $\varphi$ for describing interface modulations, the gradient of which provides a wavevector $\mathbf{k} = \nabla\varphi$. Then, in absence of creation or destruction of bulges, the phase dynamics is conservative: $D\varphi/Dt = 0$ where $D/Dt = \partial/\partial t + V_\tau\tau.\nabla$ denotes the total derivative obtained when co-moving with the drifted structures. Taking its gradient yields: $\partial k/\partial t + \partial/\partial l(kV_\tau) = 0$. In an averaged steady state for interface distortion, one has $\partial k/\partial t = 0$, so that $kV_\tau$ is conserved: $kV_\tau(l) = C$. This effect, which is similar to the red shift of light in an expanding universe, yields the bulge wavevector $k$ to decrease as it drifts along the dendrite.

As $Dl/Dt = V_\tau$, the net growth factor $\Gamma(l, l_0) = \ln[A(l)]/\ln[A(l_0)]$ of the bulge amplitude in between abscissa $l_0$ and $l$ may be written in the WKB approximation:

$$\Gamma(l, l_0) = \int_{l_0}^{l} \frac{\sigma(k, l)}{V_\tau(l)} dl. \tag{76}$$

Notice that the growth rate $\sigma$ will vanish in the grooves, $\theta = \pi/2$, together with the normal velocity $\mathbf{V}.\mathbf{n} = V\cos\theta$, so that the growth factor will no longer increase there. Denoting $\rho$ the curvature radius at cell tip, this stagnation begins at a distance $l \approx \pi/2\ \rho$ from the tip, at a net final growth factor that will be denoted $\Gamma$: $\Gamma = \Gamma(\pi/2\ \rho, l_0)$.

The determination of $\Gamma(\pi/2\ \rho, l_0)$ points to the choice of an abscissa origin $l_0$ for sidebranching. As sidebranches are thought to be generated at the tip, even if we cannot observe then yet, the natural choice would be $l_0 = 0$. This however raises two sharp difficulties. The first addresses the constant $C = kV_\tau(l)$ which vanishes at $l_0 = 0$ together with $V_\tau$. This would yields $V_\tau(l) = 0$ elsewhere, in contradiction with evidence. The caveat lies in the assumption that $\partial k/\partial t = 0$ at $l_0 = 0$, a thing which cannot be satisfied at the tip because of the stretch implied by the drift on either side of the dendrite. The second difficulty addresses the expression of the net growth factor which diverges as $l_0$ vanishes. This is because $V_\tau(l)$ reduces to $(V - V_c)\theta(l)$ where $\theta = (\mathbf{e}_z, \mathbf{n})$, i.e. to $V_\tau(l) \approx (V - V_c)\ l/\rho$, so that the integral diverges logarithmically.

Both difficulties trace back to the fact that ponctual bulges undergo no drift at the dendrite tip. They are then abusively considered as undergoing no stretch. In the present linear modeling, they are thus liable to grow at the same place without limit. To cure divergence, it has been proposed to cut the integral at a distance of the order of the critical wavelength $\lambda_c$ of the primary instability.[68,69] This, corresponds to a phenomenological ansatz that should deserve a legitimation from a stochastic analysis of the conditions prevailing at the cell tip. As it introduces the characteristic scale $\lambda_c$ in addition to the curvature radius $\rho$, it is no surprise that detailed integration yields a non-dimensional growth factor $\Gamma$ proportional to the ratio $\rho/\lambda_c$: $\Gamma = 2/3\ k_c\rho = 4\pi/3\ \rho/\lambda_c$.

Another factor than interface orientation, the wavevector stretch, makes the growth factor saturates. This is because the growth rate $\sigma$ goes to zero with $k$ (49). In addition, wavevectors larger than the most amplified wavevector $k_{max}$ will be stretched toward $k_{max}$ as they drift along the dendrite. Although they are not the fastest growing mode near the tip, they will thus become so on the grooves. All wavenumbers have then to be taken into account for properly evaluating the net growth factor and the resulting width of the corresponding bulges. This determination of $\Gamma$ from wave packets[71] modifies prefactors but does not cure the above structural difficulties.

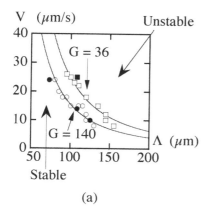
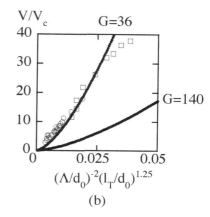

(a)  (b)

Figure 32.  Onset of sidebranching in directionally solidified succinonitrile alloy.  (a) Experimental states where sidebranching just becomes noticeable[75]. They form the frontier between a stable domain and an unstable domain regarding sidebranching. Increasing thermal gradient from 36 to 140Kcm$^{-1}$ promotes sidebranching.  (b) Confrontation between the experimental onset and the convective instability threshold in the $(V/V_c, \Lambda^{-2}G^{-1/4})$ plane[19,77]. Experimental data at onset collapse onto a line at all thermal gradients in this plane. If the convective instability is relevant, data should refer to the same net growth factor $\Gamma$. However, as $\Gamma \propto V^{1/4}\Lambda^{3/4}G^{1/2}$, adjusting the unknown prefactor so as to fit with experiment at a gradient yields strong disagreement at another.

## 7.3. *Critical surface of sidebranching*

We address the occurrence of sidebranching in the control parameter space.[75] This means that we are not interested in the *way* sidebranching occurs on definite cells but, instead, in the *domains* where it happens in the space of variables $(V, \Lambda, G)$. This corresponds to an approach that is more statistical and less phenomenological than the direct investigation of a particular dendrite. The idea behind it is that one sometimes learn more on a phenomenon by determining when it comes about rather than how it develops.

The experimental procedure has consisted in scanning the variable space for identifying the cells that display sidebranches at the weakest observable amplitude (Fig. 31-b). For this, large domains made of cells with the same spacing were achieved thanks to some experimental tricks. As all neighbors of a cell involved the same spacing, this ensured that the critical velocity measured at the sidebranching occurrence actually referred to that spacing, even if cell interactions were in order. Five thermal gradients were studied.

The onset of sidebranching is reported in Fig. 32-a for the largest and the smallest gradients. For each of them, it delimits the parameter space in two domains: the cell domain where there is no sidebranching; the dendritic domain where sidebranches are emitted. The curve in between then stands as the critical surface for sidebranching.

Increasing velocity or cell spacing at otherwise unchanged variables is found to promote sidebranching. This was expected since raising velocity strenghtens planar instability (49) while increasing cell spacing leaves more room for sidebranches to develop. Interestingly, increasing thermal gradient was also found to promote sidebranching. This is at first glance surprising since this weakens the planar instability by raising the critical velocity $V_c$. However, this also flattens cells and thus leaves more room for the development of sidebranching. This second mechanism seems to predominate over the first. Direct confirmation of this promotion of sidebranching by thermal gradient has been provided by suddenly varying thermal gradient on dendrites.[76]

To facilitate the discussion on the critical surface for sidebranching, we fit data with power laws of the variables. This is legitimated by the fact that the characteristic scales of solidification extends over an extremely large scale range, $l_T/d_0 \approx 10^{-6}$, following which a multiplicative algebra appears far more relevant than an additive algebra to express physical relationships. The onset of sidebranching then expresses easily: $V \Lambda^2 G^{1/4} = C$ where $C$ is a constant depending on the mixture variables $D, m, K, c_\infty, d_0$. If the product of variables displays values larger than the constant $C$, sidebranching sets in; if it corresponds to smaller values, cells are stable with respect to it. This criterion may be rewritten $V/V_c \, (\Lambda/d_0)^2 \, (l_T/d_0)^{-5/4} = C'$ where $C'$ is another constant. Then, performing the non linear reduction of variables $(\Lambda, G) \rightarrow (\Lambda/d_0)^{-2}(l_T/d_0)^{5/4}$ makes the two-dimensional critical surface collapse on a single line in the corresponding plane (Fig. 32-b).

Comparison of the critical surface for sidebranching with the convective instability theory makes sense because the latter being linear, it should be all the more valid that dendrites are close to the onset of sidebranching. As the onset of sidebranching has been identified at a fixed resolution $R$ for the sidebranch amplitudes whereas the noise amplitude $\eta$ is expected to be constant, it should correspond to a definite value of the growth factor $\Gamma$: $A \approx R = \exp(\Gamma)\eta; \; \Gamma \approx \ln(R/\eta)$. Accordingly, the critical surface[19,77] for sidebranching must correspond to an iso-$\Gamma$ where $\Gamma$, following section 7.2, simply expresses[69] as $\Gamma \propto \rho/\lambda_c$. In particular, it appears that the fact that the dendrite undercooling depends on velocity (Sect. 6.2.1) does not modify the form of this relationship.[78]

On the other hand, studies of cell forms[36,39] give $\rho \propto V^{-1/4}\Lambda^{3/4}G^{1/2}$ (Sect. 6.2) whereas the Mullins–Sekerka instability states that $\lambda_c$ scales as $(l_D d_0)^{1/2}$ (Sect. 5.2). This yields $\Gamma \propto V^{1/4}\Lambda^{3/4}G^{1/2}$ which conflicts with the scaling found for the critical surface. In particular, adjusting prefactors so that this surface stands as an iso-$\Gamma$ for one gradient $G = 36\mathrm{Kcm}^{-1}$ yields a large disagreement for another, e.g. $G = 140\mathrm{Kcm}^{-1}$, far beyond experimental uncertainty[19,77] (Fig. 32-b).

This discrepancy means that, according to the convective linear instability, the sidebranches of similar weak amplitude $R$ that have been observed experimentally on various dendrites should surprisingly not refer to the same growth factor, whereas no change of the fluctuation bath is in order. This indicates that the corresponding theoretical framework has missed something important in the mechanism of sidebranching. This might be non-linearities, non-locality or, as a result of both, a global mode overwhelming the convective character of the instability.

## 7.4. *Sidebranch correlations*

The level of correlations of sidebranch emission can be directly studied on signals extracted from dendrites. Doing so on a freely growing dendrite, Dougherty, Kaplan and Gollub[79] concluded to an irregular emission. Here, we report a recent study[80] performed by Georgelin, Bodéa and Pocheau on directional solidification of a succinonitrile mixture in a weak sidebranching regime, in which regular emissions have been evidenced.

Signals were extracted by cutting the interface at at given distance $z$ from the tip and by recording the corresponding abscissa $x(z,t)$ (Fig. 33-a). They show sidebranches emitted by bursts of various length and amplitude and occurring apparently without order (Fig. 33-b). However, inside each burst, the sidebranching emission seems more regular, if not periodic.

To investigate these features, we use correlation functions $C[x_1, x_2](\tau)$ of signals $x_1(t)$, $x_2(t)$. These are defined by the time integrals:

$$C[x_1, x_2](\tau) = c_{1,2}(\tau)/[c_{1,1}(\tau)c_{2,2}(\tau)]^{1/2} \tag{77}$$

$$c_{i,j}(\tau) = \int_0^{T-\tau} \tilde{x}_i(t)\tilde{x}_j(t)dt \;\; ; \;\; \tilde{x}_1(t) = x_1(t) \;\; ; \;\; \tilde{x}_2(t) = x_2(t+\tau)$$

where $T$ denotes the signal length. Here, $C[x_1, x_2](\tau)$ lies in between $-1$ and 1 and absolute values close to unity means coherence of the signals. The mean duration $T_s$ of sidebranch emission is about 4 seconds: $T_s = 4s$.

Auto-correlation of signals over almost a thousand of sidebranch emissions reveals a coherence time extended on about six sidebranches: $\tau_c \approx 6T_s$

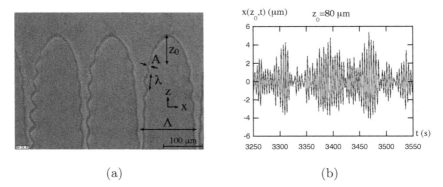

(a)                                              (b)

Figure 33.  Sidebranching emission on a dendrite in a directionally solidified succinonitrile alloy[80]. (a) Array of dendrites at growing velocity $V = 15\mu ms^{-1}$; successive inter-
sections of the interface with the x axis provides extracted sidebranching signals $x(z_0, t)$.
(b) Sidebranching signal extracted over about one hour at a distance $z_0 = 80\mu m$ from
the dendrite tip. Sidebranching occurs by bursts and shows a sidebranching period of
about $T_s = 4s$.

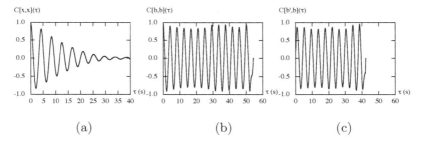

(a)                          (b)                          (c)

Figure 34.  Evidence of sidebranching coherence[80]. (a) Autocorrelation function of the
signal $x(z_0, t)$ recorded in Fig. 33-b over about 850 branch emissions. Its decrease over
about 6 oscillations reveals an unexpected large coherence time $\tau_c \approx 6T_s$ that extends
over 6 sidebranch emissions. (b) Autocorrelation function $C[b, b](\tau)$ of the central burst
$b(t)$ of Fig. 33-b ($3367s < t < 3421s$). Coherence of sidebranch emission is maintained
all over the burst length. (c) Correlation function $C[b, b'](\tau)$ of the central burst $b(t)$
($3367s < t < 3421s$) of Fig. 33-b with the following burst $b'(t)$ ($3444s < t < 3486s$).
Coherence of the two sidebranch emissions is maintained all over the burst overlap.

(Fig. 34-a). This reveals an unexpected degree of correlation which has motivated us to address the level of coherence of the burst emissions on the
whole signal, and of the sidebranch emissions on each burst.

To address the coherence of the burst emission, we consider the enve-
lope of the signal defined as the interpolation between the signal maxima.
Its auto-correlation function shows a sharp decrease over the mean burst

length $< T_b > \approx 10 T_s$. This indicates an uncorrelated character of burst occurrence. However, picking up those bursts whose length is below $6 T_s$, in between $6 T_s$ and $12 T_s$, and above, the coherence times of auto-correlation functions of the *sidebranching* signal increase from $4 T_s$ to $6 T_s$ and $10 T_s$, respectively.

This increase of coherence time with the burst length points to an intrinsic organization of branches in bursts. To point it out, we focus attention to single bursts and perform their auto-correlation functions. Interestingly, we obtain values of order unity over the whole burst length (Fig. 34-b). This reveal a large coherence inside each burst following which sidebranching refers to a non-linear oscillator instead of a noise amplifier.

To further test the stability of the sidebranch emission rate on a long time, we performed cross-correlations of sidebranching in distant bursts extracted from the signal. Here too, a large correlation is evidenced (Fig. 34-c), since the level of correlation remains at a value close to unity. Beyond the own coherence of each signal, this proves a stability of the sidebranch frequency among bursts. In particular, the mismatch of burst frequencies $\nu$ remains small enough not to induce a phase opposition during the burst overlap: $\delta\nu/\nu \approx T_s / < T_b > \approx 10^{-1}$.

Following the coherence of sidebranching inside bursts and the uncorrelation of burst occurrence, the present dendrites look similar to natural light. In particular, sidebranching signals appear to be made, as light, of intrinsically coherent wavetrains, the bursts, whose emission occurs at random regarding time, phase and amplitude. The important thing, however, is that sidebranch emission is self-organized in each of them at least, a fact that reveals an intrinsically coherent dynamical system analogous to the atoms for light emission. Understanding its origin will require non-linear or non-local modeling.[43,73,74] This constitutes a stimulating challenge since extending the coherence period of sidebranching, for instance by extending the burst periods, would open the track for achieving regular growth on a long time and on a long scale.

On a larger ground, the evidence of a coherent sidebranch emission in solidification questions the origin and the nature of similar emissions in the realm of living objects. This addresses for instance the origin of the regular emission of bulges by meristems (Fig. 2-c) which stands as a necessary precursor of phyllotaxis[66] (Fig. 2-f).

## 8.  Non-linear dynamics

The study of sidebranching has revealed a surprising self-organization that calls for a relevant role of non-linearities in growth dynamics. Here, we report two other dynamical phenomena that support this analysis. One refers to the occurrence of a limit cycle by Hopf bifurcation; the other shows evidence of spontaneous self-organization on a curved interface as far as long time scales are investigated. Both have been evidenced on directional solidification of a succinonitrile mixture.[20,82,83]

### 8.1.  *Oscillatory instability and limit cycle*

A fascinating oscillatory instability[20] takes place on deep cells at moderate velocities $V/V_c \approx 5$. It consists in coupled oscillations of cell tip position and of cell width in a nearly phase quadrature (Fig. 35-a). Cells therefore raise in the temperature gradient while thickening and go back towards colder regions while getting thinner. They are however the fattest, not at their highest position, but a quarter period later when re-descending. Similarly, they are the thinnest, not at their lowest position, but a quarter period later when raising again. Meanwhile, their two neighbors follow the same kind of dynamics, in phase between them but in phase opposition with their neighbor cell. This results in pattern oscillations whose spatial period is twice the cell spacing. They correspond to a limit cycle for cell dynamics.

Figure 35-a displays two snapshots of the oscillating pattern at times where cells involve the same height or reach their highest position. They evidence the phase opposition between neighbors. The grooves also show a change of width over a distance $VT$ that is about six cell spacings $\Lambda$. This reveals a period $T$ of about $T \approx 6$ Pe $\tau_D$ here, where $Pe$ denotes the Péclet number $Pe = \Lambda V/D$ and $\tau_D = D/V^2$ the diffusion-time. In particular, $T$ was larger than $4\tau_D$ for all oscillations here.

Measurements of the difference $\delta z_t$, $\delta w$ of cell tip position and cell width between two neighbor cells are reported in Fig. 35-b with respect to time. As neighbors cells oscillate in phase opposition, data double the oscillation amplitudes as compared to a single cell. Width amplitude, about $12\mu$m, is about thrice the tip position amplitude. Oscillation signals show a phase difference that amounts to about $100\pm10$ degrees here. This means that the modes $\delta z_t$, $\delta w$ are not slaved one to the other and thus that the dynamics is two-dimensionnal. In particular, the out-of-phase oscillations of cell tip

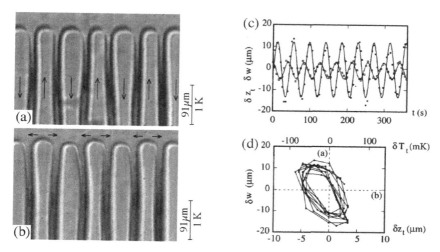

Figure 35. Oscillatory instability and limit cycle in directionally solidified succinonitrile mixture[20]. (a)(b) Cell array at different phases of the cycle. $G = 110 \text{Kcm}^{-1}$, $V = 7\mu\text{ms}^{-1}$, $V_c = 2.1\mu\text{ms}^{-1}$, $\Lambda = 55\mu\text{m}$. (a) Cells cross the mean tip position: the thickers cells are moving down to lower temperatures; the thinner ones are moving up to higher temperatures. (b) Cells reach extremal tip positions: the highest (hottest) cells are thickening; the lowest (coldest) cells are shrinking. (c)(d) Limit cycle. $G = 140 \text{Kcm}^{-1}$, $V = 13\mu\text{ms}^{-1}$, $V_c = 2.7\mu\text{ms}^{-1}$, $\Lambda = 45\mu\text{m}$. Axes refer to the differences $\delta z_t$, $\delta w$ between cell tip position and cell width of neighboring cells as a function of time. (c) periodic signals $\delta z_t(t)$ (smallest amplitude), $\delta w(t)$ (largest amplitude) (d) limit cycle in the $(\delta z_t, \delta w)$ plane. Letters (a) and (b) refer to the states shown in (a) and (b).

$\delta z_t$ as compared to cell width $\delta w$ show that the mode $\delta z_t$ is related not only to the mode $\delta w$ but also to its derivative $\partial \delta w / \partial t$: $\delta z_t \approx \psi(\delta w, \partial \delta w / \partial t)$. The objective of modeling is to identify this relationship and the resulting oscillation features.

As oscillations extend over a period of several diffusion times, they should rely on a phenomenon that is essentially diffusive and which involves scales of the order of the diffusion scale. In particular, the variations of both cell tip position and of cell width induce a vibration of the solute diffusive layer that stands ahead of the front. In turn, this vibration modifies the concentration and the flux at the cell tip, and thus the cell tip position and the cell width. This results in a feed-back loop that is suitable to induce oscillations provided that some delay is in order.

An explicit modeling of this kind of mechanism has been worked out by A. Karma and P. Pelcé to recover the oscillations of a single cell in directional solidification.[81] Its objective consists in reducing the cell dynamics to

that of a point source representing the cell tip. The analysis is performed in
the low Péclet number limit where cell tails do not affect cell tips and for a
single cell. Then, cell tips can be modeled in analogy with viscous fingering
(Sect. 6.1.3), surface tension effects being taken into account (Sect. 6.3.2).
This provides the expression of both their width $\lambda$ and the net solute flux
$J_t$ emitted by their tip with respect to tip position $z_t$, tip velocity $V_t$, $D$,
$G$, $\Lambda$, $m$ and the capillary length $d_0$:

$$\lambda = f[\frac{\Lambda^2}{Dd_0}\{c_t V_t(1 - K) + DG/m\}] \tag{78}$$
$$J_t = -D\nabla c|_t = \lambda[c_t V_t(1 - K) + DG/m] - DG/m \tag{79}$$

where $f(.)$ is a function numerically determined by McLean and Saffman[48]
and which provides the selection of the finger width by surface tension
(Sect. 6.3.2). The expression of the flux leaving the cell tip region, $J_t$,
follows from the integration of relation (62) along the interface within the
approximation $c_I \approx c_t$. Notice that it corresponds to the quickest branch of
solutions among the discrete set compatible with the existence of a surface
tension.

On the other hand, following the Gibbs–Thomson relation and the uniformity of the thermal gradient, the cell tip position $z_t$ is linearly related
to the tip undercooling $\Delta_t$ (66): $\Delta_t = 1 - z_t/l_T$, for an origin of the $z$-axis
on the solidus line.

Finally, the advection-diffusion equation couples the net flux $J_t$ and the
concentration $c_t$ by the dynamics of the diffusive layer. This feed-back is
suitable to induce a delay between the dynamics of variables that might
result in a tip position $z_t$ not in phase with the tip velocity $V_t$, and then
to oscillations. This is actually what happens in a range of parameters
compatible with observations.

## 8.2. *Pattern dynamics on a curved front*

We investigate here the implications of a long range curvature on the spatio-
temporal organization of a cellular pattern.[82,83] The radius of curvature
of the interface, $\rho$, is large compared to the scale of its modulations, the
cell spacing $\Lambda$: $\rho \approx 70\Lambda$. In addition, all cells, except the unstable ones,
have reached the saturation regime for their amplitude. They thus stand
far beyond the linear growth regime that prevails at weak modulation amplitudes. This, together with the large relative size of the curvature radius,
provides two important differences with the convective instability model
proposed for dendrites (Sect. 7.2).

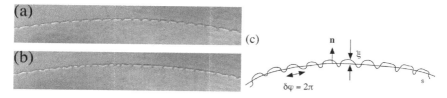

Figure 36. Repetitive cell splittings and phase modeling[82,83]. (a,b) Cellular front just before (a) and after (b) cell splitting as a result of the stretch produced by the inhomogeneous cell drift on the curved interface. Image width is $2.2mm$ and $V = 1.6\mu m.s^{-1}$. (c) Sketch of the modeling of the cellular pattern with a phase variable varying by $2\pi$ on a cell. The smooth curve refers to the interface before cellular instability and the modulated curve to the actual cellular interface.

In practice, the origin of this large scale curvature may consist in perturbations of temperature or of solute concentration at the millimeter scale. These are actually the rule in natural growth where no dedicated device imposes the uniformity of thermal gradient $\mathbf{G}$ or of solute concentration $c_\infty$. From this point of view, considering growth on a large curved interface makes modeling closer to the actual growth conditions encountered in industry or in nature.

We restrict attention to the low velocity regime $V/V_c < 2$ in which rounded cells with short grooves are displayed. No steady state is found. Cells first display a drift along the mean interface at a velocity $V_\tau = V(\mathbf{e}_z.\tau)$. Its origin is similar to that observed on dendrites (Sect. 7.2) and traces back to curvature. However, cell width also shows a permanent increase that yields cells to split in two when they reach a critical width $\Lambda_s$ (Fig. 36-a). The same scenario then resumes on the daughter cells until they involve a splitting too. This yields a dynamical pattern displaying cell spacing in between the critical width $\Lambda_s$ and its half.

This cell stretch results from the gradient of drift velocity $V_\tau = V(\mathbf{e}_z.\tau)$ along the interface. In particular, as $\mathbf{e}_z.\tau = \sin(\theta) \approx l/\rho$, where $\theta$ denotes the angle of the interface tangent with the $z$-axis, the drift velocity writes $V_\tau(l) = l\,V/\rho = l/T$ with $T = \rho/V$. It thus involves a uniform gradient $\partial V_\tau/\partial l = 1/T$ which drives a cell stretch on the interface. In particular, points located at abscissa $l$ evolve according to $Dl/Dt = V_\tau(l) = l/T$ so that $l(t) = l(0)\exp(t/T)$. The stretch factor $\sigma(t) = \exp(t/T)$ is thus, as the velocity gradient, uniform, so that all lengths are stretched the same way whatever their size or locations.

Interestingly, as time goes on, cell splits are found to occur by groups that eventually extend to the whole pattern[82,83] (Fig. 37-a). Cells then

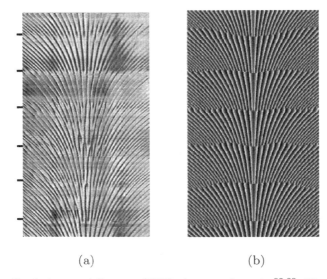

(a)                                                    (b)

Figure 37.   Spatio-temporal diagrams (STD) of pattern dynamics[82,83]. Time is reported on ordinates and space on abscissa. The lines corresponds to the trajectories of cell grooves. The occurrence of new lines reveals a cell splitting. Ticks on the left of STD indicate cell splitting avalanches. (a) Experimental STD in a window of width $2.2mm$ and over a time $13280s$. (b) STD of phase dynamics simulation with slightly inhomogenous initial conditions.

display the same width, the same stretch and the same time of splitting so that pattern dynamics has reached a self-organized state involving uniformity and periodicity.

To recover this surprising feature, we consider a phase model of cell organization on the interface. We thus introduce a phase $\varphi(l,t)$ and an amplitude $A(l,t)$ so that the normal deviation $\xi(l,t)$ of the actual interface from its averaged shape writes: $\xi = A\exp(i\varphi) + \bar{A}\exp(-i\varphi)$ (Fig. 36-b). Here the amplitude $A(l,t)$ varies on a scale large compared to the characteristic scale $\Lambda_c$ of cells, whereas $\varphi$ changes of $2\pi$ on a cell. On the other hand, $A$ represents a fast mode that has yet reached its saturation value while $\varphi$ corresponds to a slow variable that is still evolving.

Pattern analysis then shows that amplitude $A$ is slaved to phase $\varphi$ in the sense that the saturated value of $A$ slowly varies according to the evolution of the phase gradient: $A \equiv A(\partial\varphi/\partial l)$. Meanwhile the evolution of $\varphi$ is driven by the phase gradients, $\partial\varphi/\partial t = \mathcal{F}(l,\partial^n\varphi/\partial l^n), n \in \mathcal{N}$, which, at the dominant order in a phase gradient expansion, results in a diffusive equation: $D\varphi/Dt = D(k)\partial^2\varphi/\partial l^2 + o(k)$ where $k = \partial\varphi/\partial l$ and $D(k)$ a

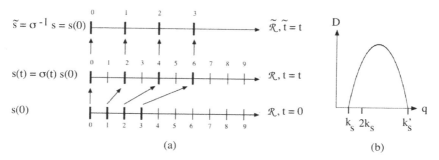

Figure 38. Reference frames and phase diffusion coefficient. (a) Sketchs of the stretched frame $\mathcal{R}$ and of the "co-stretched" frame $\tilde{\mathcal{R}}$. Numbers correspond to curvilinear abscissa $s(0)$, $s(t)$ or $\tilde{s}$. The arrows show the effect of stretch in frame $\mathcal{R}$ and the correspondence to the co-stretched coordinates $\tilde{s}$ in frame $\tilde{\mathcal{R}}$ at time $t$. It then appears that, despite the stretch, curvilinear abscissa are the same in frame $\tilde{\mathcal{R}}$ and in the frame $\mathcal{R}$ at time 0. Frame $\tilde{\mathcal{R}}$ thus corresponds to an inertial frame for the stretching dynamics. (b) Sketch of the concave shape of the diffusion coefficient $D(q)$. The experimental range lies in between $k_s$ and $2k_s$.

phase diffusion coefficient. Here $D/Dt$ denotes the total derivative obtained when comoving with cells: $D/Dt = \partial/\partial t + V_\tau(l)\partial/\partial l$ with $V_\tau(l) = l/T$. In the fixed reference frame of the interface, phase dynamics then write:

$$\frac{\partial \varphi}{\partial t} + \frac{l}{T}\frac{\partial \varphi}{\partial l} = D(k)\frac{\partial^2 \varphi}{\partial l^2} + o(k). \tag{80}$$

Interestingly, the advection-diffusion equation (80) satisfies a Liapunov dynamics that enables a conclusion to be drawn on the evolution of the pattern organization.[82,83] To express it, we introduce the non-gallilean frame $\tilde{\mathcal{R}}$ obtained when co-moving with interface points at their drift velocity $V_\tau(l)$ (Fig. 38-a). From a mechanical point of view, it corresponds to an inertial frame for the stretching dynamics. It is thus analogous to the free-falling frames used to study gravitational dynamics in general relativity. Viewed from the fixed interface frame $\mathcal{R}$, frame $\tilde{\mathcal{R}}$ corresponds to a uniformly stretched frame (Fig. 38-a). In particular, calling $(l,t)$ the coordinates of frame $\mathcal{R}$ and $(\tilde{l},\tilde{t})$ those of frame $\tilde{\mathcal{R}}$, we get: $\tilde{t} = t$ and $\tilde{l} = \sigma^{-1}l$ where $\sigma(t) = \exp(t/T)$ denotes the exponential stretch factor. As, for a drifting point, $l(t) = \sigma l(0)$, we then notice that $\tilde{l} = l(0)$: frame $\tilde{\mathcal{R}}$ keeps memory of the metrics of frame $\mathcal{R}$ at the initial time. It thus cancels out the effects of stretch on lengths and is therefore suitable for evidencing the improvement of self-organization.

We introduce in frame $\tilde{\mathcal{R}}$ the functional:

$$\mathcal{L}(\tilde{t}) = \frac{1}{2}\int_0^L (\frac{\partial \tilde{k}}{\partial \tilde{l}})^2 d\tilde{l} \tag{81}$$

where $\tilde{k}$ denotes the phase gradient $\partial \varphi / \partial \tilde{l}$ in $\tilde{\mathcal{R}}$.

Using the change of variables:

$$\left(\frac{\partial}{\partial \tilde{l}}\right)_{\tilde{t}} = \sigma \left(\frac{\partial}{\partial l}\right)_t \tag{82}$$

$$\left(\frac{\partial}{\partial \tilde{t}}\right)_{\tilde{l}} = \left(\frac{\partial}{\partial t}\right)_l + \frac{l}{\sigma}\frac{d\sigma}{dt}\left(\frac{\partial}{\partial l}\right)_t \tag{83}$$

one then obtains straightforwardly[83]:

$$\frac{d\mathcal{L}}{d\tilde{t}} = \left[\left(\frac{\partial k}{\partial \tilde{l}}\right)\left\{\frac{2}{3}D'(k)\left(\frac{\partial k}{\partial \tilde{l}}\right)^2 + D(k)\left(\frac{\partial^2 k}{\partial \tilde{l}^2}\right)\right\}\right]_0^L$$
$$- \int_0^L D(k)\left(\frac{\partial^2 k}{\partial \tilde{l}^2}\right)^2 d\tilde{l} + \frac{1}{3}\int_0^L D''(k)\left(\frac{\partial k}{\partial \tilde{l}}\right)^4 d\tilde{l} \tag{84}$$

Stability of the cellular array for $k_s < k < k'_s$ with $k_s = 2\pi/\lambda_s$ implies $D(k) > 0$ in this range. In one dimensional systems, the classical expression of the variation of phase diffusion coefficient $D$ with wavenumber $k$ is concave: $D(k) = D\,[1 - 3(k - k_c)^2]/[1 - (k - k_c)^2]$ (Fig. 38-b). This concavity is mandatory in average since $D$ is positive in the range $k_s < k < k'_s$ and vanishes at its ends. Assuming it here yields $D''(k) \leq 0$ in the working range.

In (84), the boundary terms vanish if the cellular structure is uniform at its ends. In addition, in a sufficiently large domain, they are dominated by the integral terms, unless heterogeneity is concentrated at the boundaries. Accordingly, the sign of the time-derivative of $\mathcal{L}$ should be given by those of the integral terms. These are negative since $D(k) \geq 0$ and $D''(k) \leq 0$ so that: $\forall \tilde{t}, d\mathcal{L}/d\tilde{t} \leq 0$.

The functional $\mathcal{L}$ is thus a Liapunov functional which ever decreases in time unless wavenumber gradients vanish, i.e. unless pattern uniformity is achieved. This property would not be conclusive if $\mathcal{L}$ relied on the wavenumber gradients evaluated in frame $\mathcal{R}$ since stretch does make them decrease without significance regarding pattern organization. However, $\mathcal{L}$ is based on wavenumber gradients evaluated in frame $\tilde{\mathcal{R}}$ in which stretch effects are removed. Its decrease then states that, back to the frame $\mathcal{R}$ at time 0, cells have moved so as to reduce wavenumber gradients in average: homogeneity is improved.

As stretch proceeds, cell splitting will generate new cells that are not taken into account above. This will locally raise wavenumbers, $k_s \rightarrow 2k_s$, and thus the Liapunov functional, so that the net evolution of the averaged homogeneity rate is indeterminate. However, further analysis[83] and simulation[82,83] succeed in showing that the whole pattern will stabilize at a low level of averaged phase gradient (Fig. 37-b). Then, in quasi-homogeneous patterns, cells undergo the same history and display quasi-simultaneous splittings. This recovers experimental observations.

Bending a mean interface in a non-linear, saturated, growth regime has thus given rise to drift and stretch but, more unexpectedly, to a spatio-temporal organization. On a fundamental view point, this shows that non-linearity can bring about self-organization in circumstances where linear instability is thought to amplify disorder. On a practical view point, this provides a mean for controlling from a macroscopic level the growth organization at a meso-scale.

## 9. Geometrical model of growth

Up to now, our investigation of growth interface has been restricted to transition sheets (Sect. 2.1). In particular, interfaces deserved no specific physical modeling by themselves and were only used to infer step conditions or flux conditions. Here, we draw attention on interfaces that display a nature different than the media that surround them, the so-called separation zones (Fig. 4-b). In particular, two examples stemming from biophysical growth will be considered: alga development and rhizoïd growth. Whereas their detailed modeling is out-of-reach, it will appear that geometrical models of their boundary viewed as a moving line will only require kinematic information on some physical constraints. One will refer to the elongation rate of alga cell walls and the other to the flux of incorporation of new vesicles by rhizoïd boundaries.

### 9.1. *Kinematic of a curve*

We introduce the kinematics of geometrical models of interfaces viewed as moving curves and express the evolution of some relevant geometric features with time. Care will be taken to distinguish the extrinsic properties that depend on parametrization from the intrinsic properties that are independent of its definite choice.[40,84]

### 9.1.1. Normal velocity

Introducing a field $G(M,t)$ such that the curve $\mathcal{C}$ stands at all times as a prescribed iso-$G$, $\mathcal{C}(t) = \{M; G(M,t) = 0\}$, the dynamical equation for $G$ shows that only the normal velocity $\mathbf{V}_n = (d\mathbf{M}/dt.\mathbf{n})\mathbf{n}$ of the curve matters for setting its evolution (1). The remaining tangential velocity $\mathbf{V}_t = d\mathbf{M}/dt - \mathbf{V}_n$ only makes the curve glide on itself without inducing any global change of it. It thus bears no importance for the curve dynamics and simply corresponds to a reparametrization of the curve (Fig. 5-a). By contrast the normal velocity which drives the curve dynamics appears independent of any parametrization: it is gauge invariant.

### 9.1.2. Geometrical features

We label $l$ the curvilinear abscissa on the curve and $\theta$ the angle of the tangent $\tau$ to the curve with a fixed direction $\mathbf{e}_x$: $d\mathbf{M} = dl\ \tau$; $\theta = (\mathbf{e}_x.\tau)$ (Fig. 39-a). Identifying the plane $\mathcal{P}$ with the complex plane $\mathcal{C}$ and $(\mathbf{e}_x, \mathbf{e}_y)$ with $(0,1)$ yields: $\tau = \exp(i\theta)$, $\mathbf{n} = -i\exp(i\theta)$ and:

$$\frac{\partial \tau}{\partial l} = -\kappa\mathbf{n} \ ; \ \frac{\partial \mathbf{n}}{\partial l} = \kappa\tau \tag{85}$$

where $\kappa = \partial\theta/\partial l$ denotes the curve curvature.

### 9.1.3. Evolution equations for the geometrical features

We call $U$ and $V$ the normal and tangential velocities on the curve:

$$\frac{d\mathbf{M}}{dt} = U\mathbf{n} + V\tau \tag{86}$$

Determining the evolution rate of angle increments $\delta\theta = \kappa\delta l$ straightforwardly yields:

$$\frac{d\kappa}{dt} = \frac{\partial}{\partial l}\left(\frac{d\theta}{dt}\right) - \kappa\frac{1}{\delta l}\frac{d\delta l}{dt} \tag{87}$$

where $d\theta/dt$ stands as the local rotation rate of the curve and $1/\delta l\ d\delta l/dt$ as its local elongation rate.

To explicitly determine each of these variation rates, we consider the evolution $d\delta\mathbf{M}/dt$ of the increment $\delta\mathbf{M}$ between two points of the curve. Writing it either $d(\delta l\ \tau)/dt$ or $\delta(U\tau + V\tau)$ and identifying the resulting expressions, one obtains:

$$\frac{1}{\delta l}\frac{d\delta l}{dt} = U\kappa + \frac{\partial V}{\partial l} \tag{88}$$

$$\frac{d\theta}{dt} = -\frac{\partial U}{\partial l} + \kappa V \tag{89}$$

and, from (85) (87):

$$\frac{d\kappa}{dt} = -(\kappa^2 + \frac{\partial^2}{\partial l^2})U + V\frac{\partial\kappa}{\partial l}. \tag{90}$$

Among the above variables, some are independent of the choice of parametrization and some depend on it. The former are gauge-invariant and the latter are not. Gauge-invariant quantities are: $\mathbf{n}, \tau, l, \theta, \kappa, U$. Gauge-dependent quantities are $\delta l$, $V$, $d/dt$ and therefore all the evolution rates (88) (89) (90).

An easy identification of the gauge-invariant terms of relations (88) (89) (90) may be obtained by considering the total derivative $d/dt = (\partial/\partial t)_{V=0} + V\partial/\partial l$. Here, terms of the form $V\partial/\partial l$ are explicitly gauge-dependent since they depend on the gauge-dependent variable $V$. In contrast, the remaining terms $(\partial/\partial t)_{V=0}$ are explicitly gauge-invariant since they refer to the only non-arbitrary parametrization for which any point of the curve moves normally to it. Gauge-invariant terms of relation (90) are thus $-(\kappa^2 + \partial^2/\partial l^2)U$. Noticing that $\kappa V$ also writes $V\partial\theta/\partial l$, the gauge-invariant part of relation (89) reads $-\partial U/\partial l$. Now, writing $dl(M)/dt = dH[l(M)]/dt = dl/dt\, \delta[l(M)]$ where $H(.)$ denotes the Heaviside function and $\delta(.)$ the delta-function, the elongation rate $d\delta l/dt$ reads $d/dt\{H[l(M+\delta M)]-H[l(M)]\}$. On the other hand, $Vd/dl\{H[l(M+\delta M)]-H[l(M)]\}$ writes $V(\delta[l(M+\delta M)])-V(\delta[l(M)])$, i.e $V[l(M+\delta M)]-V[l(M)]$ or $\delta l\, \partial V/\partial l$. Accordingly, the derivative $\partial V/\partial l$ appears as the convective part of the total derivative $\delta l^{-1}d\delta l/dt$ so that the remaining term $U\kappa$ of relation (88) stands as its gauge-invariant term. Finally, the terms $\partial V/\partial l$, $\kappa V$ and $V\partial\kappa/\partial l$ of relations (88) (89) (90) correspond to the additional variation rates brought about by gliding at velocity $V$ along the curve. They thus depend on the parametrization and are therefore gauge-dependent.

An easy interpretation of gauge-invariant terms may be obtained by considering simple growths in the normal gauge $V = 0$ in which points move normally to the curve. This actually makes all non-gauge invariant terms vanish, leaving only gauge-invariant contributions to the rate of change of stretch, tangent orientation and curvature. Consider for instance a growing circle of radius $R$ parametrized by the polar angle $\theta$ (Fig. 39-b). Points

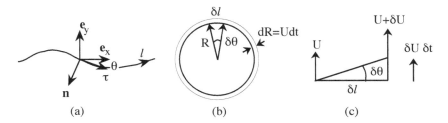

Figure 39. Moving line variables and geometrical interpretation of gauge-invariant terms. (a) Definition of variables on a moving line. (b) (c) Specific growths for which the normal gauge $V = 0$ is chosen so as to make any non-gauge invariant contribution vanish: (b) Growing circle. The stretch of an arc length writes $d\delta l = \delta\theta dR = \kappa\delta l U dt$, in agreement with (88). The evolution rate of curvature simply reads: $d\kappa/dt = -\kappa^2 dR/dt = -\kappa^2 U$, in agreement with (90). (c) Growing plane with inhomogeneous normal velocities. The plane rotation rate writes $\delta\theta/\delta t = -\delta U/\delta l$ in agreement with (89). Its gradient, $-\partial^2 U/\partial l^2$, states the additional evolution rate of curvature brought about by inhomogeneities, in agreement with (90).

moving normally to the circle undergo an elongation rate which simply corresponds to the increase rate of arc length: $1/R\, dR/dt = \kappa U$, in agreement with (88). Moreover, as $dR/dt = d\kappa^{-1}/dt = U$, the term $-U\kappa^2$ represents the evolution rate of curvature, in agreement with (90). Finally, the last gauge-invariant terms, $-\partial U/\partial l$ and $-\partial^2 U/\partial l^2$, may simply be understood by considering a plane growing with a non-homogeneous normal velocity $U(l)$ (Fig. 39-c). Its rotation rate then writes $-\partial U/\partial l$, in agreement with (89), whereas its gradient, $-\partial^2 U/\partial l^2$, corresponds to the evolution rate of inhomogeneous curvature in agreement with (90).

## 9.2. Strain mediated dynamics: the alga Micrasterias

Unicellular algae are plant cells which develop somewhat radially in a plane while showing repetitive lobe splittings that ultimately finely decorate their shape (Fig. 40). We are interested in modeling their early stages of development including lobe formation and first splits.[84]

Unicellular algae involve an inner protoplasmic medium limited by a first inner layer, the plasmalemna, which regulates ionic fluxes and macromolecules transport and a second layer, the cell wall, which supports the mechanical stress implied by the difference between the internal turgor pressure and the external ambiant pressure.

Here the cell wall corresponds to a material separation interface which brings about a definite reference frame for the dynamics: the barycentric frame of its material elements (Sect. 2.1). In particular, for growth

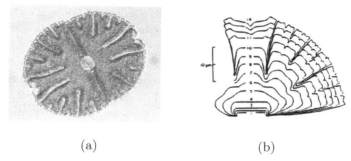

(a) (b)

Figure 40. Growth of the alga Micrasterias. (a) Alga Micrasterias Denticulata during its growth. Alga dimensions $225 \times 190\mu$m. Reprinted figure from Protist Information Server, URL: http://protist.i.hosei.ac.jp/, courtesy of Dr Y. Tsukii, Hosei University, Japan. (b) Successive snapshots of the growth of a semicell of the alga Micrasterias Rotata. Stages are separated by $20mn$ from one another[85]. Reprinted figure from T.C.Lacalli, *J.Embryol.Exp.Morph.* **33**, 95-115 (1975) with kind permission from the Company of Biologists, Cambridge, UK.

to proceed, cell wall has to incorporate new materials corresponding to microfibrils generated by vesicles in the plasmalemna. This allows it to elongate while preserving the internal structures required to support mechanical stress. Accordingly, whatever their complexity, the mechanisms of cell wall dynamics specify a particular frame: the frame $\mathcal{W}$ in which the cell wall is locally at rest.

Needless to say, the whole alga dynamics involves a number of complex biochemical interactions whose global description is still beyond the present modeling capability. However, forgetting the inner complexity of this system, one may simply reduce attention to the dynamics of its boundary, i.e. the cell wall, and model it in a coherent way.[84] This will require coefficients depending on the inner alga dynamics, whose determination will still stand beyond available capacities but whose values could be considered as an input of the model.

The modeling addresses the evolution rates of the geometrical features of the cell wall: strain (88), rotation (89) and curvature (90). As these are gauge-dependent quantities, the gliding velocity $V(l)$ of the reference frame $\mathcal{W}$ in which they are stated has to be expressed explicitly. The objective of the modeling will thus be to specify it so as to infer the normal velocity $U(l)$ of the curve and then, its dynamics.

Experiments[85,86] show that the elongation rate $\delta l^{-1}d\delta l/dt$ strongly depends on curvature. In particular, it has been recognized to be larger at cell tips where curvature is actually the largest. A counter-experiment for this has been performed by Lacalli[85] who succeeded in stopping growth of the Micrasterias denticulata alga by locally destroying its lobe tips with a laser beam, whereas destroying other less curved regions where the extension rate is expected to be weaker yielded no dramatic effect.

This strong correlation motivates us to assume that the relative elongation rate viewed in the reference frame $\mathcal{W}$ of the cell wall is a local quantity, only depending on local geometrical variables. In particular, we shall consider that the biophysical laws governing the cell wall elongation are self-contained in the cell wall frame $\mathcal{W}$ in the sense that they only depend on quantities defined in $\mathcal{W}$, with no necessary reference to another kind of frames. This in particular states that the gliding velocity of the cell wall, $V(l)$, does not enter the expression of the relative elongation rate in $\mathcal{W}$. This rate can thus only depend on local gauge-invariant quantities, as: $l$, $\theta$, and its derivatives $\kappa$, $\partial^n \kappa / \partial l^n, n \geq 1$. However, assuming rotational and translational invariance of the relative elongation rate law along the cell wall, i.e. the absence of preferred orientation or location, yields to reject both $l$ and $\theta$. This makes the relative elongation rate modeled as a function $\mathcal{F}(\kappa, \partial/\partial l)$ of curvature and its derivative. Invoking finally the absence of preferred direction on the cell wall, i.e a reflection invariance $l \to -l$, rejects even derivatives of curvature. Expanding $\mathcal{F}$ at the dominant order in curvature gradient then yields:

$$\frac{1}{\delta l}\frac{d\delta l}{dt} = H(\kappa) + \gamma \frac{\partial^2 \kappa}{\partial l^2} \tag{91}$$

where $H(.)$ is a biophysical function, $\gamma$ a biophysical coefficient and where the implicit gliding velocity $V(l)$ is that of the reference frame $\mathcal{W}$.

A second relationship is necessary to specify $V(l)$. It is obtained by considering the gradient of the rotation rate of the cell wall $\partial/\partial l(d\theta/dt)$, i.e. the way the incorpation of microfibrils modifies the rotation rate.

The necessity of considering a non-homogenous rotation rate may be shown on rhizoïds (Fig. 41-a). These are tubular forms displayed by unicellular algae and fungi and which grow at a constant velocity $V_0 \mathbf{e}_z$ while maintaining the same shape. In addition, resine spheres deposited on their surface and attached to the cell wall are observed to move back from the cell tip to its grooves, the distance between consecutive spheres increasing with time.[87] Accordingly, rhizoïds show both a non-uniform gliding of the cell wall frame $\mathcal{W}$, $\partial V/\partial l \neq 0$, and an elongation rate, $d\delta l/dt \neq 0$. Let us show that this implies a non-homogenous rotation rate $\partial/\partial l(d\theta/dt) \neq 0$.

By symmetry, the rotation rate $d\theta/dt$ must vanish at the cell tip. If it is assumed to be uniform, it then vanishes on the whole shape. Then,

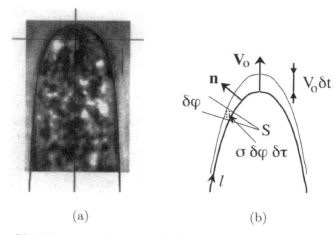

(a) (b)

Figure 41. Rhizoïd growth. (a) Image of a rhizoïd of the fungus Rhizoctonia solani[92]. The rhizoïd shape closely corresponds to the hyphoïd curve with the origin placed on the Spitzenkörper. Scale bar is 2.3$\mu m$ and parameter $d$ is 1.3$\mu m$. Reprinted figure from *Experimental Mycology* **19**, S. Bartnicki-Garcia, D.D. Bartnicki, G. Gierz, R. López-Franco and C.E. Bracker, Evidence that Spitzenkörper behavior determines the shape of a fungal hypha: a test of the hyphoïd model, 153-159, Copyright (1995), with permission from Elsevier. (b) Incorporation model for hypha growth. The rhizoïd translates itself at velocity $V_0$ when growing. It is assumed that the resulting increase of its area $d\delta S$ is driven by the incorporation of vesicles isotropically released by the Spitzenkörper in the fungal frame: $d\delta S = \alpha\Phi\delta\varphi dt$ where $\alpha$ is a constant. However, for kinematic reasons, $d\delta S = U\delta l dt$ where $U$ denotes the normal velocity $U = V_0 \mathbf{e}_z.\mathbf{n}$. This yields relation (96) whose solution is the hyphoïd curve (98).

following (89), the gliding velocity would write $V = \kappa^{-1}\partial U/\partial l$ with $U = V_0\mathbf{e}_z.\mathbf{n}$ here, and thus, $V = \kappa^{-1}V_0\mathbf{e}_z.(\kappa\tau) = V_0\mathbf{e}_z.\tau$. Accordingly, points that are fixed in the cell wall frame $\mathcal{W}$ would move with velocity $U\mathbf{n}+V\tau = V_0\mathbf{e}_z$ in the laboratory frame. They would thus all move together with the growing form. However, as the rhizoïd shape is steady, they would then involve no change of their mutual distance. Their elongation rate would thus vanish as directly confirmed by relation (88). This conflicts with experiments: non-zero elongation rates imply non-homogeneous rotation rates.

For reasons similar to those raised for the modeling of the elongation rate (91), we now expand at the dominant order the gradient of rotation rate with respect to curvature and its odd gradients. The example of a circular growth is instructive since it enables by symmetry an easy identification of its reference frame $\mathcal{W}$ as the normal growth frame: $V(l) = 0$ (Fig. 39-b). In this frame, the rotation rate then vanishes whereas curvature does not.

This shows that no term proportional to curvature is in order in the sought expansion. One then obtain, at the dominant order:

$$\frac{\partial}{\partial l}\left(\frac{d\theta}{dt}\right) = -\alpha\frac{\partial^2\kappa}{\partial l^2} \tag{92}$$

where $\alpha$ stands as a biophysical coefficient.

Both relations (91) (92) provide a complete modeling of the alga dynamics. As, by symmetry, a uniform rotation rate cannot exist, relation (92) yields $d\theta/dt = -\alpha\partial\kappa/\partial l$. This, with relation (89), gives $\kappa V = \partial(U-\alpha\kappa)/\partial l$ which serves to eliminate $V$ in (88). Equating this with relation (91) finally yields the equation determining the normal velocity $U(l)$ as a function of curvature $\kappa(l)$:

$$U+\kappa^{-2}\frac{\partial^2 U}{\partial l^2} = \kappa^{-1}H(\kappa)+\kappa^{-2}(\alpha+\gamma\kappa)\frac{\partial^2\kappa}{\partial l^2}-\alpha\kappa^{-3}\left(\frac{\partial\kappa}{\partial l}\right)^2+\kappa^{-3}\frac{\partial\kappa}{\partial l}\frac{\partial U}{\partial l}. \tag{93}$$

Denoting $\varphi$ the polar angle, we now introduce a small perturbation $\delta r = \zeta(\varphi,t)$ of the distance $r(\varphi,t)$ of the cell wall to the origin. At first order in $\zeta$, the corresponding perturbations of normal velocity and of curvature write $\delta U = \partial\zeta/\partial t$, $\delta\kappa = -\kappa^2(\zeta + \partial^2\zeta/\partial\varphi^2)$ and expansion of relation (93) around the uniform circular state yields:

$$\frac{\partial\delta\kappa}{\partial t} = -\kappa^2\frac{d(\kappa^{-1}H)}{d\kappa}\delta\kappa - (\alpha+\gamma\kappa)\kappa^2\frac{\partial^2\delta\kappa}{\partial\varphi^2}. \tag{94}$$

Notice that, following (93), $-\kappa^2 d(\kappa^{-1}H)/d\kappa = d\bar{U}/dr$ where $\bar{U}$ denotes the normal velocity of a uniformly growing alga: $\partial\bar{U}/\partial l \equiv 0$, $\partial\kappa/\partial l \equiv 0$.

Introducing normal modes, $\delta\kappa(\varphi,t) = \exp(\sigma t)\exp(im\varphi)$, one finally obtains the dispersion relation:

$$\sigma(m,\kappa) = \frac{d\bar{U}}{dR} + (\alpha+\gamma\kappa)\kappa^2 m^2. \tag{95}$$

Here $\alpha$ is positive according to experiments[87] but the sign of $\gamma$ is indeterminate.

If $d\bar{U}/dr$ is positive, the form is unstable for $m = 0$, i.e. for circular growth. This follows from the fact that increasing radius increases normal velocity too, so that the circular growth inflates. For $\alpha + \gamma\kappa$ positive too, this exponential growth is decorated by the small scale distortions at large $m$.

However, experiments show that $d\bar{U}/dr$ is negative so that radial expansion slows down with time. For positive $\alpha + \gamma\kappa$, i.e. positive $\gamma$ or low enough $\kappa$, the form is nevertheless unstable to short wavelength perturbations for $m$ above a critical wavenumber $m_c$. This corresponds to lobe formations (Fig. 40). In particular, for large radius, the perturbation wavenumber along the circular form writes $k = \kappa m$. It is then unstable for $k < k_c \approx (\alpha^{-1}|d\bar{U}/dR|)^{1/2}$.

### 9.3. *Incorporation mediated dynamics: the rhizoïd*

A number of tubular forms called rhizoïds are spontaneously generated by unicellular organisms as pollen tubes or fungi (Fig. 41-a). In particular, in the latters, a number of hypha show long fingered growing forms that seem to refer to the same kind of geometry. These include for instance Pythium aphantdermatum, Polystictus versicolor, Amillaria mellea and many others with a so distinguished spelling. Experiments with surface markers[88,89] or radioisotopes[90] have revealed an expansion of the cell surface in their apex and, in most of them, a vesicle aggregate, the Spitzenkörper. Both observations seem to be correlated since surface increase requires absorption of vesicles that the Spitzenkörper releases. It appears in particular that the position of the Spitzenkörper in the apical dome is of large importance for the resulting form of the hypha.

This ground led Bartnicki-Garcia[91] to propose a simple modeling of hypha growth according to which surface extension results from the incorporation of vesicles emitted isotropically by a moving supplied center: the Spitzenkörper. This simple ansatz enables the form of the resulting object to be derived[91,92] from both the net flux of emitted vesicles and the supplied center velocity $V$ (Fig. 41-a). It may be recovered in the following way.

In the reference frame $S$ of the Spitzenkörper, we introduce polar coordinates $(r, \varphi)$, the direction $\varphi = 0$ being the symmetry axis $\mathbf{e}_z$ of the hypha (Fig. 41-b). The hypha grows at velocity $V_0$ in the $\mathbf{e}_z$ direction while keeping the same form $r(\varphi)$. The corresponding curve thus follows a translation motion at velocity $V_0 \mathbf{e}_z$. Choosing a parametrization for which all of its points $M$ move on the $z$-axis yields $V_M = U\mathbf{n} + V\tau = V_0\mathbf{e}_z$.

We now consider the flux density $\Phi(\varphi)$ of vesicles emitted by the Spitzenkörper in the angular cone $\delta C = [\varphi, \varphi + \delta\varphi]$. When received by the cell wall, this flux induces a surface expansion of the cell wall at a rate which, viewed from the reference frame $S$ of the Spitzenkörper, is proportional to $\Phi\delta\varphi$. Within the cone $\delta C$, the hypha involves a boundary length $\delta l$, a surface $\delta S$ and a rate of surface increase $d\delta S/dt = U\delta l$. As $U = V_M.\mathbf{n} = V_0\mathbf{e}_z.\mathbf{n}$, the law of surface increase then simply writes:

$$\alpha\Phi \, \delta\varphi = V_0 \, \mathbf{e}_z.\mathbf{n} \, \delta l \qquad (96)$$

where $\alpha$ is a constant prefactor.

In the reference frame $\mathcal{S}$, the cartesian coordinates $(x, z)$ of the hypha boundary write $(-r \sin \varphi, r \cos \varphi)$. Accordingly, $\delta l = [r^2 + (\partial r/\partial \varphi)^2]^{1/2} \delta \varphi$ and $\mathbf{e}_z . \mathbf{n} = (r \cos \varphi + \partial r/\partial \varphi \ \sin \varphi)[r^2 + (\partial r/\partial \varphi)^2]^{-1/2}$. From (96), this gives:

$$\alpha \Phi = V_0 \ (r \cos \varphi + \partial r/\partial \varphi \ \sin \varphi) = V_0 \ \frac{d(r \sin \varphi)}{d\varphi}. \qquad (97)$$

Integration for an isotropic flux density, $d\Phi/d\varphi = 0$, and a curve symmetric with respect to the $z$-axis then provides the shape of hypha, the so-called hyphoïd curve:

$$r(\varphi) = \frac{\varphi \, d}{\sin \varphi} \ ; \ z = x \cot(\frac{x}{d}) \qquad (98)$$

where $d = \alpha \Phi/V_0$ denotes the distance $r(0)$ between the origin of coordinates, i.e. the Spitzenkörper, and the apical wall on the $z$-axis.

As shown in Fig. 41-b, the hyphoïd curve agrees remarkably well with both the hypha shapes and the location of their Spitzenkörper.[92] This corroborates the relevance of the model.

## 10. Conclusion

At the epilogue of this guided tour, it is useful to synthesize the main features revealed by interfacial growth phenomena and to put them into a global perspective.

Interfacial growth phenomena have first been identified as one of the main source of heterogeneity, and thus, of diversity, in our macroscopic physical world. In particular, owing to instabilities, they appear to considerably restrict in practice interface planarity and compositional homogeneity. In our environment, this results in a large variety of composition and distribution of minerals, in spite of pre-existing mixing processes (Sect. 4.2), in a macro- or micro-segregation of the solid objects we use in every day life (Sect. 5.4) and in the widespread occurrence of modulated interfacial forms (Sect. 5) from familiar snow-flakes to elementary organisms (Sect. 9.2, 9.3).

Being close to a Laplacian system, interfacial growth phenomena usually involve a tremendously large range of scales, spreading from macroscopic scales imposed by geometry or by extrinsic fields to microscopic scales mainly driven by capillarity. However, neglecting the latter because of their smallness provides not enough scales to obtain something else than similarity and, therefore, a degenerate family of solutions (Sect. 6.1). In practice, this means that tiny phenomena that might be overlooked at first time actually set the definite forms, their morphological scales and their growing velocities. In this sense, non-linear features of interfacial growth are thus mainly controlled by singular perturbations, as is the selection of forms by surface tension and anisotropy (Sect. 6.3) or that of the planar interface velocity by kinetic corrections (Sect. 3.1.3). This particularity complexifies the asymptotic matching of the partial solutions of interface geometry found in specific regions. Achieving a global understanding of growth forms (Sect. 6.2) thus still requires these matching procedures to be developed.

Another specificity of interfacial growth dynamics, as compared to other instability driven systems, refers to the long time of evolution of structures as compared to their instability time scales. This is so on the steady growth forms that result from a primary instability (Sect. 6), on the dendritic forms that emit sidebranches (Sect. 7) and, more generally, on the long-term development of cellular arrays (Sect. 8) or of monocellular organisms (Sect. 9). In this sense, interfacial growth is thus an intrinsically non-linear science that is suitable to develop self-organized states, as displayed on weakly curved interfaces (Sect. 8.2). However, understanding this ability in making a spatio-temporal organization emerge from a noisy unstructured environment requires a modeling over scales large compared to the advective or the diffusive time scales. This still remains a challenging task to achieve. For instance, for sidebranching, recovering the coherent emissions evidenced experimentally (Sect. 7.4) and their actual onset (Sect. 7.3) calls for going beyond the convective instability approach in which analysis is restricted to the gliding time of perturbations to the grooves (Sect. 7.2).

To conclude in analogy with interface growth itself, our understanding of interfacial growth phenomena appears to have grown like an unstable interface: easier things ahead, harder things behind. Further efforts should thus be put on the important features recalled above that have resisted analysis, so as to avoid them to be screened and, finally, overlooked. Among them, the ability of growth in structuring media on large or small scales stands as one of the most fascinating feature amenable to valuable applications.

## Acknowledgments

These lecture notes compile a course given at the Master "Mécanique, Physique et Modélisation" at the University of Provence with a lecture given at the "Rencontres Non-Linéaires de Peyresq". Many recent experimental

results refer to works performed in collaboration with Marc Georgelin at the
"Institut de Recherche sur les Phénomènes Hors-Equilibre". It is a pleasure
to acknowledge him for these several years of joint efforts in understand-
ing growth dynamics. Two studies have also been performed with Sabine
Bottin-Rousseau and Simona Bodéa and some are presently underway with
Julien Deschamps on his Ph.D thesis.

## References

1. M.C. Flemings, Solidification processing, *Series in Materials Science and Engineering, McGraw-Hill* (1974).
2. P.G. Saffman and G.I. Taylor, *Proc.Roy.Soc.A* **245**, 312 (1958).
3. D. Bensimon, L.P. Kadanoff, S. Liang, B.I. Shraiman and C. Tang, *Rev.Mod.Phys* **58**, 977-999 (1996).
4. Y. Couder, Viscous Fingering as an Archetype for Growth Patterns, *Perspectives in Fluid Dynamics, Cambridge University Press*, 53-98 (2000).
5. J.S. Langer, *Rev.Mod.Phys.* **52**, 1-28 (1980).
6. M.G. Worster, Solidification in fluids, *Perspectives in Fluid Dynamics Cambridge University Press*, 394-444 (2000).
7. P. Pelcé, Théorie des Formes de Croissance, *Savoirs Actuels, CNRS Editions & EDP Sciences*, (2000).
8. M. Gündünz and R. Çadirli, *Materials Science and Engineering* **A327**, 167-185 (2002).
9. W.G. Pfann, *Trans. AIME* **194**, 747 (1952)
10. R. Curry and C. Mauritzen, *Science* **308**, 17721774 (2005).
11. C.F. Vermaak, *Economic Geology* **71**, 1270-1298 (1976).
12. D.E. Jackson, *U.S.Geol.Surv.Prof.Pap.* **358**, 106 (1961).
13. S.R. Tait and C. Jaupart, in Interactive Dynamics of convection and solidification, S.H.Davis et al; (eds.), *Kluwer Academic Publishers*, 241-260 (1992).
14. La Physique de la Terre, H-C. Nataf and J. Sommeria Eds, *Belin & CNRS Editions*, (2000).
15. K.D. Collerson et al., *Nature* **349**, 20 (1991). C. Benett et al., *Planet.Sci.Lett.* **119**, (1993).
16. B. Bourdon, *La Recherche* **286**, 5457 (2005).
17. W.W. Mullins, R.F. Sekerka, *J.Appl.Phys.* **35**, 444-451 (1964).
18. A. Pocheau and M. Georgelin, *J.Phys.IV France* **11**, 169-178 (2001).
19. M. Georgelin and A. Pocheau, Experimental study of sidebranching in directional solidification, in "Branching in Nature", Ed. V. Fleury, J-F. Gouyet, M. Léonetti, *EDP Sciences, Springer*, 409-415 (2000).
20. M. Georgelin and A. Pocheau, *Phys.Rev.Lett.* **79**, 2698-2701 (1998).
21. M.A. Eshelman, V. Seetheraman and R. Trivedi, *Acta Metall.* **36**, 1165-1174 (1988).
22. B. Billia and R. Trivedi, Handbook of Crystal Growth, Vol.1 Chapter 14, *Elsevier Science Publishers*, (1993).

23. J.D. Weeks, W. Van Saarloos and M. Grant, *J.Crystal.Growth* **112**, 244 (1991).

24. R.S. Rerko, H.C. de Groh III, C. Bekermann, *NASA report* **2000-210020**, 1-108 (2000).

25. H.C. de Groh III and V. Laxmanan, Solidification processes of eutectic alloys, D.M. Stefanescu, G.J. Abbaschian and R.J. Bayuzick (eds), *The Metallurgical Society, Inc., Warrendale, Pennsylvania*, 229-242 (1988)

26. W. Kurz and D.J. Fischer, Fundamentals of solidification, *Trans tech Publications Ltd*, Uetikon-Zurich, Switzerland, (1998).

27. A. Papapetrou, *Z.Krist.* **92**,89 (1935).

28. G.P. Ivantsov, *Dokl.Akad.Nauk SSSR* **58**, 56 (1947).

29. P. Pelcé and A. Pumir, *J.Cryst.Growth* **73**, 337 (1985).

30. T. Dombre and V. Hakim, *Phys.Rev.A* **36**, 2811 (1987).

31. E. Scheil, *Z. Metallk.* **34**, 70 (1942).

32. A. Pocheau and M. Georgelin, *J.Cryst.Growth* **206**, 215-229 (1999).

33. V. Seetharaman and R. Trivedi, *Metall.Trans.A* **19**, 2955 (1988).

34. T.F. Bower, H.D. Brody and M.C. Flemings, *Trans.Met.Soc. AIME* **236**, 624 (1966)

35. P. Kurowski, C. Guthmann, and S. de Cheveigné, *Phys.Rev.A* **42**, 7368 (1990).

36. M. Georgelin and A. Pocheau, *J.Cryst.Growth* **268**, 272-283 (2004).

37. B.J. Spencer and H.E. Huppert, *Acta Mater.* **45**, 1535-1549 (1997) .

38. B.J. Spencer and H.E. Huppert, *J.Cryst.Growth* **200**, 287-296 (1999).

39. A. Pocheau and M. Georgelin, *Phys.Rev.E* **73**, 011604-18 (2006).

40. R.C. Brower, D.A. Kessler, J. Koplik and H. Levine, *Phys.Rev.Lett.* **51**, 1111-1114 (1983).

41. R.C. Brower, D.A. Kessler, J. Koplik and H. Levine, *Phys.Rev.A* **29**, 1335-1342 (1984).

42. D.A. Kessler, J. Koplik and H. Levine, *Phys.Rev.A.* **30**, 3161-3174 (1984).

43. D.A. Kessler, J. Koplik and H. Levine, *Phys.Rev.A.* **31**, 1712-1717 (1985).

44. D.A. Kessler and H. Levine, *Phys.Rev.Lett.* **57**, 3069-3072 (1986).

45. D.A. Kessler and H. Levine, *Phys.Rev.B* **33**, 7867-7870 (1986).

46. D.A. Kessler, J. Koplik and H. Levine, *Adv. Phys.* **37**, 255-339 (1988).

47. J.S. Langer, Chance and Matter, lectures on the Theory of pattern Formation, Les Houches, Session XLVI, J. Souletie, J. Vannimenus and R. Stora (eds), North Holland, Amsterdam, 629-711, 1987.

48. J.W. McLean and P.G. Saffman, *FJ.Fluid.Mech.* **102**, 455-469 (1981)

49. J.M. Vanden-Broeck, *Phys.Fluid* **26**, 2033-2034 (1983).

50. S. Tanveer, *Phys.Fluid* **30**, 2318-2329 (1987).

51. M. Rabaud, Y. Couder, N. Gerard, *Phys.Rev.A.* **37**, 935-947 (1988).

52. E. Ben-Jacob, R. Godbey, N.D. Goldenfeld, J. Koplik, H. Levine, T. Mueller, L.M. Sander, *Phys.Rev.Lett* **55**, 1315-1318 (1985).

53. Á Buka, T. Börzsönyi, N. Éber, T. Tóth-Katona, D. Reguera, L.L. Bonilla and J.M. Rubí eds LNP 567, 88-318 (2001).

54. E. BenJacob, N. Goldenfeld, B.G. Kotliar and J.S. Langer, *Phys.Rev.Lett.* **53**, 2110-2113 (1984).

55. J.S. Langer, *Phys.Rev.A* **33**, 435-441 (1986).
56. A. Barbieri and J. Langer, *Phys.Rev.A* **39**, 5314 (1989).
57. M. Ben Amar and E. Brener, *Phys.Rev.Lett.* **71**, 589 (1993).
58. D.I. Meiron, *Phys.Rev.A* **33**, 2704-2715 (1986).
59. M. BenAmar and B. Moussallam, *Physica D* **25**, 155-164 (1987).
60. J.S. Langer and H. Müller-Krumbhaar, *Acta Metall.* **26**, 1681-1687 (1977).
61. M.E. Glicksman, R.J. Schaefer and J.D. Ayers, *Metall.trans* **A7**, 1747-1759 (1976)
62. M. Muschol, D. Liu and H.Z. Cummins, *Phys.Rev.A* **46**, 1038-1050 (1992)
63. S. Akamatsu, G. Faivre and T. Ihle, *Phys.Rev.E* **51**, 4751 (1995).
64. T. Ihle and H. Müller-Krumbhaar, *Phys.Rev.E* **49**, 2972 (1994).
65. P. Huerre, Open shear flow instabilities, *Perspectives in Fluid Dynamics Cambridge University Press*, 159-229 (2000).
66. A. Douady and Y. Couder, *J.Theor.Biol.* **178**, 255-274 (1996).
67. Ya. B. Zel'dovich, A. G. Istratov, N. I. Kidin and V. B. Librovich, *Combust.Sci.Technol.* **24**, 1-13 (1980).
68. P. Pelcé and P. Clavin, *Europhys.Lett.* **3**, 907-913 (1987).
69. S.K. Sarkar, *Phys.Lett.A* **117**, 137-140 (1986).
70. R. Pieters and J.S. Langer, *Phys.Rev.Lett.* **56**, 1948-1951 (1986).
71. J.S. Langer, *Phys.Rev.A* **36**, 3350-3358 (1987).
72. E. Brener and D. Temkin, *Phys.Rev.E* **51**, 351-359 (1995).
73. W. Van Saarloos, B. Caroli and C. Caroli, *J.Phys.I* **3**, 741-751 (1993).
74. O. Martin and N. Goldenfeld, *Phys.Rev.A* **35**, 1382-1390 (1987).
75. M.Georgelin and A.Pocheau, *Phys.Rev.E.* **57**, 3189-3203 (1998).
76. M. Georgelin and A. Pocheau, *Eur.Phys.J.B.* **4**, 169-174 (1998).
77. A. Pocheau and M. Georgelin, *J.Cryst.Growth* **250**, 100-106 (2003).
78. A. Pocheau and M. Georgelin, *Eur. Phys. J. B* **21**, 229-240 (2001).
79. A. Dougherty, P.D. Kaplan and J.P. Gollub, *Phys.Rev.Lett.* **58**, 1652-1655 (1987).
80. M. Georgelin, S. Bodea and A. Pocheau, *Europhys. Lett.* **77**, 46001 (2007).
81. A. Karma and P. Pelcé, *Phys.Rev.A* **39**, 4162-4169 (1989).
82. S. Bottin and A. Pocheau, *Phys.Rev.Lett.* **87**, 076101 (2001).
83. A. Pocheau and S. Bottin-Rousseau, *Chaos* **14**, 882-902 (2004). and in: *Virtual Journal of Biological Physics Research* **8** (7), 2004
84. P. Pelcé and A. Pocheau, *J.Theor.Biol* **156**, 197-214 (1992).
85. T.C.Lacalli, *J.Embryol.Exp.Morph.* **33**, 95-115 (1975).
86. O. Kiermayer, Cytoplasmic basis of morphogenesis in Micrasterias, *Cytomorphogenesis in plants*, O. Kiermayer ed., Springer Verlag, Berlin, 147 (1981)
87. Z. Hejnowicz, B. Heinemann and A. Sievers,*Z.Pflanzenphysiol.Bd.* **81**, 409-423 (1977)
88. E.S. Castle, *J.Gen.Physiol.* **41**, 913-926 (1958)
89. S. Bartnicki-Garcia, C.E. Bracker, G. Gierz, R.López-Franco and H.Lu, *Biophysical Journal* **79**, 2382-2390 (2000).
90. G.W. Gooday, *J.Gen.Microbiol.* **67**, 125-133 (1971).
91. S. Bartnicki-Garcia, F. Hergert and G. Gierz, *Protoplasma* **153**, 46-57 (1989).

92. S. Bartnicki-Garcia, D.D. Bartnicki, G. Gierz, R. López-Franco and C.E. Bracker, *Experimental Mycology* **19**, 153-159 (1995).
93. Y. Couder, J. Maurer, R. González-Cinca and A. Hernández-Machado, *Phys.Rev.E* **71**, 031602 (2005).
94. P.P. Trigueros, F.Sagués and J. Claret, *Phys.Rev.E* **49**, 4328-4335 (1994).